TRENDS IN HAZARDOUS MATERIALS RESEARCH

TRENDS IN HAZARDOUS MATERIALS RESEARCH

EDWARD C. BOOKING
EDITOR

Nova Science Publishers, Inc.
New York

For permission to use material from this book please contact us:
Telephone 631-231-7269; Fax 631-231-8175
Web Site: http://www.novapublishers.com

Library of Congress Cataloging-in-Publication Data

Trends in hazardous materials research / Edward C. Booking (editor).
p. cm.
Includes index.
ISBN 978-1-60021-335-9
1. Hazardous substances. 2. Hazardous wastes. I. Booking, Edward C.
T55.3.H3T747 2006
628.4'2--dc22

2006021086

Published by Nova Science Publishers, Inc. ✦ New York

CONTENTS

PREFACE

Hazardous waste is a waste with properties that make it dangerous or potentially harmful to human health or the environment. Hazardous waste generally exhibits one or more of these characteristics: ignitability, corrosivity, reactivity or toxicity. The universe of hazardous wastes is large and diverse. Hazardous wastes can be liquids, solids, contained gases, or sludges. They can be the by-products of manufacturing processes or simply discarded commercial products, like cleaning fluids or pesticides. One major type is radioactive waste. This new book brings together the latest research in this diverse field.

Chapter one describes how plating sludges generally contain a mixture of metal compounds. Often due to high concentrations of mobile/leachable metallic species - particular transition metals like chromium and nickel, which can rise up to 20 and 30%, respectively (w/w, dry weight) - these sludges are classified as hazardous waste and they are disposed of in special landfills involving enormous costs to the waste producers. This option undoubtedly leads to serious disadvantages, as it contributes to a great build up of environmentally hazardous materials on the earth's crust. On the other hand, it does not consider the recovery of the heavy metals for re-use, which might represent an economy of raw materials and potential profits.

Taking into account the above considerations and also that metal deposits of Zn, Cu and Ni may be depleted within 30, 36 and 50 years, respectively [1], a plating sludge treatment using the hydrometallurgical route is proposed herein. Using simple, cheap and ecofriendly processes, the sludge could be valued if some economical value metals were extracted from it and transformed into commercially pure grade raw materials. At the end, the remaining sludge would be a non hazardous waste, suitable for landfilling in available sites at a much lower cost, or be incorporated into different construction materials such as ceramics, cements or bituminous aggregates.

A new process envisaging value-adding through the recycling of plating sludge materials has been developed to recover some heavy metals (Ni, Cu, Zn and Cr).

The plating sludge samples used for this work were collected from different plating shops spread out in Portugal. Their characterizations involved the determination of relevant physical and chemical properties, and then they were grouped according to their typology.

To accomplish the hydrometallurgical value-adding process, the plating sludge selected was first digested using a sulfuric acid leaching. Hence, the leaching solution had its acidity neutralized by employing the same kind of sludge (fresh one) as a conditioner agent. When an adequate pH value was reached, a copper cementation step was applied using zinc dust as a precipitating agent, from which a copper cement with an appreciable purity grade was obtained. The resulting solution was subsequently treated with a calcium carbonate suspension, and a selective chromium precipitate was achieved. The remaining solution, free from copper and chromium, was sequentially subjected to solvent extraction using D2EHPA in kerosene as organic solvent. The chosen extractant successfully promoted the separation of zinc from nickel. In the last step nickel and zinc sulfate salts were produced through the drying of both final solutions, by simulation of an evaporative crystallization process.

The overall results collected allow the proposal of an integrate hydrometallurgical process to promote a plating sludge value-adding through the production of a copper cement, chromium oxide and, finally, nickel and zinc sulfate salts. The data picked up from a sequential experiment is presented, which simulates results obtained at a pilot scale.

Chapter two states that Polyacrylonitryl (PAN) and bentonite (B)/zeolite (Z)-PAN composites were prepared by direct polimerization of pure AN and AN saturated suspensions of B and Z. PAN and the composites were subjected to amidoximation procedure to obtain PAO, B-PAO and Z-PAO. FT-IR and XRD were employed to characterize their structures. The sorption dependency of the materials on ion concentration and temperature were then investigated for Pb^{2+} and UO_2^{2+}. The reusability and storage ability of the adsorbents were also considered for Pb^{2+} and UO_2^{2+} adsorption.

All isotherms were L and H type of the Giles classification. For both ions, the adsorption capacities of B and Z-PAO composites were higher than that of pure PAO, when the PAO contents of composites were normalized to pure PAO. The introduction of B and Z in to PAO significantly increased the Langmuir equilibrium constants (L mol^{-1}), so as 353, 2180 and 149100 for Pb^{2+} 1980, 25900 and 284200 for UO_2^{2+} adsorption onto PAO, B-PAO and Z-PAO respectively. The adsorption was enthalpy controlled. For the use of 1 mol L^{-1} HCl as a regenerative effluent, the structure of PAO completely changed to be gel in the first use, whilst the composites were reusable for five sequential treatments without any change in their structures. Storage foregoing had also no effect on the structural stability.

The studied features of the composites suggest that these materials should be considered amongst the new adsorbents. It is envisaged that the use of B-PAO and Z-PAO composites will provide practicality and effectiveness for separation and removal procedures involving di/trivalent cations.

Chapter three informs us that certain inorganic species exert toxic effects when present in excess of threshold concentrations. Unlike toxic organics, inorganics are not biodegradable and so builds up in the environment. The adverse effects of heavy metals such as cadmium, lead, chromium, arsenic, mercury, etc at very low levels is well known and these are often regarded as highly toxic. In recent past, it is established that uranium and its compounds, like lead are highly toxic which cause progressive or irreversible renal injury and in acute cases, may lead to kidney failure and death. Hence, at this juncture, there is a need for exploring novel strategies for the removal of toxic heavy metals and uranium from hazardous environmental, biological and geological samples (particularly in view of the increasing realization of "Green Science and Technology").

The need for removal of toxic heavy metals and uranium from various hazardous materials is highlighted in the introductory part. Subsequently, the basic principles of different techniques employed for the removal of toxic inorganic species has been described. Finally, the removal strategies reported in literature by several researchers for toxic inorganic species viz. heavy metals, radio isotopes, uranium and binary/multivalent heavy metal mixtures from environmental, biological and geological samples have been discussed. The chapter ends with a concluding note highlighting the exciting and rapidly developing Solid Phase Extraction (SPE) technique for removal of micro and macro amounts of toxic inorganic species in a variety of hazardous materials.

Chapter four explores the colloidal particles that have a few micro-scale or nano-scale diameters and are suspended as charged particles in aqueous solution. The geochemical environment and engineering process generate colloidal particles and their suspension exposed to aqueous and terrestrial environments. These colloidal particles are very reactive due to large surface area and then highly contaminated with inorganic or organic contaminants. The colloidal stability in the aqueous solution is significant for its deposition condition. Due to the long suspension time of the particles it is difficult to precipitate them. Evidently, the colloidal phenomenon is applicable for the remediation process of soil and groundwater environment. The practical design in many remedial processes was not suitable for the application of nano-scale particles. This study presents the principle of colloidal control and their case studies of lab-scale and pilot-scale the remedial design using fine particles. The reactive barrier using nano-scale particles is effective as colloid flushing or colloid barrier formation for the removal through adsorption and degradation. Particularly, the deposition of nano-scale particles was highlighted for the removal of various contaminants. Soil washing with surfactant removes inorganic and organic contaminants through solubilization, mobilization, detachment and emulsion by formation of colloidal micelle above CMC (critical micelle concentration).

In chapter five, experimental tests of different techniques for contaminated soils and groundwater remediation are presented and discussed. The contaminant of concern is the MtBE, a common antiknock additive in the production of unleaded fuel. MtBE's great water solubility, low partitioning into soils and low tendency to be degraded by microbial activity, are responsible for a considerable mobility that allows rapid dispersion and strong persistency in groundwater.

To remove or/and destroy this contaminant, three *in situ* technologies are evaluated and compared: Air Sparging (AS), soil flushing (ISSF), and chemical oxidation (ISCO). AS, which mainly involves the stripping of the contaminant by injected air in the subsurface, is currently one of the most commonly used, due to the simplicity of the plant installation, operation and maintenance and, consequently, to its low associated costs. Other systems, like ISSF and ISCO can be also successfully implemented in MtBE contaminated soils and groundwater treating. Both ISSF and ISCO technologies deal with the chemical oxidation of the pollutant, with the difference that for ISSF the oxidant does not come into contact with the soil matrix. In the experimentation presented in this chapter, Fenton oxidation was performed in either technologies used.

For each technology after the description of the basic principles and environmental applications, the results of lab scale experiments conducted on samples of saturated soil artificially contaminated, are discussed.

AS tests were performed both in a cylindrical column (small scale test) and in a large tank of 1 m^3 (intermediate scale test). These two testing set-ups were used to study the optimal air flow rate for MtBE transfer to the gas phase and the influence of the geometry around the injection well.

ISSF tests were performed by flushing the soil with an aqueous solution of a surfactant agent (ethanol). The pH, the ethanol concentration and the flow rate of this extracting solution were optimized to achieve the complete removal of MtBE from the matrix. The recovered solution was then oxidized with Fenton's reagent to destroy the pollutant. The effect of ethanol concentration towards MtBE oxidation was also evaluated, considering the non selectivity of the oxidant.

ISCO tests were performed by injecting Fenton's reagent directly into the soil, both at the natural pH of soil and under acidic conditions.

The most common formulation, which has been used to preserve wood, is chromated copper arsenate (CCA) discussed in chapter six. Based on the average wood service life of 20–50 years, it is expected that enormous amount of CCA-treated wood is now being, or will soon be, removed from service, which raises an environmental issue of how to deal with this wood waste management problem. Both legislation and drawbacks of the traditional disposal method (land-filling) is motivating studies of alternatives for treatment and possible reuse of CCA-treated wood waste. Besides the economic and technological issues, the method of choice is expected to be environmentally beneficial.

This paper is designed to provide an overview of the main membrane-assisted techniques that can be used for removal of heavy metals from CCA-treated wood waste, the main advantage of which is the combination of leaching of the target contaminant(s) by suitable agents, with the complete isolation of the treated material by membranes, thus forming separate compartment(s), in which the contaminant(s) can be accumulated and recovered. By a proper selection of the membrane material and operating conditions, the transfer of the target charged pollutants through the membranes can be regulated. Furthermore, the metal removal degrees can be controlled by appropriate pre-treatment of the wood waste, by adjusting the flow rates of the solutions, contacting the membranes, by introducing electrodes into the compartments, etc.

The authors have studied comprehensively both the dialytic and electrodialytic removal of Cu, Cr and As from CCA treated timber waste. The dialytic method is solely based on the Donnan equilibrium principle while the electrodialytic method uses a low-level direct current as the "cleaning agent", combining the electrokinetic movement of ions in the matrix, with the principle of electrodialysis. The electrodialytic treatment, performed under appropriate direct current conditions, proved to efficiently remediate CCA-treated timber waste in the form of sawdust (Ribeiro et al., 2000a; 2000b; Moreira *et al.*, 2005) or wood chips (Velizarova *et al.*, 2002; 2004; Virkutyte *et al.*, 2005) in laboratory scale, as well as in pilot scale (Pedersen *et al.*, 2005; Christensen *et al.*, 2005).

The principles and the applicability of these ion exchange membrane-assisted techniques for the removal of heavy metals from CCA-treated wood waste is discussed and critically compared with the electrokinetic remediation technique throughout the paper. The design of more efficient and cost-effective operating conditions, especially in long-term operations without or with minimal membrane fouling, reduction of the specific energy consumption requirements and development of efficient recovery or disposal options for the contaminants, in order to prevent their adverse impact on the ecosystems, are the major challenges.

Chapter seven describes how flammability limits data are essential for a quantitative risk assessment of explosion hazard associated with the use of combustible gas. To obtain the fundamental flammability data for prevention of the hazards in the practical applications, experiments were conducted in a constant volume combustion bomb, for promising alternative gaseous fuels, such natural gas, liquefied petroleum gas, mixtures in air with or without diluent gas. Following the investigation on the flame initiation and its early development, the role of spark energy on the flame initiation of combustible mixtures had been investigated experimentally. Results showed that, a stable flame kernel yields easily with higher spark energy, but generally, spark energy well above the minimum ignition energy, has relatively weak influence on flame propagation, as identified by the fact that the flame propagation speeds become independent upon ignition energy after flame radii greater than a criterion value. The effect of residence time on flammability limit was investigated because of the possibility of preoxidation reactions during the waiting time from mixture adding to spark ignition. On the basis of global chemical kinetics mechanism, a simple method was proposed to determine the preoxidation rate of fuel. With the growing implementation of exhaust gas recirculation in reducing NOx emissions of engine, it is necessary to perform a fundamental research on the flammability characteristics of fuel-air-diluent mixtures. The influences of diluent gas on the flammability region of these alternative gaseous fuels in air were then experimentally explored. It is known that, the flame burning velocity is one of the most relevant parameters of flame chemistry. Thereby, an assumption of critical burning velocity at flammability limit was proposed to approximate the flammability region of these mixtures, as a result, the diluent limits of studied mixture was achieved.

Chapter eight states that the current Population Growth Rate (PGR) for Sub-Sahara Africa (3 percent) is by far the highest for any region in the world. This high population growth is particularly ominous regarding food demands if the rapid rate of movement of people from the rural area, the site of food production, to urban areas is considered. In view of the need for self-sufficiency in food production, and the expansion of export agriculture, several countries in Sub-Sahara Africa have been aggressively promoting the use of agricultural inputs such as fertilizers and pesticides.

Agricultural pesticides are, however poisons by design as these chemicals are intended for use in *destroying* any pest that may interfere with the production or the processing of agricultural products. The capacity of pesticides to *destroy* is however not limited to pests but may result in serious risks to human health and to the environment especially. While occupational agricultural pesticide poisoning may be universal, the incidence of such poisoning or contamination is much higher in developing countries.

This paper which provides an overview of past research on the use/misuse of agricultural pesticides in Sub-Sahara Africa includes the following: the upsurge of the use of agricultural pesticides in the region; factors of pesticide poisoning or contamination; adverse health effects of pesticide poisoning including self-reported effects and the extent of these adverse effects, and the adverse effects pesticide contamination on the environment In contrast to the zero-risk approach favored by some researchers, this paper emphasizes the need to reduce agricultural pesticide poisoning in Sub-Sahara Africa through pesticide risk communication and education programs, provision of government subsidy for the purchase of less toxic but usually more expensive pesticides, discontinuation of calendar-based pesticide application, delimiting the importation of highly toxic agricultural pesticides, and reassessing the need for *special* pesticide protective clothing for farmers working in the *hot tropical climate* of Sub-Sahara Africa.

In: Trends in Hazardous Materials Research
Editor: Edward C. Booking, pp. 1-58
ISBN 1-60021-335-9

Chapter 1

PLATING SLUDGE VALUE-ADDING BY APPLICATION OF HYDROMETALLURGICAL PROCESSES

J. E. Silva[a], A. P. Paiva[b], D. Soares[a], A. Labrincha[c] and F. Castro[a]

[a] Mechanical Engineering Department, University of Minho, Guimarães, Portugal
[b] Department of Chemistry and Biochemistry, Centre of Chemistry and Biochemistry, Faculty of Sciences, University of Lisbon, Lisbon, Portugal
[c] Ceramics and Glass Engineering Department, University of Aveiro, Aveiro, Portugal

ABSTRACT

Plating sludges generally contain a mixture of metal compounds. Often due to high concentrations of mobile/leachable metallic species - particular transition metals like chromium and nickel, which can rise up to 20 and 30%, respectively (w/w, dry weight) - these sludges are classified as hazardous waste and they are disposed of in special landfills involving enormous costs to the waste producers. This option undoubtedly leads to serious disadvantages, as it contributes to a great build up of environmentally hazardous materials on the earth's crust. On the other hand, it does not consider the recovery of the heavy metals for re-use, which might represent an economy of raw materials and potential profits.

Taking into account the above considerations and also that metal deposits of Zn, Cu and Ni may be depleted within 30, 36 and 50 years, respectively [1], a plating sludge treatment using the hydrometallurgical route is proposed herein. Using simple, cheap and ecofriendly processes, the sludge could be valued if some economical value metals were extracted from it and transformed into commercially pure grade raw materials. At the end, the remaining sludge would be a non hazardous waste, suitable for landfilling in available sites at a much lower cost, or be incorporated into different construction materials such as ceramics, cements or bituminous aggregates.

A new process envisaging value-adding through the recycling of plating sludge materials has been developed to recover some heavy metals (Ni, Cu, Zn and Cr).

The plating sludge samples used for this work were collected from different plating shops spread out in Portugal. Their characterizations involved the determination of relevant physical and chemical properties, and then they were grouped according to their typology.

To accomplish the hydrometallurgical value-adding process, the plating sludge selected was first digested using a sulfuric acid leaching. Hence, the leaching solution had its acidity neutralized by employing the same kind of sludge (fresh one) as a conditioner agent. When an adequate pH value was reached, a copper cementation step was applied using zinc dust as a precipitating agent, from which a copper cement with an appreciable purity grade was obtained. The resulting solution was subsequently treated with a calcium carbonate suspension, and a selective chromium precipitate was achieved. The remaining solution, free from copper and chromium, was sequentially subjected to solvent extraction using D2EHPA in kerosene as organic solvent. The chosen extractant successfully promoted the separation of zinc from nickel. In the last step nickel and zinc sulfate salts were produced through the drying of both final solutions, by simulation of an evaporative crystallization process.

The overall results collected allow the proposal of an integrate hydrometallurgical process to promote a plating sludge value-adding through the production of a copper cement, chromium oxide and, finally, nickel and zinc sulfate salts. The data picked up from a sequential experiment is presented, which simulates results obtained at a pilot scale.

1. Introduction

1.1 Relevance of the Study

Although the first concept of the waste management sector is the prevention and minimization of waste by the optimizing productive processes to zero discharge, it is known that there are many cases of productive processes that even if the waste producer invests in the construction of a modern productive line, some wastes will never be extinct. In this way, the R&D of technological processes to treat wastes, in order to recycle or re-use those effluents, are fully justified; they are now a reality and will certainly grow in the future.

Currently there are new proposals to redefine waste and waste management, and those assume concepts closer to sustainable management of secondary resources, in order to achieve a full and clear property transfer (property = waste herein) between the producer and who acquires the property [2].

In many surface treatments of metals and plastics - such as aluminum in anodizing, chromium, nickel and zinc plating - aqueous effluents containing high concentrations of metallic elements are normally produced. The physicochemical treatment of those aqueous effluents, made with the objective of purifying and recycling process waters (spent baths and rinsing waters) generates plating sludge, whose composition consists basically on metallic hydroxides, sulfates, phosphates, chlorides, etc. [3]. Since sometimes the concentration of some metals might exceed the legal limits, the plating sludge receives a hazardous classification.

Once the plating sludge is precipitated, there is commonly no further suitable treatment, the disposal in landfills being their usual destination. However, the use of landfills as final destination for industrial hazardous waste is being replaced by other practices, such as mitigation of waste generation, re-use of waste, physical / chemical / biological treatments, incineration and stabilization / solidification methods [4].

According to Crowson [1], the estimated lives for ore deposits that are now source of metals such as zinc, copper and nickel might deplete in 30, 36 and 50 years, respectively, assuming global growth rates for demand similar to those of the past decade and also considering that there is no discovery and development of new sources of supply through new base metal reserves. An evident signal of these estimates is the evolution of some product prices such as the non-ferrous metals in the world market. An evolution of the market prices for some metals, taking into account the annual averages alluded to 16 years ago, is shown in Figure 1.1 (1995-2005) [5,6].

It can be observed that nickel is the most valuable metal, followed by chromium, copper, aluminum, zinc and lead.

Bearing these data and the estimation done by Crowson [1] in mind, Ni, Cu and Zn are considered the most economical valuable metals along this chapter.

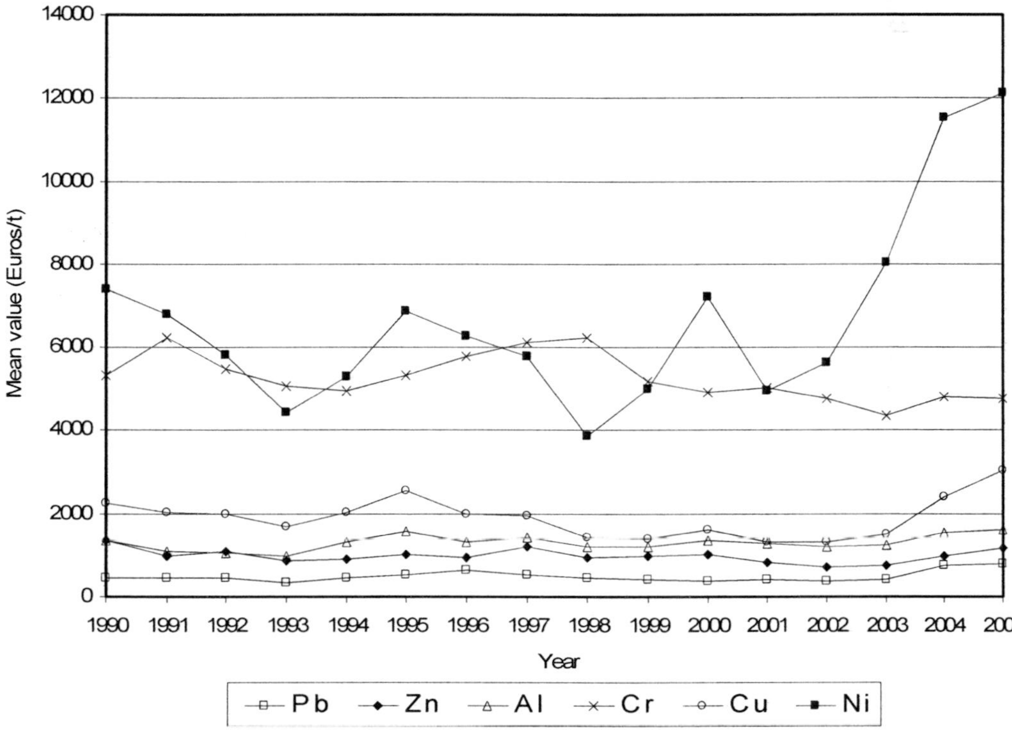

Figure 1.1. Evolution of market prices for some non-ferrous metals, considering the annual averages recorded for the last 16 years [5,6].

1.2 Plating Sludge as Secondary Source

The characteristics of plating sludges depend on the treated workpiece nature (e.g., ferrous and non-ferrous materials, plastics, etc.) and on the implemented productive process. In fact, Magalhães et al. [3] have established the existence of a high variability in the chemical composition of plating sludges from different installations, and even in sludges originated from the same company.

In R&D projects such as ECOCERAM I, ECOINERT, ECOCERAM II [7-9], an inertization process for some plating sludge types was successfully studied and developed, through their incorporation in clay-based ceramics. However, some compositional complex sludges may show high immobilization resistance. For example, the accumulation of metallic elements inside waste floccules that were inserted in the ceramic matrix may become soluble after firing, since their reaction towards the inertization material was found to be very incipient [10-12].

On the other hand, some plating sludges might contain significant amounts of valuable elements, and viable processes for the recovery of such species as secondary source should be implemented, in alternative to the inertization method.

Hydrometallurgical techniques might be used for this purpose, since they are well established in the ores treatment (primary sources) since the Middle Ages. However, their use is nowadays also extended to the minimization of the environmental impact, lowering the hazardous wastes build up on the earth crust and recovering heavy metals for re-use.

Taking into account the above issues, a plating sludge treatment using the hydrometallurgical route is described in this chapter. Using simple, cheap and eco-friendly processes, the sludge could be valued if some economical value metals were extracted and transformed into commercially pure grade raw materials.

1.3 Specific Objectives of the Study

The present study was based on the following specific objectives:

1. to carry out the physicochemical characterization of plating sludges collected from different plating shops;
2. to select the most adequate plating sludge for the hydrometallurgical treatment;
3. to choose an efficient leaching process to dissolve the valuable elements present in the sludge;
4. to identify fortuitous applications for the products generated from this treatment;
5. to establish a flowsheet for the treatment of the plating sludge.

Characterization

This section has been divided in two parts. In the first one, data achieved from plating sludge samples, collected at various Portuguese plating shops, is used to group those samples according to their typology. Relevant physical and chemical parameters was evaluated for that purpose. The second part summarizes the characteristics of the plating sludge that has been chosen for the proposed herein recovery process.

2.1 Characterization of the Plating Sludges

2.1.1 Survey and Collection of Plating Sludge Samples

An inquiry about different plating shops spread out in Portugal was firstly conducted, and then several companies were selected, in an attempt that they could represent the following sectors:

(i) aluminum anodizing;
(ii) decorative chromium plating;
(iii) zinc plating;
(iv) complex / mixed surface treatments (zinc, chromium, brass, copper plating, etc.).

Thirty nine plating sludge samples were collected, with mass values ranging from 10 to 30 kg (humid basis).

2.1.2 Chemical Characterization

Each plating sludge sample (about 1 g) was submitted to acid digestion and the quantitative analysis (Al, Ca, Cu, Cr, Fe, Ni and Zn) of the corresponding liquor was obtained by atomic absorption spectrometry (AAS - GBC 904 AA). Table 2.1 describes the technical conditions used for the chemical analysis of the above elements.

Thus, plating sludge samples analyzed were grouped according to the metal elements contents for Al, Ca, Cu, Cr, Fe, Ni and Zn. Tables 2.2 to 2.7 show the refined groups.

Group 1 – Table 2.2 presents plating sludges containing a high average iron content (Fe = 39.3%). They also present a significant amount of zinc (Zn = 11.7%) and are those having the highest average total percentage of analyzed metals (55.97%), mainly due to the Fe contribution.

Group 2 – Table 2.3 assembles a collection of sludges generated from a typical zinc plating process, easily characterized by a high amount of zinc (Zn = 28.4%), and simultaneously by a low average amount of iron (Fe = 6.67%) when compared to the samples belonging to Group 1.

Table 2.1. Technical conditions adopted for the chemical analysis of aluminum, calcium, copper, chromium, iron, nickel and zinc by AAS

Element	λ (nm)	I (mA)	F.T.	S.W. (nm)	W.R. ($mg.L^{-1}$)	S ($mg.L^{-1}$)
Al	396.2	10.0	NO-Ac.	0.5	25 – 110	0.55
Ca	422.7	10.0	NO-Ac.	0.5	1 – 4	0.02
Cu	324.7	3.0	Air-Ac.	0.5	1 – 5	0.025
Cr	359.3	6.0	NO-Ac.	0.2	4 – 20	0.09
Fe	248.3	7.0	Air-Ac.	0.2	2 – 9	0.05
Ni	232.0	4.0	Air-Ac.	0.2	1.8 – 8	0.04
Ni	341.5	4.0	Air-Ac.	0.2	6 – 25	0.12
Zn	213.9	5.0	Air-Ac.	0.5	0.4 – 1.5	0.008

λ – wavelength; I – lamp current; F.T. – flame type; NO-Ac. – nitrous oxide/acetylene; Air-Ac. – air/acetylene; S.W. – slit width; W.R. – work range; S – sensitivity.

Table 2.2. Group 1: chemical composition of Fe-rich plating sludges

Sample	Content in metals (wt. %)							
	Al	Ca	Cu	Cr	Fe	Ni	Zn	Total
F1	0.12	0.28	0.71	nd	43.70	6.40	2.14	53.32
F2	0.11	10.16	0.51	3.27	31.89	0.01	0.48	46.43
F3	0.01	0.04	nd	nd	33.01	nd	16.60	49.62
F4	0.83	1.00	0.01	0.09	58.22	0.05	7.10	67.30
F5	0.41	0.24	nd	0.53	29.83	nd	32.20	63.19
Average	0.30	2.34	0.25	0.78	39.33	1.29	11.70	55.97
Standard deviation	0.33	4.38	0.34	1.41	11.85	2.85	13.06	8.93

nd = under the detection limit.

Table 2.3. Group 2: chemical composition of sludges generated from zinc plating processes

Sample	Content in metals (wt. %)							
	Al	Ca	Cu	Cr	Fe	Ni	Zn	Total
Z1	1.59	6.56	0.08	2.38	6.34	nd	18.3	35.25
Z2	0.08	0.33	0.03	0.78	6.93	0.01	37.5	45.66
Z3	0.53	0.63	0.04	6.66	0.07	0.22	15.4	23.55
Z4	0.13	2.19	0.64	2.17	12.3	0.57	26.9	44.9
Z5	0.05	0.22	0.67	3.02	13.4	1.12	32.4	50.88
Z6	0.18	0.22	3.62	14.5	4.83	2.04	47.4	72.79
Z7	0.16	0.13	nd	0.08	2.8	nd	20.6	23.77
Average	0.39	1.47	0.73	4.23	6.67	0.57	28.36	42.40
Standard deviation	0.55	2.36	1.3	4.99	4.81	0.77	11.50	17.16

nd = under the detection limit.

Table 2.4. Group 3: chemical composition of sludges generated from aluminum anodizing processes

Sample	Content in metals (wt. %)							Total
	Al	Ca	Cu	Cr	Fe	Ni	Zn	
A1	40.20	1.07	nd	0.49	0.21	0.08	0.01	42.06
A2	38.40	1.22	0.06	0.07	0.07	0.06	0.02	39.90
A3	29.50	0.01	nd	3.23	0.25	0.13	0.01	33.13
A4	30.50	0.16	nd	0.87	0.18	0.01	0.02	31.74
A5	38.50	0.99	nd	0.47	0.09	nd	0.01	40.06
Average	35.42	0.69	0.01	1.03	0.16	0.06	0.01	37.38
Standard deviation	5.01	0.56	0.03	1.26	0.08	0.05	0.01	4.62

nd = under the detection limit.

Group 3 – Table 2.4 describes the composition of samples arising from aluminum anodizing processes, this element being distinctively dominant (Al = 35.42%). When compared to the previous groups, the total amount of analyzed metals (37.38%) is lower.

Group 4 – Table 2.5 shows the composition of plating sludges originated from mixed/complex surface treatments. This is the largest group and obviously the one involving samples with a vast range of chemical composition differences (maximal heterogeneity).

Tables 2.6 and 2.7 present data related with samples separated in two groups (5 and 6) based on the relative amount of the main components, chromium and nickel. Group 5 comprises samples that are typical of the chromium plating treatments, showing high Cr contents (Cr = 22.76%), whereas Ni exists in low amounts (Ni = 12.16%). On the contrary, group 6 includes samples that are poorer in Cr (Cr = 5.28%) and richer in Ni (Ni = 36.03%), typically generated from nickel plating processes.

Table 2.5. Group 4: chemical composition of sludges generated from complex/mixed surface treatments

Sample	Content in metals (wt. %)							Total
	Al	Ca	Cu	Cr	Fe	Ni	Zn	
M1	0.19	16.40	nd	1.16	2.25	0.28	2.14	22.42
M2	0.11	18.60	0.01	1.97	0.42	1.36	0.22	22.69
M3	0.12	0.49	0.01	0.04	1.25	0.03	0.01	1.95
M4	0.79	0.83	0.04	17.50	1.50	0.21	0.06	20.93
M5	nd	0.02	nd	nd	1.76	nd	2.95	4.73
M6	0.25	9.33	5.85	2.75	0.34	5.84	4.01	28.37
M7	8.94	0.04	0.32	0.38	1.27	0.02	0.02	10.99
M8	0.20	0.40	0.51	6.37	8.13	1.01	2.58	19.20
M9	0.08	12.50	0.02	nd	3.62	nd	0.08	16.30
M10	18.90	0.35	1.11	15.30	0.85	1.33	0.60	38.44
M11	0.10	0.06	2.24	11.50	0.25	2.64	1.54	18.33
M12	0.19	11.60	nd	0.03	0.91	0.01	0.21	12.95
M13	15.40	8.69	nd	2.93	0.41	nd	0.25	27.68
M14	0.76	0.40	3.59	nd	0.21	0.01	1.91	6.88
Average	3.29	5.69	0.98	4.28	1.66	0.91	1.18	17.99
Standard deviation	6.35	6.88	1.76	6.06	2.09	1.62	1.33	10.02

nd = under the detection limit.

Table 2.6. Group 5: chemical composition of sludges coming from chromium plating processes

Sample	Content in metals (wt. %)							
	Al	Ca	Cu	Cr	Fe	Ni	Zn	Total
C1	0.11	1.12	2.59	23.10	0.40	10.60	1.91	39.83
C2	0.14	0.80	4.24	25.40	0.46	6.86	2.42	40.32
C3	4.37	0.05	3.42	22.20	0.59	9.14	0.29	40.06
C4	0.28	0.29	15.80	22.50	1.85	11.80	8.13	60.65
C5	0.06	0.11	5.75	20.60	1.83	22.40	3.01	53.76
Average	0.99	0.47	6.36	22.76	1.03	12.16	3.15	46.92
Standard deviation	1.89	0.47	5.40	1.74	0.75	6.01	2.96	9.70

In all groups a reasonable high dispersion in the content of certain elements is observed. For some extreme cases, the standard deviation supplanted the average value. These results confirm that significant differences exist among productive processes expectably similar and/or among effluent treatment flowsheets.

Table 2.7. Group 6: chemical composition of sludges generated from nickel plating

Sample	Content in metals (wt. %)							Total
	Al	Ca	Cu	Cr	Fe	Ni	Zn	
N1	0.18	0.05	1.99	3.65	0.90	41.00	1.57	49.34
N2	0.94	0.10	1.51	5.42	0.17	40.90	1.01	50.05
N3	0.08	0.09	8.24	6.75	0.22	26.20	4.21	45.79
Average	0.40	0.08	3.91	5.27	0.43	36.03	2.26	48.39
Standard deviation	0.47	0.03	3.75	1.56	0.41	8.52	1.71	2.28

2.1.3 Physical Characterization

Some relevant characteristics of plating sludge samples were estimated, namely the moisture content (%), average particle size (μm), density ($g.cm^{-3}$) and specific surface area of dried and fired sludge samples ($m^2.g^{-1}$).

Moisture content (M) was determined in a special balance (Adam Co.—model AMB 310). Particle size distribution (PS) of dried and previously aqueous dispersed samples was obtained by a laser technique (Beckman Coulter LS 230). Density (ρ) was determined using a helium picnometer (Quantachrome Multipicnometer). The specific surface area (SSA) of dried and calcined samples was determined by the BET method (Gemini II 2370).

Table 2.8 shows the most relevant physical characteristics of plating sludge samples, grouped according to their chemical composition.

Table 2.8. The most relevant physical characteristics of plating sludge samples, grouped according to their chemical composition

Group	Sample	M (wt. %)	PS (μm)	ρ ($g.cm^{-3}$)	SSA ($m^2.g^{-1}$)	
					dried sludge	calcined sludge
	F1	90.94	228	2.84	150.71	0.02
	F2	70.81	526	2.65	48.88	1.42
1	F3	33.99	3	2.44	51.32	0.39
	F4	42.48	4	3.32	68.86	2.25
	F5	36.42	5	3.14	85.67	0.56
	Z1	74.76	428	2.61	43.81	2.38
2	Z2	70.73	263	2.29	65.71	0.22
	Z3	71.11	452	2.53	39.01	0.01
	Z4	68.11	280	2.58	73.94	0.06
	Z5	75.92	415	2.69	87.61	0.10
	Z6	62.31	438	2.82	79.27	0.21
	Z7	41.35	364	3.07	82.62	1.98
	A1	82.64	406	2.12	90.77	9.22
	A2	80.26	383	2.22	85.09	2.36

Table 2.8. The most relevant physical characteristics of plating sludge samples, grouped according to their chemical composition (Continued)

Group	Sample	M (wt. %)	PS (μm)	ρ ($g.cm^{-3}$)	SSA ($m^2.g^{-1}$)	
					dried sludge	calcined sludge
3	A3	79.85	410	2.17	92.44	1.42
	A4	80.47	480	2.21	153.51	53.06
	A5	79.11	400	2.14	73.82	1.74
	M1	45.61	117	2.51	27.74	0.01
	M2	44.72	120	2.50	14.09	1.91
	M3	55.99	34	1.55	32.87	8.45
	M4	49.55	376	2.49	82.86	16.51
	M5	23.31	58	1.47	158.17	0.07
4	M6	60.90	206	2.49	41.29	0.93
	M7	60.52	32	1.98	29.62	4.98
	M8	53.77	337	2.40	51.72	1.38
	M9	53.22	389	1.72	50.27	2.83
	M10	83.19	407	2.27	49.81	29.23
	M11	35.09	122	2.44	17.40	0.97
	M12	75.16	24	2.39	43.88	0.17
	M13	77.74	397	2.22	106.53	0.98
	M14	59.83	330	2.00	17.71	0.22
	C1	70.94	286	2.67	38.88	9.18
	C2	74.89	262	2.47	54.06	3.15
5	C3	79.48	268	2.54	18.68	7.81
	C4	69.93	448	2.62	81.66	1.54
	C5	89.31	288	2.73	23.84	4.62
	N1	87.85	63	2.24	32.19	1.49
6	N2	80.02	265	2.68	12.43	3.67
	N3	89.88	213	2.52	11.95	0.92

M = moisture (wt. %); PS = particle size = 90% (vol.%) of particles are under the value shown (in μm); ρ = density ($g.cm^{-3}$); SSA = specific surface area ($m^2.g^{-1}$).

2.1.4 Physical and Chemical Characteristics of the Grouped Sludges

It is important to identify what properties may distinguish samples from different group(s). Such identification is shown in Table 2.9.

2.1.4.1 Total Amount of Metals

The maximum content in metals is observed in samples from group 1, with an average value of 55.97%, mainly due to Fe and Zn contributions. On the contrary, sludge samples of group 4 show the lowest total metal content (average = 17.99%) and a high dispersion (standard deviation = 10.02%). This dispersion seems to characterize installations that carry out mixed/complex surface treatments.

Group 1 is the one composed by samples containing maximum average metal contents; however, the economically valuable species (such as Ni, Cu and Zn, according to the description in section 1.1 in this chapter) are mainly concentrated on samples from groups 5 and 6.

Table 2.9. Physical and chemical characteristics of the sludges, detached by sample groups

Characteristic detached	Group(s)
Higher average total content (wt.%) in metals	1
Lower average total content (wt.%) in metals	4
Higher average moisture value (wt.%)	6
Lower average moisture value (wt.%)	1 and 4
Lower particle average size (vol.%)	1
Higher particle average size (vol.%)	3
Higher average density value	1
Lower average density value	3 and 4
Higher average SSA value (m^2/g) – dried samples	3
Lower average SSA value (m^2/g) – dried samples	6
Higher average SSA value (m^2/g) – calcined samples	3
Lower average SSA value (m^2/g) – calcined samples	1 and 2
Higher homogeneity of chemical and physical characteristics	3

2.1.4.2 Moisture (M)

As expected, almost all sludges have very high moisture levels, over 60%. Group 6, which comprises samples originated from nickel plating shops, shows the highest average value (~86 wt. %), whereas samples pertaining to groups 1 and 4 present the lowest moisture contents. However, these both latter groups show moisture average values over 50%, which still represents a great amount of water being disposed together with sludges.

2.1.4.3 Particle Sizes (PS)

Approximately 97% of sludge samples are composed by particles whose size is under 600 μm (90% in volume are under 500 μm), confirming the high fineness level of this kind of waste. Group 1 has the finest sludge samples (certain samples with particle sizes under 5 μm), probably as a result of mechanical grinding and polishing techniques applied prior to the surface treatment of the workpiece. Sludge samples of group 1 show a maximum dispersion (with a standard deviation of 230 μm and an average of 153 μm). By contrast, samples from group 3 show the highest average particle size (416 μm) but a lower dispersion (standard deviation = 37 μm).

2.1.4.4 Density (ρ)

The sludges present roughly similar density values (average = 2.43; standard deviation = 0.38). Since sludges of group 1 show the maximum levels of metallic species, their density is obviously higher (average = 2.88). On the contrary, the lowest density values are shown by sludges from groups 3 and 4 (average = 2.17). It can then be concluded that the powder density of the dried material directly reflects the ratio between metallic and non-metallic components.

2.1.4.5 Specific Surface Area (SSA)

In general, a great dispersion of specific surface area (SSA) values was observed, both in dried and fired samples. In all cases, samples from group 3 show the maximum values (average = 99 $m^2.g^{-1}$ and 14 $m^2.g^{-1}$ for the dried and calcined samples, respectively). Dried samples from group 6 show the lowest values (average = 19 $m^2.g^{-1}$), while calcined samples having minimum SSA values pertain to groups 1 and 2 (average of about 1 $m^2.g^{-1}$).

2.1.4.6 Homogeneity Level of Chemical and Physical Characteristics

Group 3 comprises sludges that seem to be more homogeneous, that is, those showing a low dispersion of the measured chemical and physical characteristics. It then seems that the aluminum anodizing processes are more reproducible or are conducted under strictly controlled operating conditions.

2.2 Plating Sludge Selected for the Study

A sample from group 6 was selected, since it contains maximum levels of the highest valuable metals, namely nickel.

2.2.1 Collection and Characterization of a Plating Sludge Sample

The sample was randomly collected from several containers stored in the company. After a careful mixing, a single batch of about 60 kg was obtained.

Before drying, small aliquots were extracted to determine their pH according to U.S. EPA 9045 C method [13].

The physical characterization of this sludge was made as already described in section 2.1.3.

After drying at 105 °C, the sludge was milled (with balls in a porcelain jar) and then sieved at 1 mm. After proper homogenization, the chemical composition of several small samples was estimated by XRF (X Unique II Philips). Table 2.10 shows these results, and confirms their homogeneity.

Table 2.10. Chemical composition of the selected sludge (ten dried samples, wt. %)

Sample	Al_2O_3	CaO	Cu	Cr	Fe	Ni	Zn	S	Cl	Pb	P_2O_5	SiO_2
1	2.62	0.34	4.14	7.01	1.55	9.45	2.94	5.67	0.13	0.36	5.72	60.08
2	2.54	0.34	4.18	7.05	1.54	9.41	2.98	5.72	0.13	0.37	5.89	59.86
3	2.60	0.37	4.22	7.05	1.51	9.49	2.98	5.70	0.10	0.36	6.41	59.21
4	2.50	0.38	4.22	6.96	1.58	9.41	2.98	5.72	0.13	0.36	5.68	60.08
5	2.50	0.34	4.18	7.05	1.54	9.45	2.94	5.72	0.13	0.36	5.94	59.86
6	2.54	0.36	4.18	7.01	1.53	9.45	2.98	5.84	0.13	0.36	5.99	59.64
7	2.34	0.34	4.22	7.05	1.47	9.57	2.98	5.72	0.12	0.36	6.18	59.64
8	2.42	0.34	4.18	7.05	1.50	9.45	2.98	5.65	0.11	0.36	5.68	60.29
9	2.52	0.34	4.18	7.01	1.53	9.49	2.94	5.72	0.13	0.36	5.93	59.86
10	2.50	0.35	4.18	6.96	1.55	9.45	2.94	5.74	0.11	0.36	5.78	60.08
Average	2.51	0.35	4.19	7.02	1.53	9.46	2.96	5.72	0.12	0.36	5.92	59.86
Standard deviation	0.081	0.014	0.025	0.037	0.032	0.047	0.021	0.051	0.011	0.003	1.17	0.306

Table 2.11 summarizes the most relevant physical and chemical characteristics of the plating sludge selected for the recovery study. As expected, the relative amounts of valuable

metals (Ni = 9.46%; Cu = 4.19% and Zn = 2.96%) are interesting. However, silica is the most abundant species, representing almost 60% of the total weight. Silica may be originated from dusts of the foundry, as well as from contamination of silicates used in pretreatment steps such as mechanical polishing and pickling.

Table 2.11. Main physical and chemical characteristics of the selected sludge

M	[%]	87.6
ρ	[g.cm^{-3}]	2.24
SSA	[m^2.g^{-1}] dried sludge, ground to <1mm	19.73
SSA	[m^2.g^{-1}] calcined sludge, ground to <1mm	0.77
pH		6.8

Chemical composition of the dried sludge with grain size <1mm

	Cr	**Cu**	**Fe**	**Ni**	**Pb**	**Zn**	**SiO_2**	**Al, Cl, P, S,...**
(wt.%)	7.02	4.19	1.53	9.46	0.36	2.96	59.86	Rest

3. LEACHING

In any hydrometallurgical process, the first step comprising the dissolution of the constituents is probably the most important. This process is called leaching, a solid-liquid mass transfer process. Leaching may be carried out at ambient conditions or at elevated temperatures and/or under pressure, making use of many kinds of solvents, including water. Processing conditions will depend on the chemical reactions taking place during the leaching step.

Different mineral acids, namely sulfuric, nitric and hydrochloric acid, alone or combined with suitable oxidizing agents, may dissolve the majority of metals and other materials present in solid fractions. However, leaching stages carried out with alkaline agents are also common in hydrometallurgy. Important examples of alkaline reagents are ammonia, sodium hydroxide and sodium carbonate solutions [14].

Since both sulfuric acid and ammoniacal leaching (this latter involving a mixture of ammonia and ammonium salt) are reagents that may offer a good alternative to the dissolution of interesting metal elements when tested and optimized under certain conditions, these two leaching media were tested herein, in order to choose the most convenient for the plating sludge. The subsequent steps of separation, extraction and recovery of metals were also considered when this choice was carried out.

3.1 Leaching Experiments

Leaching experiments with sulfuric acid and ammoniacal solutions were carried out at room temperature in semi-closed, cylindrical 2000 mL glass beakers for 24 hours. A stirring speed of a 700 r.p.m., with a magnetic stirrer, was set up to ensure the full suspension of particles and to improve the liquid-solid contact. Pure grade chemicals were used.

Based on the knowledge of the plating sludge composition (see section 2.2) and assuming that the main reactions involving Cr-, Cu-, Fe-, Ni- and Zn-hydroxides with sulfuric acid are of the type:

$$Ni(OH)_2 + H_2SO_4 \rightarrow NiSO_4 + 2H_2O$$

a simple calculation of the stoichiometric amount of sulfuric acid needed to neutralize the hydroxides was done aiming to obtain a 20 $g.L^{-1}$ Ni as the final concentration in the leaching solution (a slight margin of confidence of about 10% was assumed). As previously mentioned (section 1.1), the sludge is a very complex multi-component-system consisting of a few major components and a lot of minor/trace constituents. Therefore, it should not be forgotten that several interactions eventually occurring between ions in solution had to be neglected. On this basis, a ratio of about 100 g H_2SO_4 per 100 g of dried sludge was calculated. This determination was used to estimate the sulfuric acid consumption and the L/S ratio on the leaching experiments.

In the ammoniacal leaching tests, the L/S ratio was kept constant at 10:1. Aqueous solutions were prepared by varying the dissolution of known amounts of reagent grade NH_3 and $(NH_4)_2CO_3$ in distilled water, in accordance with data mentioned in literature [15-20].

During the leaching tests, samples were withdrawn at regular intervals using a pipette (10 mL) and, after filtration, the resulting solution was analyzed for the metal elements by atomic absorption spectrometry (AAS). The remaining cake was analyzed by X-ray fluorescence spectrometry (XRF, Philips X'UNIQUE II).

The conditions adopted for each set of experiments are shown in Tables 3.1 and 3.2 {I} for sulfuric acid and ammoniacal leaching, respectively.

Table 3.1. Sulfuric acid leaching conditions adopted {I}

Condition	Test number					
	1	2	3	4	5	6
H_2SO_4 [$g.L^{-1}$]	100	100	100	200	200	200
L/S ratio [$g.L^{-1}$]	5: 1	10: 1	20: 1	5: 1	10: 1	20: 1

Table 3.2. Ammoniacal leaching conditions followed in the different tests {I}

Reagent	Test number					
	1	2	3	4	5	6
Ammonia [$g.L^{-1}$]	0	50	100	100	150	200
Amm. carbonate [$g.L^{-1}$]	100	100	0	100	150	200
pH	8.54±0.64	9.63±0.13	11.32±0.41	9.61±0.17	9.75±0.15	9.85±0.19

Results given in Tables 3.3 and 3.4 {I} show the concentrations ($g.L^{-1}$), weight percentages of each element dissolved, and also the [Ni+Cu+Zn]/[Cr+Fe+Pb] ratio, calculated from the metal element concentrations, for a set of six experiments conducted under different conditions, as detailed in Tables 3.1 and 3.2. This latter ratio denotes the selectivity achieved for the most economical value metals in relation to those considered as impurities. The leaching efficiency is defined as the ratio of the leached amount of a given element over the total amount of this element in the plating sludge sample. Moreover, a mass balance was calculated taking into account the fraction of unreacted sludge, in order to confirm the percentages of metal elements of interest leached at the end of the tests.

Table 3.3. Effect of different leaching conditions for the dissolution of Cu, Cr, Fe, Ni, Pb and Zn by sulfuric acid {I}

Leaching test	Leach metal ions [g.L^{-1}] (%)						[Ni+Cu+Zn]/ [Cr+Fe+Pb]
	Cu	Cr	Fe	Ni	Pb	Zn	
1	7.45 (88.6)	13.83 (98.2)	1.13 (36.8)	18.59 (98.0)	0.04 (5.6)	5.87 (99.2)	2.13
2	3.72 (88.5)	6.47 (91.9)	0.57 (37.1)	8.60 (90.8)	0.01 (3.0)	2.83 (95.5)	2.15
3	1.77 (84.1)	3.48 (98.8)	0.27 (35.8)	4.70 (99.2)	0.002 (2.5)	1.43 (96.3)	2.10
4	6.52 (77.6)	13.97 (99.2)	1.53 (50.0)	18.08 (95.4)	0.04 (4.9)	5.31 89.8)	1.93
5	3.28 (78.0)	5.97 (84.8)	0.70 (45.9)	8.80 (92.9)	0.02 (5.5)	2.83 (95.7)	2.23
6	1.66 (79.0)	2.90 (82.3)	0.34 (44.2)	4.19 (88.4)	0.01 (4.0)	1.31 (88.6)	2.21

Leaching conditions: t = 24 h, stirring speed = 700 rpm, values in brackets give the percentage of dissolution for each metal species.

Table 3.4. Effect of different leaching conditions for the dissolution of Cu, Cr, Ni, Pb and Zn in ammoniacal media {I}

Leaching test	Leach metal ions [g.L^{-1}] (%)					[Ni+Cu+Zn]/ [Cr+Pb]
	Cu	Cr	Ni	Pb	Zn	
1	1.24 (30.0)	0.07 (1.2)	0.98 (10.3)	nd	0.36 (23.0)	35.86
2	1.11 (28.0)	0.13 (1.8)	1.19 (20.6)	0.009 (25.0)	0.45 (18.0)	20.24
3	0.74 (18.6)	0.02 (0.5)	0.46 (5.6)	nd	0.13 (8.5)	83.44
4	1.88 (44.9)	0.55 (7.7)	2.94 (30.4)	0.003 (7.5)	0.89 (42.0)	10.22
5	2.59 (61.4)	1.09 (14.3)	4.25 (42.3)	0.009 (25.0)	1.21 (47.5)	7.29
6	2.90 (69.7)	1.52 (19.7)	4.75 (49.4)	0.004 (12.5)	1.37 (51.5)	5.92

Leaching conditions: t = 24 h, L/S ratio = 10:1, stirring speed = 700 rpm, nd = values under the detection limit, values in brackets give the percentage of dissolution for each metal species.

3.1.1 Effect of Reagent Concentration

Concerning the sulfuric acid leaching, as can be concluded from the analysis of Tables 3.1 and 3.3, the increase in the ionic strength of the solution beyond the calculated stoichiometric value (100 g.L^{-1} H_2SO_4) did not improve the concentration of the valuable metals in solution. In Table 3.3, the comparison of results obtained by using fixed L/S ratios (pairs {1,4}, {2,5} and {3,6}) shows that only the concentrations of the undesirable elements Fe and Pb significantly increased with the increase of the sulfuric acid concentration. In test 1 (see Table 3.3), in which the exact acid concentration required for the calculated stoichiometric amount was used, maximum simultaneous conversions of Cu (88.6%), Ni (98.0%) and Zn (99.2%) were reached.

By looking to the ammoniacal leaching, the comparison of results obtained on tests 1 and 3 (see Table 3.4) led to the conclusion that the presence of ammonium carbonate has a stronger influence than ammonia in the metal ion dissolution. As a whole, however, it can be considered that the presence and the combined effect of both reagents (ammonia and ammonium carbonate) are fundamental to assure an effective dissolution of the metal ion elements [14,21]. This assumption can be confirmed if Tables 3.2 and 3.4 are directly compared: for example, the extraction of Ni sharply rises by increasing the ionic strength of the solution containing both ammonia and ammonium carbonate, particularly for tests 4 to 6. Burkin [21] has established that Cu could not be dissolved in solutions of ammonium salts containing no free ammonia. However, the addition of ammonium ions increased the rate of solution in ammonia until a certain ammonia-ammonium ratio is reached; from that point, further ammonium addition had no apparent effect.

For the general conditions tested, one can say that sulfuric acid proved to be an efficient leaching medium for the most valuable metals present in the sludge, although with a quite low selectivity, whereas the inverse situation occurred when ammoniacal media was used instead.

3.1.2 Effects of Liquid/solid Ratio and pH

For the sulfuric acid leaching, the comparison of results obtained from the tests performed by using a fixed acid concentration, that is {1,2,3} or {4,5,6} groups - Table 3.3 - shows that no significant variations were detected on the dissolution of the valuable metals with the increase in the L/S ratio. This trend might indicate that equilibrium leaching conditions were reached even by using the lower L/S ratio. Accordingly, test 1 seems to be the best in conjugating a higher relative dissolution of valuable metals with a lower sulfuric acid consumption.

In the ammoniacal leaching experiments, the L/S ratio was kept constant in all cases. However, the pH control in all these experiments played a fundamental role. As can be inferred from the analysis of Tables 3.2 and 3.4, the dissolution of Cu, Ni and Zn was higher for average pH values closer to 10. In tests 1 and 3, pH values were somewhat distant from that value, and results were not so constant (high standard deviations were found). It is known that ammonium carbonate plays an important role in buffering the leaching solution, thus preventing a pH rise to the point where nickel and copper can precipitate as their hydroxides (pH beyond 10). In addition, NH_4^+ ions, supplied by the ammonium salt, can favour the occurrence of amine complex reactions by removing OH^- ions [14,22]. Therefore, a critical compromise between the ionic strength and pH should be found, in order to enhance the leaching efficiency of ammoniacal media for the valuable metals contained in the plating sludge.

3.1.3 Selectivity

The selectivity values obtained for the sulfuric acid leaching tests were quite close, however a relatively high [Ni+Cu+Zn]/[Cr+Fe+Pb] ratio (2.23) was achieved in test 5 – Table 3.3 - thus representing the maximum selectivity ratio attained for a final leach liquor. However, this test involved a higher sulfuric acid consumption and therefore, balancing the experimental conditions and overall results obtained for both tests 1 and 5, the latter was considered the lesser favourable.

In the ammoniacal leaching, as can be inferred from results of Table 3.4, Cu was the most readily leached component, whereas Cr seemed to be the main element rejected within the system, since the amount of Cr is relatively abundant in the feed sludge when compared to the Pb and Fe contents. Richardson [23] arguments that ammoniacal leach liquors reject iron and offer excellent kinetics with respect to Cu metal dissolution, and the results found in this work for the [Cu+Ni+Zn]/[Cr+Pb] ratio indicate that test 3 displayed in Table 3.4 assured optimum selectivity conditions. Nevertheless, very low concentrations of valuable metal elements were found in the leach solution obtained in test 3, when ammonia was used as a single reagent, and this is obviously an important drawback. Hence, the simultaneous use of ammonia and ammonium carbonate in similar proportions seemed to enhance the efficiency of the leaching process, despite the decrease of the system selectivity towards the valuable elements. In any case, if the balance between reagent consumption and effective metal ion dissolution efficiency is taken into account, test 6 corresponds to the optimum conditions, since the dissolution of Cu, Ni and Zn were the highest ones while the presence of Cr and Pb was minimum.

3.1.4 Leaching Time

The dependence of the metal ion dissolution by sulfuric acid (expressed as wt. % of their initial concentration) on the leaching time is shown in Figure 3.1 (a-f) {I} for different testing conditions (tests 1 to 6, respectively). It is possible to conclude that the contact/reaction time between the solid and liquid phases did not play a relevant role, 1 hour seeming to be enough for the systems to reach equilibrium. In fact, for almost all the metal elements of interest, their dissolution by sulfuric acid solutions is completed within about 1 hour or less.

In general, the extent of dissolution of Cr, Ni and Zn is comparable, whereas the dissolution of Cu is systematically lower. This latter result is quite obvious, as sulfuric acid is not the most suitable solvent to dissolve Cu. On the other hand, the effect of the leaching time on the dissolution of the metal ion elements (expressed as wt.% of the initial concentration) by ammoniacal media is shown in Figure 3.2 (a-f) {I} for different testing conditions (tests 1 to 6, respectively).

Some important trends are easily visible:

(i) firstly, the total dissolved amounts of all species were relatively low after 24 hours, in contrast with the results obtained with sulfuric acid;
(ii) secondly, with the ammonia/ammonium carbonate leaching solutions, Cu was always the best dissolved metal, while Cr dissolution was systematically lower.

Bell et al. [20] studied the leaching of chalcopyrite with ammoniacal solutions and established that temperature, ammonium salt concentration and the combined effect of these two factors were the relevant and influent parameters on the extraction of Cu. In particular, the breakdown of ammonium hydroxide into water and ammonia was found to be a crucial step, since ammonia is known to be an active contributor to the Cu leaching reaction. Blanco et al. [24], in their work about leaching of Zn-based wastes by ammonia/ammonium carbonate solutions, also concluded that temperature should be as high as possible, but below the boiling point of the ammonia/ammonium carbonate solution used. Therefore, an increase of temperature for the systems under study would probably be a good option, in order to promote the leaching reaction kinetics and to increase the extracted amounts of the metallic elements of interest. However, this route would involve extra costs with higher energy consumptions, currently an unwise way.

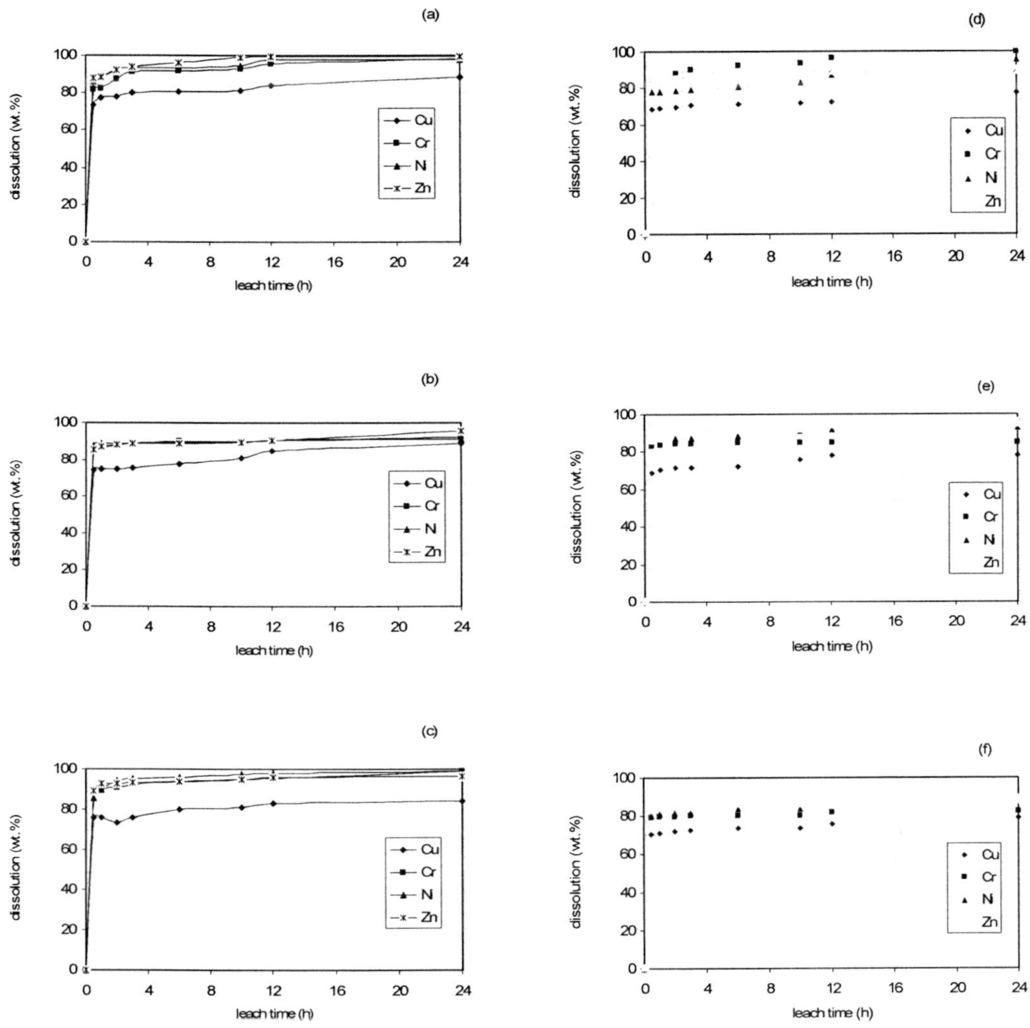

Figure 3.1. Effect of leaching time on the dissolution of the major metal elements contained in a plating sludge by sulfuric acid: (a) L/S ratio 5:1; [H_2SO_4] 100 g.L^{-1}; (b) L/S ratio 10:1; [H_2SO_4] 100 g.L^{-1}; (c) L/S ratio 20:1; [H_2SO_4] 100 g.L^{-1}; (d) L/S ratio 5:1; [H_2SO_4] 200 g.L^{-1}; (e) L/S ratio 10:1; [H_2SO_4] 200 g.L^{-1}; (f) L/S ratio 20:1; [H_2SO_4] 200 g.L^{-1} {I}.

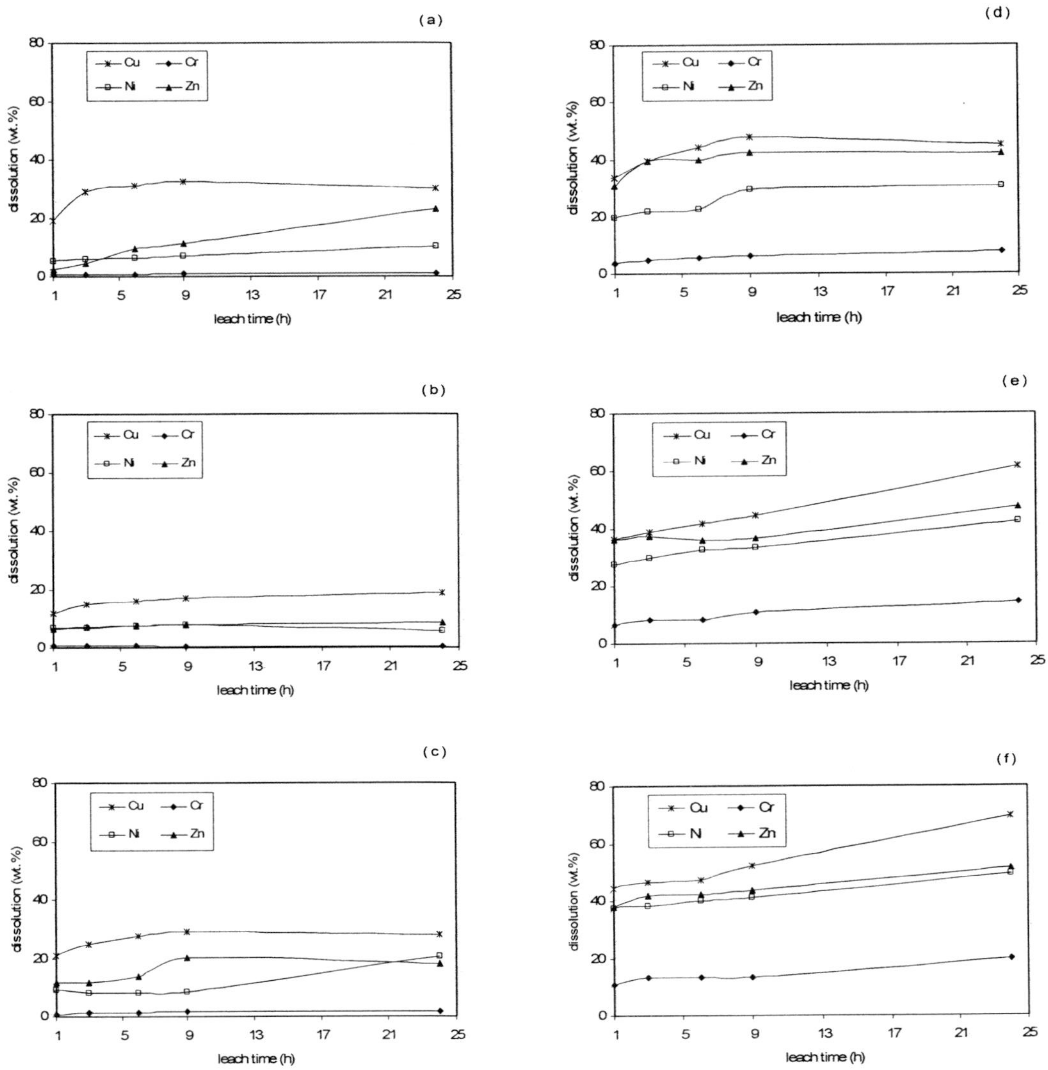

Figure 3.2. Effect of leaching time on the dissolution of the major metal elements contained in a plating sludge by ammoniacal media: (a) L/S ratio 10:1; $[NH_3]$ 0 $g.L^{-1}$; $[(NH_4)_2CO_3]$ 100 $g.L^{-1}$; (b) L/S ratio 10:1; $[NH_3]$ 50 $g.L^{-1}$; $[(NH_4)_2CO_3]$ 100 $g.L^{-1}$; (c) L/S ratio 10:1; $[NH_3]$ 100 $g.L^{-1}$; $[(NH_4)_2CO_3]$ 0 $g.L^{-1}$; (d) L/S ratio 10:1; $[NH_3]$ 100 $g.L^{-1}$; $[(NH_4)_2CO_3]$ 100 $g.L^{-1}$; (e) L/S ratio 10:1; $[NH_3]$ 150 $g.L^{-1}$; $[(NH_4)_2CO_3]$ 150 $g.L^{-1}$; (f) L/S ratio 10:1; $[NH_3]$ 200 $g.L^{-1}$; $[(NH_4)_2CO_3]$ 200 $g.L^{-1}$ {I}.

3.2 Evaluation of the Remaining Solid Fraction from the Sulfuric Acid Leaching

From the leaching process a remaining solid fraction has resulted, corresponding to an average of 60% of the plating sludge initial mass and containing significant amounts of metal elements in its composition. Since sulfuric acid leaching was selected as the most suitable to reach the predetermined objectives, the characterization of the solid fraction, here called cake 1, is shown in Table 3.5.

Table 3.5. Chemical composition of cake 1 (wt. %)

	Cr	Cu	Fe	Ni	Pb	Zn	SiO_2	Al, Ca, Cl, P, S,...
(wt.%)	0.91	1.90	2.21	3.38	2.04	0.25	79.20	Rest

As expected, the amount of metals in cake 1 is still very significant. Thus, it was decided to promote a second sulfuric acid leaching step in order to recover the remaining valuable metals. The remaining solid fraction was called cake 2.

Operational conditions were the same as those used in the first leaching step, in particular those reported for test 1 (see Table 3.1): acid concentration (H_2SO_4) of 100 $g.L^{-1}$ and S/L ratio 1:5.

Table 3.6 compares the concentration of some metals detected in liquors 1 and 2, that is, the liquors obtained after the first and the second leaching steps, respectively.

Table 3.6. Comparison between liquors 1 and 2, resulting from the leaching of the initial sludge and cake 1, respectively

Metal	Liquor 1 ($g.L^{-1}$)	Liquor 2 ($g.L^{-1}$)
Cr	10.70	0.92
Cu	5.50	2.15
Ni	18.0	3.35
Pb	0.04	0.05
Zn	2.70	0.03

The adoption of a second leaching step seems to be important, since there is further significant dissolution of copper and nickel.

Table 3.7 shows the cake 2 chemical composition, expressed as the wt. % of the solid fraction remaining from the sulfuric acid leaching of cake 1.

Table 3.7. Chemical composition of cake 2 (wt. %)

	Cr	Cu	Fe	Ni	Pb	Zn	SiO_2	Al, Ca, Cl, P, S,...
(wt.%)	0.71	0.62	1.29	0.93	2.21	0.09	80.19	Rest

It is concluded that only 8% of the initial mass of cake 1 was dissolved after leaching, hence cake 2 is 92% of cake 1. Prosser [25], in its extensive review about leaching (quoting several authors), points out that, sometimes, some portions of elements to be dissolved may, at some point, assume a refractory behavior.

4. Treatment of the Leach Liquor

From the preceding section, a metal-rich (Ni, Cu, Zn, and Cr) liquor was obtained. The present section describes some attempts to find optimum extraction conditions for some of those valuable metals from the leach liquor, namely nickel, copper and zinc.

Preliminary studies on the use of solvent extraction techniques have shown the unfeasibility of this technique to produce solutions with high purity metal levels, due to the complexity of the involved liquors in terms of their chemical composition. As will be soon understood, hydrometallurgical methods such as solvent extraction or crystallization should not be directly applied to aqueous solutions with chemical compositions so complex as those obtained immediately after the leaching step.

4.1 Solvent Extraction – First Approach

Currently, solvent extraction is a very well-established process in hydrometallurgy. It is used for the hydrometallurgical processing of copper, nickel, cobalt, zinc, uranium, molybdenum, tungsten, vanadium, rare earths, zirconium, hafnium, niobium, tantalum, indium, gallium, germanium, the platinum group metals, boron, reprocessing nuclear fuels, purification of wet process phosphoric acid, nitric acid recovery, etc. [26]. Nowadays, a very large number of stable extractants is available for use in hydrometallurgy, showing excellent selectivity for particular metal ions, coupled with advances in the engineering and increasing demands for higher purity products and more environment-friendly processing routes [27].

A literature survey covering the most recent information on the above metal separations reveals that organophosphorus acids are often the chosen extractants. Hence, di-(2-ethylhexyl)-phosphoric acid (D2EHPA) allows an efficient recovery of Zn and Cu from sulfuric acid solutions of low pH (2.5–3.5). At this pH, if Fe is present in the leach solution, it would be totally extracted [28-30]. Additionally, D2EHPA has also been employed to recover Ni from aqueous sulfate solutions (a pH of about 6–7 is necessary) [28, 30-33]. Some works report an efficient separation of Ni from Co at lower pH values by the use of D2EHPA and Acorga M5640 (a hydroxyoxime derivative) [34], and D2EHPA and LIX 860 (a β-hydroxyaryloxime compound) [35]. Cr extraction from aqueous sulfuric acid solutions by D2EHPA has also been investigated [36], and it was found that a pH varying between 3 and 4 is necessary to reach good extraction efficiencies. For an effective Cr recovery, however, the use of a modifier and a stripping mixture composed by NaOH and H_2O_2 is compulsory [36], whereas for all the other metals, suitable H_2SO_4 or sulfate salt solutions have always resulted in good stripping efficiencies [28-33]. An additional work deserving a mention [37] refers to the use of D2EHPA or Ionquest 801 (a phosphonic acid derivative) to previously extract Fe and Zn from spent electroless nickel plating baths at pH 4 and then LIX 84I (a β-hydroxyaryloxime) to finally recover Ni from the raffinate (for pH values higher than 6).

Figure 4.1 {II} illustrates the extraction of some metal ions from sulfate solutions by D2EHPA and their dependence on the pH values involved.

Another organophosphorus acid also reported to be quite suitable for these separation purposes is Cyanex 272 (its active ingredient being bis-(2,4,4-trimethylpentyl)-phosphinic acid). Several works report the adequacy of Cyanex 272 to extract, for instance, Fe, Zn, Cr,

Cu and Ni from sulfuric and/or sulfate solutions (their average $pH_{0.5}$ being about 1.6, 3.0, 4.5, 4.6 and 7.05, respectively) [38-42]. A remarkable separation of Co from Ni using Cyanex 272 was also reported: for a pH of about 6, 99.4% Co was extracted together with only 3% of Ni [43].

Extraction isotherms involving Cyanex 272 and sulfate solutions, for the first-row transition metals, are shown in Figure 4.2 {III}.

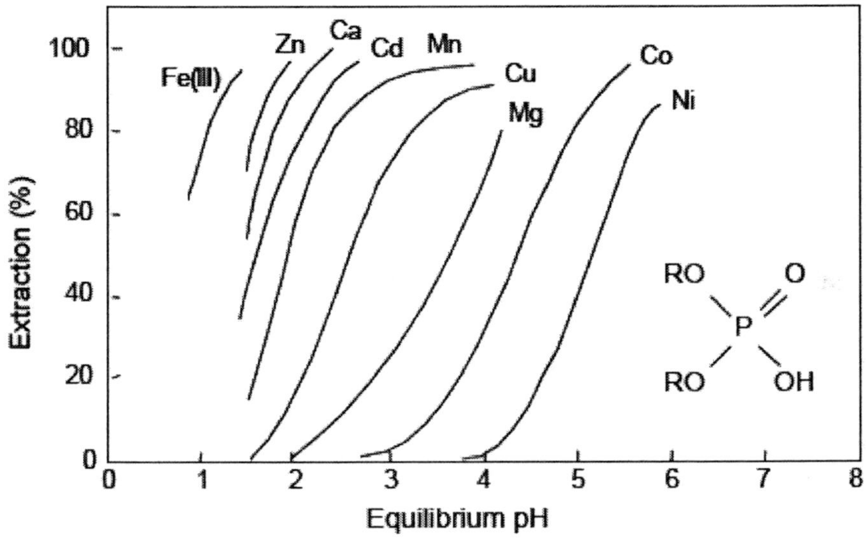

Figure 4.1. pH dependence of the extraction of some metal ions from sulfate solutions with D2EHPA {II}.

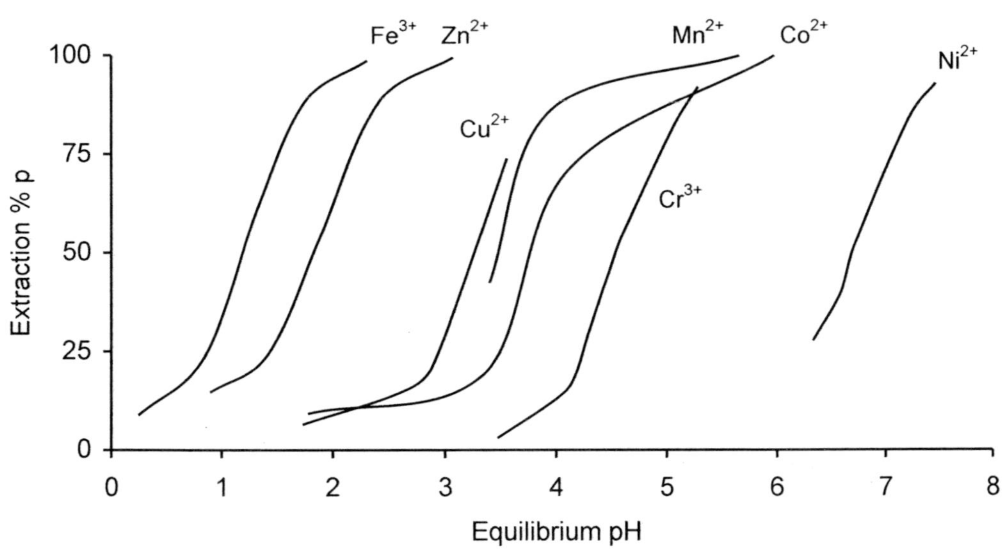

Figure 4.2. Extraction isotherms of some first-row transition metals, from sulfate solutions, with Cyanex 272 {III}.

Generally it can be pointed out that Cyanex 272 needs higher pH values than D2EHPA to extract the same metal ions, this behavior being explained by the fact that phosphinic acid derivatives are weaker acids than the phosphoric ones [28]. This acidity property is determinant if one takes into account that the metal ion extractions occur by their exchange with the acidic hydrogen atoms of the extractants, these latter being consequently released to the aqueous solution, decreasing therefore the equilibrium pH after extraction [28].

Thus, some liquors obtained from plating sludge leaching were subjected to some solvent extraction tests, using D2EHPA and Cyanex 272 as extraction reagents. The initial idea was the promotion of a sequential separation of Zn, Cu, Cr and Ni, as suggested by Figures 4.1 and 4.2, that could allow the recovery of those metal species under the metallic or compound form.

4.1.1 Leach Liquors as Feed Solutions in Solvent Extraction Tests

Liquor samples were previously neutralized through pH adjustment by the same kind of sludge (fresh, dried and ground <1 mm), used in the leaching stage, and filtered afterwards. The pH control of the aqueous samples, whenever necessary, was made with a WTW InoLab pH meter.

D2EHPA – Di-(2-ethylhexyl)-phosphoric acid (93% purity) was gently provided by Albright & Wilson Americas Inc., and used as received. Cyanex 272 – bis-(2,4,4-trimethylpentyl)- phosphinic acid, 80–87% purity – was kindly supplied by Cytec Canada Inc., and also used without further purification. Both extractants were diluted in kerosene (Fluka, purum, boiling point range 190–250°C, without aromatics content).

Unless otherwise stated, the following conditions were generally adopted in the solvent extraction tests, carried out in stoppered flasks:

(i) extractant concentrations in the organic phase of 1M, equal volumes of organic and aqueous solutions (A/O = 1);
(ii) phase dispersion achieved by a stirring speed of a 800 r.p.m. for a period of 60 minutes;
(iii) A one molar sulfuric acid solution was generally used to regenerate the loaded organic phases;
(iv) stripping conditions were similar to the ones adopted in the extraction stage.

After equilibration, the aqueous phases were always filtered to minimize organic contaminations. All the tests and analytical measurements were made at room temperature (25°C).

Analyses of the metal contents in the aqueous solutions before and after extraction, and also of the stripping ones, were carried out by flame atomic absorption spectrometry (AAS) on a "GBC 904 AA" model. Metal concentrations in the organic solutions were calculated by mass balance.

4.1.2 Results

4.1.2.1 Sample A

The feed liquor corresponding to sample A contained Cr = 13.5 $g.L^{-1}$, Cu = 9.1 $g.L^{-1}$, Ni = 40 $g.L^{-1}$ and Zn = 7.6 $g.L^{-1}$. Single-stage extraction tests at an initial pH 3.0, using different contact times with 1M D2EHPA, were carried out. Figure 4.3 shows the extraction percentages observed for the metal ions.

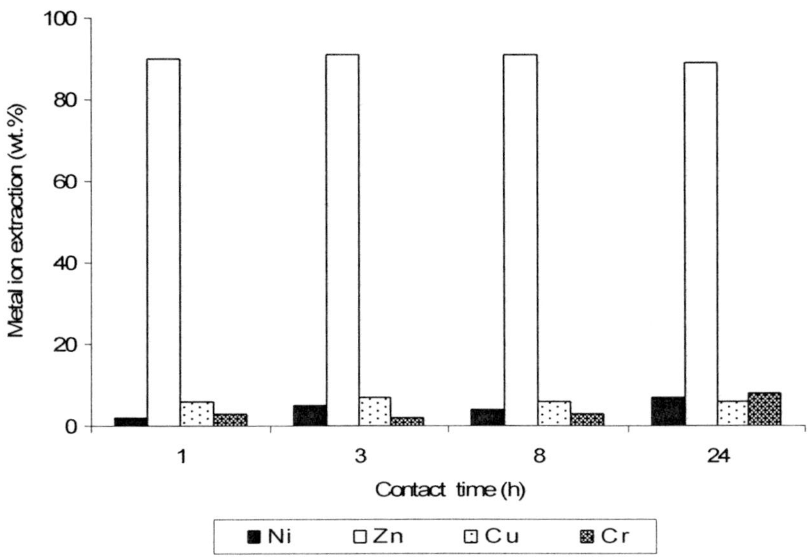

Figure 4.3. Different tests carried out with sample A and extraction percentages found for the metals of interest as a function of contact time.

Under the adopted experimental conditions, Figure 4.3 shows that only Zn was extracted in a significant extent. In addition, the contact time between both phases is not a relevant parameter, since 1 hour is clearly enough for the system to reach equilibrium. Also, based on literature data [28,36], it was decided to reduce the equilibration period of extraction and stripping stages to 30 minutes in the subsequent solvent extraction tests.

4.1.2.2 Sample B

The second feed liquor (sample B) contained Cr = 18 $g.L^{-1}$, Cu = 12 $g.L^{-1}$, Ni = 39 $g.L^{-1}$ and Zn = 10.5 $g.L^{-1}$, at an initial pH = 3.5. Keeping in mind the good results presented by D2EHPA to extract Zn from sample A, under the adopted experimental conditions, it was decided to test the solvent extraction behavior shown by both D2EHPA and Cyanex 272 towards sample B, since this latter organic reagent was developed to extract preferentially Co to Ni from sulfate media [38,40,42,43]. Several papers in literature report that Cyanex 272 exhibits high extraction efficiencies for several transition metals provided that some parameters be controlled (pH range, concentration, etc). Stripping of metals from loaded Cyanex 272 is usually very easy [28,30,38,40,42,43]. According to Lanagan and Ibana [39], the $pH_{0.5}$ value observed for Cr(III) extraction by Cyanex 272 is about 4.5, whereas the values shown by the same organic reagent towards Zn, Cu and Ni extractions are 2.0, 3.0 and ~6.5, respectively, if the data reported by Rickelton et al. [40] is taken into account. Hence, a separation of Zn and Cu from Ni by Cyanex 272, at pH 3.0–4.0, could be expected.

Hence, for sample B, single-extraction stages performed with 1M D2EHPA and 1M Cyanex 272, with an initial pH value for the leach solution lowered to 3.0 by adding H_2SO_4 1M, were carried out, followed by another extraction stage, with the same extractant, applied to each of the raffinates after their neutralization with CuO until pH 4.0 (the copper (I) oxide salt used for neutralization was processed from scraps in order to simulate economical and environmental- friendly conditions).

Loaded organic phases from the extraction tests at pH 3.0 and 4.0, for each extractant, were stripped by equal volumes of H_2SO_4 2M.

In Table 4.1, the percentages of total extraction and stripping of metal ions obtained for sample B, at pH 3.0 and 4.0, using D2EHPA and Cyanex 272, can be observed.

Table 4.1. Extraction of the interesting metals, using Cyanex 272 and D2EHPA, before and after pH adjustment

	Extraction (%) with Cyanex 272		Stripping (%) with H_2SO_4 2M		Extraction (%) with D2EHPA		Stripping (%) with H_2SO_4 2M	
	pH 3.0	pH 4.0	pH 3.0	pH 4.0	pH 3.0	pH 4.0	pH 3.0	pH 4.0
Cr	20	3	20	10	20	5	4	10
Cu	5	5	5	20	6	20	5	50
Ni	10	1	15	5	8	20	5	10
Zn	40	80	60	95	80	95	40	60

According to the results presented in Table 4.1, only Zn was again efficiently extracted by both D2EHPA and Cyanex 272, with more emphasis on the second extraction stage. Using Cyanex 272, there is a significant extraction enhancement with the increase of the pH value, reaching 80% (pH 4.0). In tests with D2EHPA, a higher Zn extraction than with Cyanex 272 was verified. The extraction efficiency increased by enhancing the pH value, this effect being more remarkable than with Cyanex 272, reaching 95% (pH 4.0).

In general, Cu extraction by Cyanex 272 was very incipient. With D2EHPA, Cu extraction was slightly higher, especially when pH was increased to 4.0. Anyway, for both extractants, the Cu extraction efficiency was always lower than those predicted in literature (see Figures 4.1 and 4.2).

The extraction of Ni with both extractants was also very low but, even so, a bit better than those predicted in literature.

Both D2EHPA and Cyanex 272 revealed similar incipient performances for Cr extraction at pH = 3.0, a further decrease at pH 4.0, and these were unexpected results if compared with those from literature (see Figure 4.2).

After extraction with Cyanex 272, the stripping step with H_2SO_4 2M was found to be very efficient for Zn recovery (95%). However, after utilization of D2EHPA, stripping of Zn was difficult and incomplete, its recovery level not surpassing 60%, even when stronger concentrated acid solutions were used. Cu stripping efficiency was always poorer than those found for Zn. In all cases, final solutions contained always some Ni and Cr contaminations.

Bearing in mind those observations, it was easy to conclude that performing sequential and selective metal separations from such a complex sulfate solution, through the use of a common solvent extraction technique involving Cyanex 272 or D2EHPA, would not be an easy task. As an alternative, it was decided to assay the reduction of the leach liquor chemical complexity through the removal of some of its metals by precipitation techniques.

4.2 Precipitation Techniques

In an attempt to produce leaching liquors that, after purification, would result in solutions with reasonable high purity metal grades, a sequential precipitation process was established.

Three precipitation methods were tested:

(i) leach liquor neutralization;
(ii) copper cementation;
(iii) chromium precipitation.

For each case, several trials were carried out in order to optimize the most determinant operational conditions. Afterwards, the best precipitation methods were sequentially applied.

4.2.1 Leach Liquor Neutralization

Neutralization was carried out to condition the leach liquor pH to the range where precipitation reactions of interesting metals are favoured.

Two alternative ways were tested: (i) addition of NaOH, (ii) addition of plating sludge (fresh, dried and ground <1 mm).

4.2.1.1 Neutralization by NaOH 2M

To 875 mL of leach liquor, small volumes of a NaOH 2M solution were gradually added, its pH being determined under agitation (100 r.p.m. with magnetic bar and agitator). Once reached the pre-established pH value, the NaOH addition was immediately interrupted and the liquor was filtered. At pH = 1 and 2 no precipitate was formed; the liquor was just filtered once its pH value reached 3.5, for which precipitation really occurred. The liquor was analyzed for metal contents determination by atomic absorption spectrometry (AAS - GBC 904 AA), while the precipitate was plentifully washed with distilled water, dried at 105 °C, disintegrated, homogenized, sampled and then analyzed by X-ray fluorescence spectrometry (XRF, Philips X'UNIQUE II). A summary of the tests carried out with addition of NaOH 2M is shown in Table 4.2.

Table 4.2. Neutralization tests carried out with NaOH 2M

pH		Precipitate mass (g)	NaOH 2M (L) consumption	NaOH mass added (g)	Liquor volume (L)
Initial	Final				0.87
0.4	1.0	no precipitate	0.28	22.8	1.16
1.0	2.0	no precipitate	0.09	7.2	1.25
2.0	3.5	8.7	0.08	6.4	1.33
Total		8.7	0.45	36.4	1.33

4.2.1.2 Neutralization by Sludge Addition

To 1000 mL of leach liquor, kept under permanent stirring, a certain amount of plating sludge was gradually added until the pH = 1. After pH stabilization, the mixture was filtered. The retained cake was plentifully washed with distilled water, dried at 105 °C, disintegrated, homogenized, sampled and analyzed by X-ray fluorescence spectrometry (XRF, Philips X'UNIQUE II). The same procedure was repeated by adding a new batch of fresh sludge, to reach different pH values (about 2, 3, etc). After all filtrations, the resulting liquor was analyzed by atomic absorption spectrometry for metal contents determination (AAS - GBC 904 AA). Table 4.3 summarizes the most relevant neutralization conditions.

Table 4.3. Neutralization tests carried out with sludge

pH		Added sludge (g)	Precipitate mass (g)	Dissolved sludge (g)	Liquor volume (L)
Initial	Final				1.0
0.3	1.3	259.15	175.30	83.85	0.76
1.6	2.5	70.34	62.11	8.23	0.67
2.6	3.0	77.56	76.12	1.44	0.59
3.0	3.6	131.77	128.58	3.19	0.46
Total		538.82	442.11	96.71	0.46

4.2.1.3 Results

Comparing the data of Tables 4.2 and 4.3, it is easy to estimate a NaOH consumption of ~27 $g.L^{-1}$ (used to rise pH values from 0.3 to 3.5), while about 100 g of sludge were required to treat each L of leach liquor. It should also be emphasized that this latter option is an alternative re-use for this industrial waste.

Concentrations of relevant elements in the liquor, precipitate or filtered residues, obtained before and after neutralization with NaOH 2M and sludge, are shown in Tables 4.4 and 4.5, respectively.

Table 4.4. Liquor and cake chemical compositions, before and after neutralization tests carried out with NaOH 2M

	pH		Elements ($g.L^{-1}$)				
			Cu	Cr	Na	Ni	Zn
Liquor	Initial	Final	5.5	10.7	0.56	18.0	2.7
	0.4	1.0	na	na	na	na	na
	1.0	2.0	na	na	na	na	na
	2.0	3.5	3.2	6.2	9.0	10.6	1.6
cake			Elements (wt. %)				
			Cu	Cr	Na	Ni	Zn
			11.18	30.21	-	2.63	1.65

The analyzed elements are combined with sulfur and phosphorus; the vestigial presence of other compounds containing iron, lead and aluminum was also detected; na – not analyzed.

In the precipitation process of metal hydroxides using hydrolyzing agents, hydrolysis reactions of metal ions (the most common example of ionic precipitation), is basically dependent on metal ion concentrations and solution pH values (controlled by acid or alkali additions) [44]. Some typical solubility product values (K_{sp}) of a few metal hydroxides are given in Table 4.6 (selected data from Jackson [19], who, in turn, made a selection from Smith and Martell).

Table 4.5. Liquor and cake chemical compositions, before and after neutralization tests carried out with sludge

	pH		Elements ($g.L^{-1}$)				
			Cu	Cr	Na	Ni	Zn
Liquor	initial	Final	5.4	9.3	0.5	18.9	2.7
	0.3	1.3	na	na	na	na	na
	1.3	2.5	na	na	na	na	na
	2.5	3.0	na	na	na	na	na
	3.0	3.6	9.1	13.5	2.0	40.0	7.6

	pH		Elements (wt.%)				
			Cu	Cr	Na	Ni	Zn
Cake	**initial**	**Final**	-	-	-	-	-
	0.3	1.3	4.85	8.79	-	3.27	1.06
	1.3	2.5	6.72	15.73	-	8.91	0.46
	2.5	3.0	8.63	17.30	-	11.56	0.46
	3.0	3.6	8.35	13.52	-	12.70	0.47

The analyzed elements are combined with sulfur and phosphorus; the filtered mass also contains a great proportion of silica; the vestigial presence of other compounds containing iron, lead and aluminum was also detected.

Table 4.6. Solubility product values (K_{sp}) of some metal hydroxides. Data selected from Jackson [19] who, in turn, made a selection from Smith and Martell)

Metal hydroxide	K_{sp}	Log K_{sp}
$Cr(OH)_3$	$[Cr^{3+}].[OH^-]^3 = 1.58 \times 10^{-30}$	-29.8
$Cu(OH)_2$	$[Cu^{2+}].[OH^-]^2 = 4.79 \times 10^{-20}$	-19.32
$Zn(OH)_2$	$[Zn^{2+}].[OH^-]^2 = 3.47 \times 10^{-17}$	-16.46
$Ni(OH)_2$	$[Ni^{2+}].[OH^-]^2 = 6.31 \times 10^{-16}$	-15.2

Assuming as a reference the final concentrations of metal ions from the neutralization tests (Tables 4.4 and 4.5) and K_{sp} data (Table 4.6) it is possible to calculate equilibrium pH values at which precipitation reactions are thermodynamically favourable, leading to the respective hydroxides, shown in Table 4.7. The most probable precipitation order is also presented.

Table 4.7. Values of equilibrium pH (assuming the final concentrations of metal ions as a reference, from neutralization tests with NaOH and with sludge). The most probable order of metal precipitations is also given

Metal element	Equilibrium pH (NaOH)	Equilibrium pH (sludge)
Cr	4.44	4.26
Cu	4.99	4.76
Zn	6.58	6.24
Ni	6.77	6.48
Most probable order of metal precipitations: Cr > Cu > Zn > Ni		

Table 4.8 summarizes the mass variation of the interesting metal species, while Figure 4.4 presents that change as a function of pH values when sludge was used as neutralizing agent.

Table 4.8. Mass variation of the interesting metal species in the neutralization tests

pH		Mass variation of the metal elements of interest* (wt. %)							
		NaOH addition				Sludge addition			
		Cu	Cr	Ni	Zn	Cu	Cr	Ni	Zn
Initial	Final								
0.3	1.3					57.42	39.38	130.77	285.79
1.3	2.5					27.53	-28.98	138.56	374.02
2.5	3.0					-53.35	-138.27	128.38	469.59
3.0	3.6					-180.43	-253.34	101.48	631.50
0.4	3.5	-20.21	-28.07	-1.45	-6.08				

* A negative variation (-) means precipitation.

As previously mentioned, when neutralization with NaOH was carried out, no precipitation occurred for pH values under 3.5. Once this value was reached, about 8.7 g of precipitate formed, essentially composed by Cr and Cu. By the end of the process, the mass of precipitate tended to decrease (see Table 4.8), according to the following descending order: Cr > Cu > Zn > Ni. This sequence coincides with the probable order of precipitation calculated from equilibrium pH data (see Table 4.7).

Results shown in Table 4.8 and Figure 4.4, corresponding to the neutralization with sludge, suggest that : initially, the dissolution of all the metal elements of interest occurred. However, at pH values higher than 2.5, precipitation of Cr became too high and, above pH 3.0, precipitation of Cu also raised significantly. During neutralization of the leach liquor, a continuous increase of Zn mass and a decrease of Ni mass happened, whereas the increase of Cr and Cu precipitate masses were noted, despite that, apparently, in accordance with the final concentrations of these latter metal ions (Table 4.5) it seemed that no increase of their mass occurred. Due to the inopportune metals precipitation and to the marked reduction of liquor volume as sludge was added, it was imperative to limit the final pH of the solution.

Looking now to the resulting cake, it was easy to conclude that it still contained significant contents of valuable species, mainly Ni and Cu. Hence, this sludge should return to the initial stage and be mixed with the fresh plating sludge to feed the leaching process.

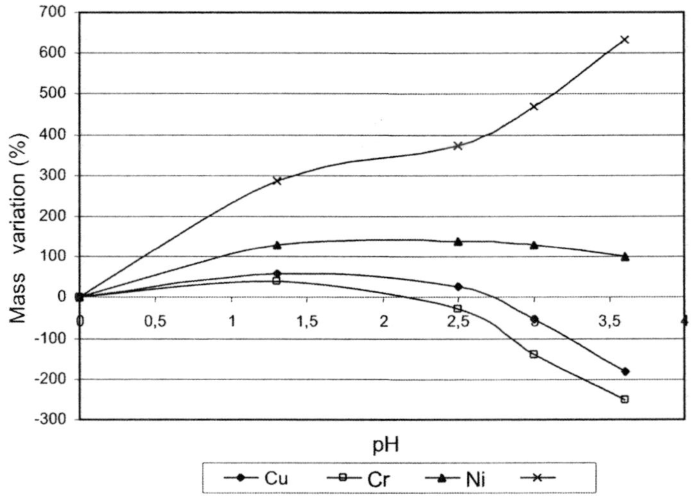

Figure 4.4. Mass percentage variation of the metal elements during the liquor neutralization with the addition of sludge, according to pH.

4.2.2 Copper Cementation

Cementation or metal displacement reactions are possibly one of the oldest hydrometallurgical techniques to recover metals, in elemental forms, from aqueous solutions [44]. This technique is still widely used nowadays in the recovery of some metals, as well as in the purification of process liquors and wastewater streams.

Regarding the application of a Cu cement, this can be used as a raw material in the aluminum industry, where it can be added as an alloy element.

Tests performed to precipitate and recover Cu from the leach liquor are described in this section. At the beginning, the effect of Fe and Zn as reducing agents for Cu was studied. The presence of other metals, besides Cu, on the precipitation process was also investigated, both in synthetic and in real solutions.

Then, in a second stage, after achieving the best precipitating agent for real solutions, the effect of some relevant parameters on the Cu cementation process, such as liquor pH and reaction time, was evaluated.

4.2.2.1 Tests with Synthetic Solutions

4.2.2.1.1 Effect of Fe and Zn as Reducing Agents

According to literature [45-55], Fe and Zn are amongst the most effective reducing agents of Cu. Hence, several tests using Zn and low carbon steel – "Fe" plates - were carried out in order to choose the most effective material. Zinc plates were previously polished, washed with a diluted solution of hydrochloric acid, subsequently with distilled water, dried with a napkin and then immediately used.

The solution was prepared from Cu-, Cr- and Ni sulfates, with the following concentrations: Cu^{2+} = 5.6 $g.L^{-1}$; Cr^{3+} = 7.3 $g.L^{-1}$ and Ni^{2+} = 20.3 $g.L^{-1}$, to simulate the composition of the leach liquor. Each Zn plate, with a high pure grade and exposed surface area of 2.56 cm^2, was dived into a 50 mL portion solution and accommodated in plastic flasks with a 100 mL capacity. Flasks labeled A to H were hermetically closed and immediately put under agitation (25 r.p.m.), allowing a permanent contact between the Zn plate and the solution at ambient temperature.

At regular intervals, each flask was withdrawn from the equipment following the labeling alphabetic order (A to H). After filtration, solutions were analyzed by AAS for Cu, Cr, Ni and Zn contents. Zinc plates were separated from the solutions, washed and dried, and finally scraped to remove the deposits. The scraped deposits and the filtered materials (the cement) were mixed and analyzed by XRF to determine their chemical composition.

Copper cementation using Fe (low carbon steel plate) was conducted under similar experimental conditions than those followed for Cu cementation by Zn plates. However, zinc sulfate was also added to the synthetic solution, whose general composition, regarding the interesting metals, was the following: Cu^{2+} = 5.6 gL^{-1}, Cr^{3+} = 7.3 gL^{-1}, Ni^{2+} = 20.3 gL^{-1} and Zn^{2+} = 6.2 gL^{-1}. Each low carbon steel plate had an exposed surface area of 3.24 cm^2.

The Cu cementation results achieved by the use of Zn and Fe plates are given in Tables 4.9 and 4.10, respectively. Curves showing the evolution of metal contents upon time are shown in Figures 4.5 and 4.6.

Table 4.9. Test identification, elapsed time, pH values and metal ion contents (in solutions and in cements), using Zn plates.

Test	Time	pH		Liquor ($g.L^{-1}$)				Cement (wt. %)			
	(min.)	Initial	Final	Cu	Cr	Ni	Zn	Cu	Cr	Ni	Zn
	0	-	-	5.60	7.30	20.30	0	-	-	-	-
A	5	1.71	1.73	4.70	6.45	20.31	1.22	91.36	2.65	1.68	4.32
B	10	1.71	1.77	4.03	6.41	19.88	2.12	91.55	1.52	0.45	6.49
C	15	1.71	1.79	2.44	6.65	20.05	4.11	94.15	0.73	0.27	4.85
D	30	1.71	1.93	0.92	6.53	19.62	5.88	96.06	0.78	0.23	2.93
E	60	1.71	2.17	1.01	6.54	19.63	7.27	93.76	1.67	0.42	4.14
F	120	1.71	2.39	0.29	6.81	19.82	7.35	95.34	1.10	0.40	3.16
G	180	1.71	2.51	0.11	6.83	19.92	7.65	89.10	3.35	0.58	6.98
H	240	1.71	2.49	0.08	6.97	20.43	8.26	94.54	1.60	0.27	3.59

Table 4.10. Test identification, elapsed time, pH values and metal ion contents (in solutions and in cements), using Fe plates

Test	Time	pH		Liquor ($g.L^{-1}$)					Cement (wt. %)				
	(min.)	Initial	Final	Cu	Cr	Fe	Ni	Zn	Cu	Cr	Fe	Ni	Zn
	0	-	-	5.60	7.30	0	20.30	6.20	-	-	-	-	-
A	5	1.85	1.84	4.51	7.31	0.11	20.35	6.21	61.62	5.49	22.72	4.58	5.59
B	10	1.85	1.85	4.13	7.29	0.43	20.41	6.21	72.97	4.39	12.33	4.92	5.40
C	15	1.85	1.87	4.03	7.21	0.34	20.19	6.19	73.98	5.85	8.45	6.33	5.39
D	30	1.85	1.91	3.98	7.19	0.79	19.89	6.1	77.77	4.74	9.34	3.87	4.28
E	60	1.85	2.05	3.85	7.23	1.53	20.34	6.23	80.46	3.96	9.59	2.77	3.22
F	120	1.85	2.02	3.56	7.18	1.89	20.31	6.34	76.51	4.92	10.29	4.61	3.67
G	180	1.85	2.07	2.23	7.10	3.13	20.35	6.13	74.42	5.71	12.82	4.53	2.52
H	240	1.85	2.17	0.34	7.15	4.32	20.22	6.10	81.93	3.31	11.73	1.85	1.18

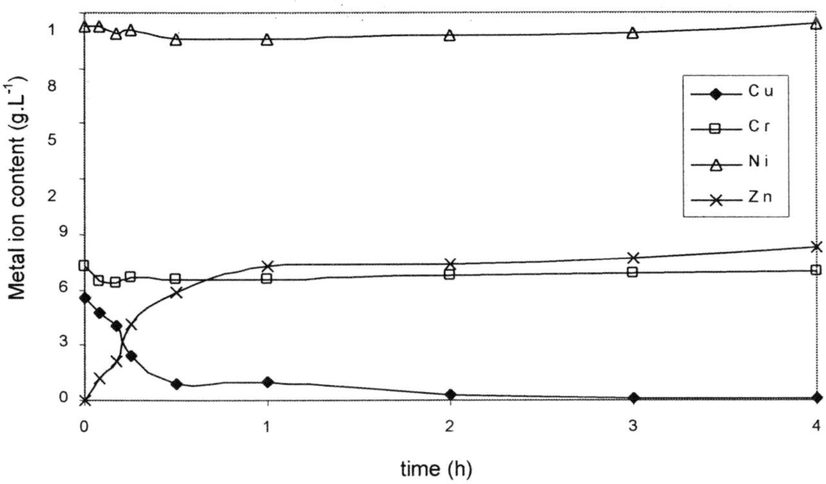

Figure 4.5. Concentration evolution of the metal contents in the liquor throughout Cu cementation time, using Zn plates.

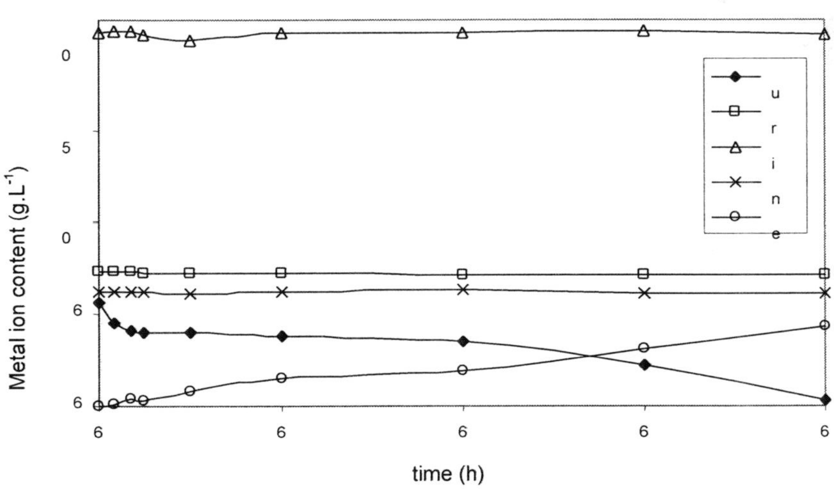

Figure 4.6. Concentration evolution of the metal contents in the liquor throughout Cu cementation time, using Fe plates.

Copper cementation with Zn as reducing agent was found to be more effective than when Fe was used. In addition, the relative amounts of Cu in the cement, after the displacement reactions with Fe, were significantly lower than those obtained with Zn, whereas the contents of others metals (Cr, Ni) in the cement were also higher. Furthermore, according to Figures 4.5 and 4.6, the intersection between the Cu-Zn curves occurs earlier than that occurring with the Cu-Fe curves, and this fact seems to confirm that displacement reactions between Cu and Zn are kinetically more favourable.

From a thermodinamic point of view, both Fe and Zn are able to precipitate Cu from aqueous solution, according to the following reactions [44]:

$$Cu^{2+} + Fe \rightarrow Cu + Fe^{2+} \; \Delta G^{o}_{298} = \text{-35,83 kCal/mol}_{Fe} \quad (4.1)$$

$$Cu^{2+} + Zn \rightarrow Cu + Zn^{2+} \; \Delta G^{o}_{298} = \text{-50,73 kCal/mol}_{Zn} \quad (4.2)$$

Comparing the ΔG^{o}_{298} values related to those reactions, Cu precipitation by Zn is predictably more favourable. It should also be reminded that Zn occupies a top position in the Electromotive Force (EMF) series when compared with Fe, the larger gap between the electrode potentials of the two half-cells (Zn-Cu) corresponding to a higher driving force for the Cu precipitation reaction [44].

4.2.2.1.2 Effect of the Presence of Other Metal Ions in Solution

To evaluate the effect of other metal ions, besides Cu, in solution, Cr and Ni were also added. Hence, some solutions were prepared in such a way that, in addition to 5 $g.L^{-1}$ Cu (by using copper sulfate), Cr and Ni contents were adjusted by adding their respective sulfate salts.

Since Zn proved to be more effective than Fe (see sub-section 4.2.2.1.1), the tests were just conducted using Zn plates only.

Each cementation test involved 50 mL of a synthetic solution accommodated in a 100 mL plastic flask, and previously prepared Zn plates were used. As referred to before, Zn plates had an exposed surface area of 2.56 cm^2. Once hermetically closed, flasks were immediately put under agitation (25 r.p.m.) for 24 hours.

At the end, Zn plates were separated from the solutions, then washed (with distilled water), dried and finally scraped to remove the deposits. The scraped deposits and the filtered material were mixed and then analyzed by XRF. After filtration, solutions were submitted to AAS analysis for Cu, Cr, Ni and Zn determination. The plan of tests and results is shown in Table 4.11.

Table 4.11. Test plans, metal ion contents in solution after Cu cementation and metal contents in the cements

N.	Tests plan ($g.L^{-1}$)			Solution after cementation ($g.L^{-1}$)				Cement (wt.%)			
	Cu	Cr	Ni	Cu	Cr	Ni	Zn	Cu	Cr	Ni	Zn
1	5	5	10	0.003	1.318	7.960	8.706	77.64	7.03	11.06	4.27
2	5	10	10	0.094	6.191	8.931	14.440	56.75	32.49	7.52	3.24
3	5	10	5	0.107	5.920	4.452	10.280	65.28	28.80	0.55	5.36
4	5	5	5	0.010	2.942	8.650	11.362	83.59	12.91	0.36	3.14

In general, Cu concentrations in the final solutions were very low, the differences found probably pertaining to the error margin of the chemical analysis. In addition, some important information can be picked up from these tests:

(i) for a 5 g.L^{-1} Ni concentration, the increase of the Cr amount from 5 to 10 g.L^{-1} in solution induced a decrease of the Cu content in the cement (18.31%), whereas for a 10 g.L^{-1} Ni concentration the same increase of the Cr content caused a stronger drop in the Cu amount in the cement (20.89%);

(ii) for a 5 g.L^{-1} Cr concentration, improving the Ni content from 5 to 10 g.L^{-1} caused only a slight decrease on the Cu amount of the cement (5.95%), whereas for a 10 g.L^{-1} Cr concentration the Cu content in the cement decreased to 8.53%.

It can hence be concluded that both Cr and Ni species in solution negativelly affect the Cu cementation process, Cr influence being the most determinant.

4.2.2.2 Tests with Real Solutions *(*Leach Liquor*)*

4.2.2.2.1 Copper *Cementation* by *Zinc Dust*

After Zn selection as a reducing agent, it was decided to use it under a metallic dust form (available with 99.6% Zn in weight and grain size < 100 μm). Zn powder is a by-product from surface treatments of metals and plastics, such as zinc plating. Moreover, its common fineness might increase the cementation reactivity, by assuring a high contact solid-liquid area.

4.2.2.2.1.1 Effect of the Solution pH

A certain volume of leach liquor was divided in three portions (A, B and C). Afterwards, they were properly neutralized with plating sludge (fresh, dried and grain sized < 1 mm) until pH values of 2.2, 2.8, and 4.7 in samples A, B and C, respectively. After neutralization and filtration, the pH of liquors was determined and the respective chemical composition was evaluated by AAS for determination of the Cu, Cr, Ni and Zn contents.

Copper cementation tests consisted in adding, under agitation (magnetic bar and agitator), a stoichiometric amount (further 10% as a margin of confidence) of Zn dust for 80 mL of liquor. Once the pre-established test time was reached (1 hour), samples were again analyzed in terms of pH and chemical composition (Cu, Cr, Ni and Zn). The cement was extensively washed with distilled water, dried and submitted to chemical analysis by XRF. The obtained results are presented in Table 4.12.

Table 4.12. Evaluation of the pH effect on Cu cementation; pH values (initial and final) of liquors and their metal contents are also provided

Test	pH		Liquor before (g.L^{-1})				Liquor after (g.L^{-1})				Cement (wt. %)			
	initial	Final	Cu	Cr	Ni	Zn	Cu	Cr	Ni	Zn	Cu	Cr	Ni	Zn
A	1.9	2.3	8.53	15.53	42.35	7.08	0.43	16.57	41.29	21.43	78.6	13.4	0.5	2.1
B	2.5	2.8	9.16	13.52	40.39	7.63	1.44	15.18	43.03	23.73	82.6	8.4	0.1	1.6
C	4.5	5.1	9.50	16.41	52.56	7.11	1.89	16.52	51.21	19.98	58.2	28.7	2.9	8.2

In the cement some traces of sulfur, phosphorus, chlorine, tin and iron were also detected.

The efficiency of Cu cementation was calculated from the initial and final concentrations in the liquor. Estimated values of 95%, 84% and 80% were obtained for tests A, B and C, respectively, leading to the conclusion that Cu recovery levels diminish with increasing pH values.

Concerning the chemical composition of the cement, Cu concentration decreased with higher solution pH values. At the highest pH level, the sum of Cr, Ni and Zn percentages in the cement reached about 40%, whereas in the other two situations that sum did not surpass 16%. In particular, the increase in the Cr content in the cement, which became relevant for higher pH values, might induce a strong contamination of the Cu cement. The unwanted occurrence of Cr/Cu co-precipitation was confirmed by SEM/EDS, conducted on a cement obtained from test "C", and it is illustrated in Figure 4.7 and Table 4.13.

Figure 4.7. SEM micrographs showing some Cr contamination, detected in two different zones of the cement sample: Z1 corresponds to a highly pure grade Cu region; Z2 zone shows Cr contamination.

Table 4.13. Chemical analysis (obtained by EDS) of zones Z1 and Z2 marked on the cement (see Figure 4.7)

Element	Zones (wt. %)	
	Z1	Z2
Cu	96.12 ± 3.54	30.52 ± 3.14
Cl	0.94 ± 0.17	0.01 ± 0.13
Cr	1.53 ± 0.28	31.07 ± 0.86
P	0.03 ± 0.16	3.04 ± 0.18
Si	0.11 ± 0.14	0.40 ± 0.10
S	0.08 ± 0.18	3.06 ± 0.21
Sn	0.52 ± 0.41	0.03 ± 0.39
Pb	0.68 ± 0.66	0.81 ± 0.68

Figure 4.7 shows SEM micrographs of two distinct zones of the Cu cement resulting from test "C", Z1 and Z2. The first one is almost composed of pure metallic Cu, while the second is seriously contamined by Cr hydroxide, as denoted by the typical green color easily detectable by the naked eye. By looking to the chemical composition shown in Table 4.13, it is easily concluded that the Cu purity level may change from more than 96% to less than 30%, depending mostly on the level of Cr contamination, which might reach 31%.

It was also observed that the cement generated from test "C" has a granular appearance, with an average particle size of about 300 μm, whereas the other cements, resulting from "A" and "B" tests, presented much finer particles. Kuntyi [55] concluded that solution acidity can increase the nucleation rate of Cu precipitates, lowering the grain growing rate. In fact, hydrogen formation ($2H^+ + 2e^- = H_2$) is a side reaction that consumes the precipitating agent and competes with the main cathodic reaction, lowering its effectiveness [23,44]; however, a minimal acidity grade of the liquor should always be used to run Cu cementation processes.

Taking into account the available solution and actual experimental constraints, the optimal pH value for Cu cementation using Zn dust as reducing agent is of about 2. Nevertheless, values reported in the literature are different. Dvořák and Jandová [56] suggested a pH 4 as the optimal value for Cu cementation, whereas Karavasteva [50] recommended the use of pH <1 in order to avoid the deposition of precipitates in the form of metal hydroxides. Similarly, Gupta [44] established that Cu cementation, using Fe and Al as precipitating agents, might form metal hydroxides, phosphates and other complex salts for solutions with pH<3. Since it is not economically feasible to purify the solution under study, the salts must be kept dissolved by a proper control of the acidity level.

According to the results described in sub-section 4.2.1.2 for the neutralization of the leach liquor by the plating sludge, the conditioning of the pH value to 2 will surely allow that a significant concentration of the interesting metals will remain in solution. In fact, taking into account the physicochemical characteristics of the sludge and the curves shown in Figure 4.4, it can easily be concluded that the enhancement of the pH values from 0 to 2 will increase the dissolved metal contents in the following manner: $\Delta Cu = 54\%$; $\Delta Cr = 10\%$, $\Delta Ni = 140\%$ and $\Delta Zn = 340\%$.

For all cases, the pH control is always critical, to avoid or reduce Cr/Cu co-precipitation and then to minimize the consequent Cu cement contamination.

4.2.2.2.1.2 Effect of Reaction Time

Liquor samples previously neutralized to pH 2 were tested. A mass of 0.8 g of Zn dust was added to each flask containing 80 mL of liquor. Immediately after, they were put under agitation. Once reached each pre-established time (1, 2, 5, 10, 60, 180 and 360 minutes, corresponding to tests 1, 2... 7, respectively), the agitation was stopped and the liquor was filtered. The pH values of the liquors were measured and the chemical contents of Cu, Cr, Ni and Zn were determined by AAS. The cement was extensively washed with distilled water, dried and submitted to chemical analysis by XRF. The observed results are presented in Table 4.14.

From Table 4.14 it is easy to conclude that the optimal reaction time is of about 5 - 10 minutes; for a 10 minutes reaction, the obtained cement contains more than 83% Cu. Reis and Carvalho [53] also postulate that Cu cementation using Zn dust is complete in just 5 minutes.

However, the cement purity grade is seriously affected by Cr co-precipitation, the presence of Zn and Ni being less significant.

It is also obvious that, after a 30 minutes reaction, the Cu concentration in the liquor started to rise, suggesting that its re-dissolution was occurring. This tendency was even more significant for longer reaction times (3 and 6 hours). Stole-Hansen [48] developed a controlling process of Cu cementation in a real Zn-producing plant (Falconbridge Kidd Creek) and showed that the optimal residence time inside the reactor goes from 30 minutes to 1 hour. Gupta [44] compared several processes and operation conditions in cementation plants and concluded that the optimal contact times vary between 3 and 80 minutes.

Table 4.14. Effect of the reaction time on the Cu cementation with Zn dust

Test	pH		Initial liquor ($g.L^{-1}$)				Time	Final liquor ($g.L^{-1}$)				Cement (wt. %)			
	Initial	Final	Cu	Cr	Ni	Zn	(min.)	Cu	Cr	Ni	Zn	Cu	Cr	Ni	Zn
1	1.9	2.1	8.53	15.53	42.35	7.08	1	0.28	15.75	41.56	16.21	79.21	12.12	0.76	3.40
2	1.9	2.3	«	«	«	«	2	0.23	15.61	42.46	16.45	78.54	13.36	0.51	2.09
3	1.9	2.2	«	«	«	«	5	0.21	15.25	42.33	16.03	82.67	12.89	0.38	1.54
4	1.9	2.5	«	«	«	«	10	0.23	14.95	42.93	16.28	83.21	13.02	0.09	2.00
5	1.9	2.3	«	«	«	«	60	0.31	15.34	42.15	16.14	79.29	14.32	1.02	3.16
6	1.9	2.3	«	«	«	«	180	0.43	14.97	44.29	17.43	72.15	17.33	0.99	2.32
7	1.9	2.6	«	«	«	«	360	1.09	15.63	44.39	16.13	67.19	16.49	2.03	2.67

In the cement traces of sulfur, phosphorus, chlorine, tin and iron were also detected.

4.2.3 Chromium Precipitation

As for Cu cementation, Cr precipitation should be conducted under suitable conditions, bearing in mind that the increase in the metal purity grades of the leach liquor is desirable, in order to facilite the subsequent Ni and Zn recovery operations.

Regarding the fortuitous application of Cr precipitate, it is possible to identify at least three important outlets: (i) as part of a raw meal in the cement production [57]; (ii) as a raw material of ceramic refractory products [58]; finally, as coloring agent of ceramic pigments [59,60]. Obviously, in some of such noble applications (the two latter mentioned), the Cr precipitate must be thermally treated, at reasonable high temperatures, to form the stable Cr_2O_3, that can then be used as a magnesium-chromitic refractory material or inserted in glasses and enamels.

At the beginning, the behavior of the Cr ion was followed by considering the effect of some experimental parameters (pH, nature of precipitating agent and temperature), in order to identify what are the factors which mostly determine its selective precipitation from the actual complex system.

References in literature about industrial precipitation of trivalent Cr by using alkaline agents, namely limestone, are scarce, so there is just a few information concerning optimum conditions for this operation [61-64].

The tests and respective results described herein refer to Cr precipitation by using the liquor that directly results from the sulfuric acid leaching step, meaning that this liquor did not suffer previous neutralization.

4.2.3.1 Selection of the Precipitating Agent

Several Cr precipitation tests using aqueous suspensions of CaO 2 M and $CaCO_3$ 1.5 M were carried out at ambient temperature.

To 1000 mL of leach liquor verted to a flask of 2000 mL capacity, small volumes of suspension were gradually added. The pH of the mixture was determined, and the suspension was kept under permanent stirring (100 r.p.m. with magnetic bar and agitator). Once each pre-established pH value was reached, the addition was immediately stopped and the liquor was filtered. A small liquor sample was collected and chemically analyzed by AAS for determination of the Cr, Cu, Ni and Zn contents. The formed precipitate was plentifully washed with distilled water, dried at 105 °C, disintegrated, homogenized, sampled and then analyzed by XRF. New small volumes of suspension were added to the filtered liquor, to reach the next pre-established pH value. This procedure was then repeated for all pH values.

Table 4.15 shows the concentration of the interesting metals in the liquor, for each pH range and after Cr precipitation with the use of either CaO 2 M or $CaCO_3$ 1.5 M solutions.

Table 4.15. Concentration of metals in the liquor for each pH range, after Cr precipitation by either CaO 2 M or $CaCO_3$ 1.5 M

Precipitating agent	pH		Liquor ($g.L^{-1}$)			
	Initial	Final	Cr	Cu	Ni	Zn
	Initial liquor →		11.0	5.0	18.0	3.0
$CaCO_3$ 1.5 M	0.4	2.2	10.80	4.98	17.91	2.98
$CaCO_3$ 1.5 M	2.0	3.5	10.10	4.83	17.89	2.96
$CaCO_3$ 1.5 M	4.0	5.5	4.57	2.16	17.38	2.32
CaO 2 M	0.4	2.2	10.48	4.80	17.78	2.92
CaO 2 M	2.0	3.5	9.44	4.63	17.73	2.89
CaO 2 M	4.0	5.5	4.50	2.12	12.85	0.95

Table 4.16 gives the total mass of precipitate and the relative amount of interesting elements in the precipitate, for each tested pH range and after Cr precipitation by either CaO 2 M or $CaCO_3$ 1.5 M solutions.

Table 4.16. Mass of precipitate and relative amount of the interesting elements found in the precipitate, obtained for each pH range, after Cr precipitation by either CaO 2 M or $CaCO_3$ 1.5 M

Precipitating agent	pH		Precipitate	Precipitate (wt.%)						
	Initial	Initial	(g)	Cr	Ni	Cu	Zn	Ca	S	P
$CaCO_3$ 1.5 M 25°C	0.4	2.2	7.2	0.09	0.03	0.01	0.01	32.47	20.39	0.08
$CaCO_3$ 1.5 M 25°C	2.0	3.5	0.5	20.93	0.53	5.54	0.84	20.93	1.04	6.65
$CaCO_3$ 1.5 M 25°C	4.0	5.5	10.34	2.10	0.20	1.51	0.39	67.45	0.26	0.13
CaO 2 M 25°C	0.4	2.2	6.9	0.25	0.08	0.12	0.05	34.21	19.76	0.14
CaO 2 M 25°C	2.0	3.5	0.25	18.14	0.67	3.53	0.69	20.77	9.59	5.76
CaO 2 M 25°C	4.0	5.5	1.93	13.33	10.18	8.23	6.46	14.76	13.92	1.65

The analyzed elements are probably involved in carbonate, sulfate or phosphate compounds.

With both precipitating agents, the increase of pH tends to induce a gradual decrease on the concentration of the interesting metals in solution. At pH 4.0-5.5, addition of $CaCO_3$ 1.5 M ensured a selective precipitation of Cr much more pronounced than that achieved by using CaO 2 M. However, Cu also precipitates in significant amounts.

Concerning the formed precipitates, the detection of reasonable amounts of Ca and S indicates the preferential precipitation of $CaSO_4$ in the pH 0.4-2.2 range. Chong and Sheikholeslami [65] studied the precipitation process of a mixture of calcium carbonate and calcium sulfate and concluded that its thermodynamic behavior is very complex. However, solubility of calcium carbonate increases by enhancing the sulfate ion concentration until a certain limit. This limit may be imposed by the presence of impurities, in addition to the expected effect of reagents concentration ($CaCO_3$ and $CaSO_4$).

In this study, pure solutions were used to investigate the co-precipitation process between $CaCO_3$ and $CaSO_4$, when $CaCO_3$ is dominant. However, it is admissible that the presence of impurities (cationic and anionic species in solution) may catalyze the precipitation of sulfate species. Probably this explanation can justify the verified order of precipitation: initially, calcium sulfate was exclusively formed, and only when a certain limit was reached the other metal precipitations started.

Comparing the total mass of precipitates and their composition (see Table 4.16) it may be concluded that:

(i) at the lower pH range (pH 0.4-2.2), the conversion of CaO or $CaCO_3$ to $CaSO_4$ was almost complete;
(ii) at an intermediate pH range (pH 2.0-3.5), the total mass of precipitate was very low; the Cr relative content was significantly higher than those of the other metals, the formation of $CaSO_4$ still persisted and P precipitation was also appreciable;
(iii) at the pH range 4.0-5.5, a much higher total mass of precipitate was produced for the tests with $CaCO_3$ 1.5 M when compared with the results obtained with CaO 2 M (about 5 times). The use of this latter reagent has promoted a general metals precipitation, with a poor selectivity. By contrast, the addition of $CaCO_3$ 1.5 M tends to favour the selective precipitation of Cr, but Cu precipitation still remained reasonably high.

Therefore, $CaCO_3$ 1.5 M proved to be the best precipitating agent under the tested experimental conditions and further experiments were then carried out with this reagent, in an attempt to optimize the Cr removal efficiency and to get a highly pure liquor.

4.2.3.2 Tests of Cr Precipitation Using the Selected Precipitating Agent

Chromium precipitation tests were planned under the following experimental conditions:

(i) addition of $CaCO_3$ 2 M at ambient temperature;
(ii) addition of $CaCO_3$ 2 M at 40 °C (liquor and reagent suspension kept at 40 °C);
(iii) direct addition of $CaCO_3$ (powder) at ambient temperature.

The experimental procedure was analogous to that described earlier, except for the tests performed at 40°C, for which the related samples were kept immersed in a conditioning temperature bath.

Table 4.17. Concentration (g.L^{-1}) of metals in the liquor for each studied pH range and for three experimental conditions

pH		$CaCO_3$ 2 M at 25°C Liquor (g.L^{-1})				$CaCO_3$ 2 M at 40°C Liquor (g.L^{-1})				$CaCO_3$ (powder) at 25°C Liquor (g.L^{-1})			
Initial	Final	Cr	Ni	Cu	Zn	Cr	Ni	Cu	Zn	Cr	Ni	Cu	Zn
-	-	11.0	18.0	5.0	3.0	11.0	18.0	5.0	3.0	11.0	18.0	5.0	3.0
0.5	2.0	10.83	17.94	4.98	2.98	10.60	18.0	4.98	3.0	10.62	18.00	5.00	3.0
2.0	3.5	10.22	17.92	4.80	2.96	4.24	17.80	1.81	2.67	9.30	17.97	4.46	2.90
3.5	4.0	8.82	17.92	4.46	2.91	3.23	17.74	1.02	2.66	8.57	17.93	4.09	2.80
4.0	4.5	4.36	17.54	1.36	2.16	2.79	17.62	0.55	2.65	7.60	17.87	3.53	2.73
4.5	5.5	3.63	17.36	1.28	1.74	2.78	17.60	0.54	2.65	7.43	17.80	3.44	2.67

Table 4.18. Total mass of precipitate and the relative concentrations of the interesting elements, obtained for three experimental conditions

Condition	pH		Precipitate	Precipitate (wt. %)						
	Initial	Final	(g)	Cr	Ni	Cu	Zn	Ca	S	P
$CaCO_3$ 2 M 25°C	0.5	2.0	19.07	0.83	0.02	0.08	0.02	29.22	29.32	0.41
$CaCO_3$ 2 M 25°C	2.0	3.5	2.54	21.38	0.35	5.00	0.66	26.46	1.12	7.42
$CaCO_3$ 2 M 25°C	3.5	4.0	13.42	11.21	1.64	7.32	1.82	29.44	14.14	2.12
$CaCO_3$ 2 M 25°C	4.0	4.5	3.13	9.23	7.85	2.60	6.86	24.56	17.24	0.76
$CaCO_3$ 2 M 25°C	4.5	5.5	2.14	0.20	19.87	0.03	11.06	34.41	6.70	0.02
$CaCO_3$ 2 M 40°C	0.5	2.0	16.15	0.95	0	0.05	0.01	30.52	23.33	0.34
$CaCO_3$ 2 M 40°C	2.0	3.5	2.54	30.47	1.53	21.31	1.33	13.59	7.66	1.94
$CaCO_3$ 2 M 40°C	3.5	4.0	28.03	14.27	0.41	6.31	1.09	38.95	5.45	2.53
$CaCO_3$ 2 M 40°C	4.0	4.5	2.51	12.32	2.95	11.51	2.77	42.81	2.30	0.41
$CaCO_3$ 2 M 40°C	4.5	5.5	0.50	0.16	5.67	0.09	7.08	55.14	5.39	1.67
$CaCO_3$ powder	0.5	2.0	15.25	0.55	nd	nd	nd	29.58	27.73	0.28
$CaCO_3$ powder	2.0	3.5	21.35	2.20	0.05	0.80	0.16	66.03	0.56	0.55
$CaCO_3$ powder	3.5	4.0	45.10	0.57	0.02	0.26	0.06	70.37	0.06	0.09
$CaCO_3$ powder	4.0	4.5	35.41	0.50	0.06	0.50	0.07	69.35	0.32	0.07
$CaCO_3$ powder	4.5	5.5	5.24	1.21	0.38	0.54	0.32	68.77	0.33	0.04

The analyzed elements are probably linked to carbonate and sulfate forms, in addition to some phosphate.

nd = under the detection limit.

The liquor concentrations of the metals under study, determined for each tested condition, are given in Table 4.17. Figure 4.8 shows the effect of the liquor pH on the accumulated amount of those metals. Table 4.18 presents the total mass of precipitate and its composition in terms of the same elements. Finally, Table 4.19 summarizes the $CaCO_3$ consumptions for each pH range and its global consumption per test.

As it may be verified from the results displayed in Table 4.17 and Figure 4.8, the addition of $CaCO_3$ 2 M at 40 °C (graph "b") gave an optimal Cr precipitation. The selectivity achieved for Cr was remarkable, at least when compared to Ni and Zn elements. Nevertheless, as demonstrated in previous tests, Cu co-precipitation still occurred and, at least under the tested experimental conditions, seemed inevitable.

Table 4.19. $CaCO_3$ consumption for each pH range and $CaCO_3$ global consumption per test

pH		$CaCO_3$ consumption ($g.L^{-1}$)		
Initial	Final	Suspension 2 M at 25°C	Suspension 2 M at 40°C	powder at 25°C
0.5	2.0	50.04	50.04	51.90
2.0	3.5	4.00	55.05	96.15
3.5	4.0	70.06	5.00	208.13
4.0	4.5	86.08	11.00	154.30
5.0	5.5	16.01	2.00	25.19
Total		226.20	123.11	535.67

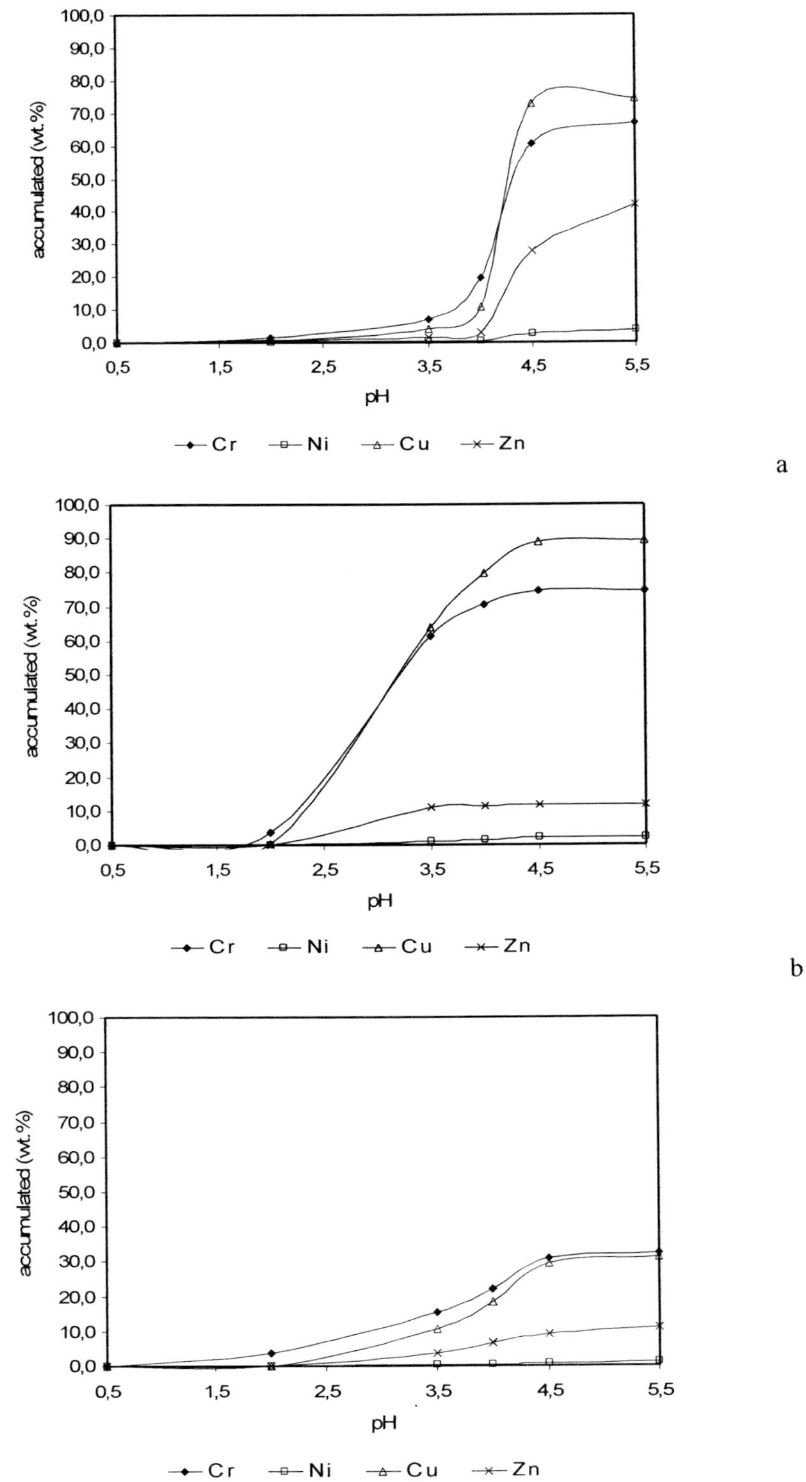

Figure 4.8 – Liquor pH dependence on the relative amount of accumulated metals for the three experimental conditions tested: (a) addition of $CaCO_3$ 2 M at 40°C; (b) addition of $CaCO_3$ 2 M at 40°C; (c) addition of $CaCO_3$ in the powder form, at 25°C.

Returning to the curves of graph "b" in Figure 4.8, the optimal pH seems to be 4.5, since a maximal Cr precipitation was achieved through the addition of $CaCO_3$ 2 M at 40 °C, and that did not compromise the permanence of Ni and Zn in the liquor.

Concerning the mass and the chemical composition of the formed precipitates, it can be concluded that:

(i) even if a heavier mass of accumulated precipitate is formed by addition of powdered $CaCO_3$ (122.35 g, which corresponds to about 3 times the value reached in the other two tested conditions), this agent did not promote any significant Cr precipitation;
(ii) as already observed, an almost exclusive $CaSO_4$ precipitation occurred in the tested conditions for low pH values (range 0.5-2.0);
(iii) the addition of $CaCO_3$ 2 M at 40 °C at an intermediate pH range (3.5-4.0) resulted in a maximal mass of precipitate, that is, the Cr richest one;
(iv) the addition of $CaCO_3$ 2 M at 40 °C assured a significant selective Cr precipitation, with a minimal consumption of reagent (see Table 4.19), and that corresponded to the optimal processing conditions.

To avoid or minimize Cu co-precipitation, the previous cementation of this element could be carried out before Cr precipitation.

4.3 Sequential Tests

After studying and optimizing each single step, a sequential plan was designed, as shown in Figure 4.9. In this sequence, the liquor resulting from one operation has been treated as feed in the next one.

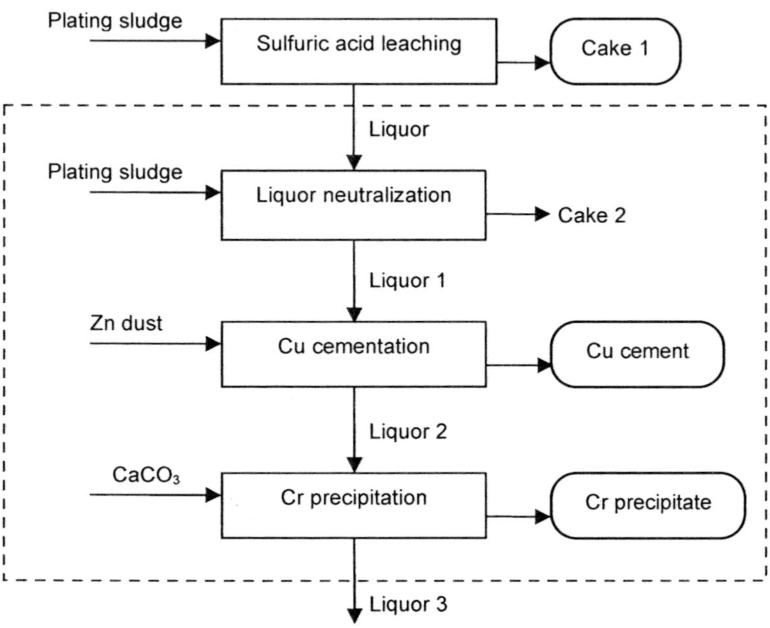

Figure 4.9. Sequence of precipitation tests to treat the leach liquor.

According to Figure 4.9, the following order is suggested: (i) leach liquor neutralization; (ii) Cu cementation; (iii) Cr precipitation. This sequence assures the natural increase of liquor pH after each step, as was experimentally verified in sub-section 4.2.

4.3.1 Experimental

The leach liquor used as feed solution for the sequential tests (ST) was previously neutralized, to minimize the acidity level of the solution in order to adjust the precipitation process to the optimal thermodynamic situation.

The adopted experimental procedure was similar to that earlier described in sub-section 4.2, with exception of the following details:

(i) the plating sludge used in the neutralization of the leach liquor was much richer in Ni and Cr, containing Ni = 40,09%, Cr = 21,97%, Cu = 3,61% and Zn = 3,16% (wt. % of dried sludge);
(ii) tests of Cu cementation with Zn dust have elapsed for 1 hour.

4.3.2 Results

Table 4.20 summarizes the most relevant parameters controlled in both sequential tests, identified as ST1 and ST2.

The chemical compositions of the liquor and precipitate, the mass of cement produced after each step, as well as the $CaCO_3$ 2 M consumption in the Cr precipitation, are all presented in Table 4.21.

Table 4.22 summarizes the mass variation found for the interesting metals after each sequential step.

Table 4.23 shows the chemical composition of precipitates produced after each sequential step.

Table 4.20. Experimental conditions tested in both sequential tests, ST1 and ST2

Sequential Test	Step	Liquor volume (L)		pH		Added sludge (g)	Added Zn dust (g)	Washing water (L)
		Initial	Final	Initial	Final			
	Initial neutralization	2.0	1.50	0	1.0	170		
ST1	Neutralization	1.20	1.05	1.0	1.2	15		
ST1	Cu cementation	0.20	0.18	1.2	1.6		2.55	
ST1	Cr precipitation	0.16	0.04	1.5	3.5			0.80
ST2	Neutralization	1.0	0.80	1.0	1.8	20		
ST2	Cu cementation	0.20	0.17	1.8	2.3		2.55	
ST2	Cr precipitation	0.15	*	2.3	4.5			2.0

* There was no collection of filtered solution since during $CaCO_3$ 2 M addition the mixture (liquor plus reagent suspension) transformed into a gel, making difficult its filtration.

Table 4.21. Results obtained for the treated liquor, and for the precipitate and cement masses produced after each sequential step. $CaCO_3$ 2 M consumption is also indicated

	Concentration of elements of interest in the liquor ($g.L^{-1}$)						Preci-pitate (g)	$CaCO_3$ 2 M use ($g.L^{-1}$)
	Cr	Cu	Ni	Zn	Pb	Ca		
ST1								
Neutralization							3.05	–
Liquor before	17.49	9.03	23.22	4.70	0.01	0.23	–	–
Liquor after	20.80	11.69	32.04	5.77	0.01	0.23	–	–
Cu Cementation							2.25	–
Liquor before	20.80	11.69	32.04	5.77	0.01	0.23	–	–
Liquor after	20.19	1.98	31.74	16.75	0.01	0.20	–	–
Cr Precipitation							16.43	54.59
Liquor before	20.19	1.98	31.74	16.75	0.01	0.20	–	–
Liquor after	0.57	0.28	39.55	29.50	0.00	0.27	–	–
Washing H_2O	0.01	0.01	4.43	1.87	0.00	0.07	–	–
ST2								
Neutralization							11.29	–
Liquor before	17.49	9.03	23.22	4.70	0.01	0.23	–	–
Liquor after	22.45	12.68	38.85	6.56	0.01	0.30	–	–
Cu Cementation							2.47	–
Liquor before	22.45	12.68	38.65	6.56	0.01	0.30	–	–
Liquor after	25.34	1.64	41.45	20.94	0.01	0.14	–	–
Cr Precipitation							13.38	69.29
Liquor before	25.34	1.64	41.45	20.94	0.01	0.14	–	–
Liquor after	(*)	(*)	(*)	(*)	(*)	(*)	–	–
Washing H_2O	0.01	0.11	3.13	1.38	0.00	0.06	–	–

* There was no collection of filtered solution since during $CaCO_3$ 2 M addition the mixture (liquor plus reagent suspension) transformed into a gel, making difficult its filtration.

Table 4.22. Mass variations found for the interesting metals, after each sequential operation

	Before neutralization (g)				After neutralization (g)				Mass variation (wt.%)			
	Cu	Zn	Cr	Ni	Cu	Zn	Cr	Ni	ΔCu	ΔZn	ΔCr	ΔNi
ST1	10.84	5.64	20.99	27.86	12.27	6.06	21.84	33.64	13.19	7.45	4.05	20.75
ST2	9.48	4.94	18.36	24.38	10.14	5.25	17.96	31.08	6.96	6.28	-2.18	27.48
	Before cementation (g)				After cementation (g)				Mass variation (wt.%)			
	Cu	Zn	Cr	Ni	Cu	Zn	Cr	Ni	ΔCu	ΔZn	ΔCr	ΔNi
ST1	2.34	1.15	4.16	6.41	0.36	3.35	3.63	5.71	84.62	-191,30	12.74	10.92
ST2	2.54	1.31	4.49	7.73	0.28	3.56	4.31	7.05	88.98	-171.80	4.0	8.80
	Before precipitation (g)				After precipitation (g)				Mass variation (wt.%)			
	Cu	Zn	Cr	Ni	Cu	Zn	Cr	Ni	ΔCu	ΔZn	ΔCr	ΔNi
ST1	0.32	2,68	3.23	5.08	0.02	2.68	0.03	5.13	93.75	0	99.07	-0.98
ST2*	0.25	3.14	3.8	6.22	0.22	2.76	0.02	6.26	12.0	12.10	99.47	-0.64

* In ST2, the concentration of metals determined after the Cr precipitation step was inferred from the analysis of the washing water, since there was no production of filtered solution due to the formation of a gel.

Table 4.23. Chemical composition of precipitates produced after each sequential step

	Ca	Cr	Cu	Fe	Ni	Pb	Zn	SiO_2	Al_2O_3	Cl	P	S
ST1												
Cake da neutralização	0.32	12.91	2.24	1.43	7.84	3.85	0.81	49.07	2.31	0.32	5.46	13.44
Cu cement de cobre	-	6.48	78.81	-	0.17	0.53	1.99	-	-	0.58	4.63	6.81
Cr precipitate de crómio	12.26	20.07	1.89	0.20	3.37	0.04	3.15	-	-	-	5.10	47.70
ST2												
Cake da neutralização	0.40	15.97	1.54	0.89	2.82	9.04	0.37	46.96	1.18	0.16	1.92	18.75
Cu cement de cobre	-	4.19	81.66	-	0.24	2.01	2.55	-	-	0.80	2.12	6.43
Cr precipitate de crómio	6.83	30.30	0.46	0.26	4.18	0.13	3.56	-	-	-	9.29	44.99

In Cr precipitates, carbonates, phosphates and sulfates are common element-containing compounds.

In the second sequential test – ST2 - there was no filtered solution after Cr precipitation because, when trying to optimize Cr removal, the mixture had its pH elevated above that achieved in ST1 (pH 4.5 against pH 3.5). Immediately after the mixture reached pH 4.5, a gel has widely formed, which seriously affected its efficient filtration. Anyway, the gel was transferred to a filter paper, extensively washed with distilled water, filtered, and the water was collected and analyzed.

During the neutralization step, a Ni enrichment was observed, mainly in ST2, whereas the Cr content changed inversely in the liquor (Table 4.22). However, the changes in concentration of the analyzed metals were not very significant, as expectable from the low pH range considered. The chemical composition of cakes formed after sludge addition still revealed the presence of significant contents of valuable metals, recommending the re-use of the sludge as feed for the leaching process, to be mixed with the fresh plating sludge.

Cu cementation in the ST2 test, carried out at pH values closer to 2, was found to be very effective, confirming previous observations (see in Table 4.22, Cu recovery of ~89% at pH 1.8 against ~85% at pH 1.2). Table 4.23 allows the confirmation that cement resulting from ST2 was slightly richer in Cu than the one formed in ST1.

Finally, Cr precipitation was also found to be very effective, the recovery level being close to 100% in both ST1 and ST2 tests (see Table 4.22). The overall performance at pH = 3.5 was higher than at 4.5, due to the mentioned risk of gel formation at higher pH values. Thus, it seems critical to limit herein the pH to values under 3.5.

4.4 Solvent Extraction – Final Approach

Once reduced the compositional complexity of the sludge, through Cu and Cr removal, it was decided to revisit the solvent extraction technique.

Starting from the Cu and Cr almost free liquors, efforts were concentrated in Zn and Ni separation and recovery. The D2EHPA extractant was chosen, since it showed technical and economical advantages over Cyanex 272. Table 4.24 {IV} shows relevant characteristics of the samples C, D and E used in these tests.

Table 4.24. Initial pH values and contents of the interesting metals in liquor samples before the solvent extraction treatment {IV}

Sample	pH	Concentration ($g.L^{-1}$)			
		Cu	Cr	Ni	Zn
C	5.5	0.003	0.003	23.81	12.95
D	4.3	0.008	0.015	4.0	0.80
E	3.4	0.016	0.03	8.0	1.60

Feed solutions of C and D samples had their pH adjusted to 3.0 by adding H_2SO_4 1 M.

4.4.1 Sample C

The extraction tests for zinc and nickel recovery were performed in two sequential stages: (i) first Zn at pH 3.0, adjusted as described earlier; (ii) then Ni, after adjusting the pH value of the raffinate to about 6 by adding NaOH 1 M. In fact, it is known that the extraction of those species by D2EHPA is optimal by using those equilibrium pH values [30,31,40]. The extraction mechanism of Zn and Ni by acidic organophosphorus-based extractants obviously shows an inverse dependence on the acidity, since hydrogen ions from the extractant are released during the extraction process. The loaded Zn and Ni ions were stripped from D2EHPA organic phases by H_2SO_4 2 M.

Table 4.25 {IV} presents the total extraction and stripping levels for Zn and Ni initially present in sample C.

Table 4.25. Removal efficiency of Zn and Ni from sample C, by D2EHPA extraction and H_2SO_4 2 M stripping {IV}

	Extraction (%) with D2EHPA 1 M		Stripping (%) with
	pH 3.0	pH 6.0	H_2SO_4 2 M
Zn	96	-	88
Ni	-	63	5

With the adopted procedure, an excellent Zn recovery level was achieved, actively helped by a high efficient stripping stage. As expected, Ni was only extracted at pH = 6 but the extraction efficiency was much lower (63%). Moreover, Ni stripping from the organic phase was also very ineffective. Sarma and Reddy [33] suggested that the presence of sodium salts in the feed solutions causes a significant decrease of the Ni extraction efficiency when D2EHPA is used, and this can probably be the explanation for the actual poor performance found for its solvent extraction.

4.4.2 Samples D and E

After evaluation of the results obtained with sample C, it was decided to optimize the Zn recovery by solvent extraction by trying to maintain as much as possible the Ni dissolved in the original solution. At the same time, the reduction of the required D2EHPA amount was envisaged. Therefore, the pH of sample D was lowered to 3.5 by adding H_2SO_4 1 M and both solutions D and E were treated with 0.5 M D2EHPA through single-extraction stages. The loaded organic phases were then stripped with H_2SO_4 2 M to reextract Zn ions.

Table 4.26 {IV} presents Zn and Ni extraction and stripping levels from samples D and E.

Table 4.26. Removal efficiency of Zn and Ni from samples D and E, by D2EHPA extraction and H_2SO_4 2 M stripping {IV}

	Extraction (%) with D2EHPA 0,5 M		Stripping (%) of Zn with H_2SO_4 2 M	Stripping (%) of Ni with H_2SO_4 2 M
	D	E	D	E
Zn	100	88	100	100
Ni	14	~0	1	1

Results on Table 4.26 confirm that Zn recovery was almost complete, even from less concentrated liquors. The Zn stripping efficiency was remarkable, since it was obtained with a half concentrated extractant solution.

At the end, two solutions were obtained, one essentially composed of zinc sulfate while the other was rich in nickel sulfate.

5. Crystallization from Solution

Crystallization is the physical process of salts separation, under the form of crystals, from aqueous solutions. Crystallization and precipitation processes are responsible for 70% of all solid materials produced by the chemical industry [66]. Crystallization of a metal salt from an aqueous solution can be achieved by evaporation until the solute concentration surpasses the saturation limit. By knowing that solubilities of most metal salts drastically decrease beyond about 200 °C, this technique is normally conducted at reasonably high temperatures and/or pressures [14].

Currently, certain sulfate salts represent an important niche market. Nickel sulfate can be applied in the agrochemical industry or can return to plating and metal finishing plants. Zinc sulfate, produced by crystallization, has been applied in the animal food industry and in agricultural blends (as nutrient) [67-70].

This sub-section describes some preliminary tests of evaporative crystallization of salts, that is, simple drying operations to remove the solvent, aiming to produce Ni- and Zn sulfates.

5.1 First Test

At the beginning, it was decided to firstly use the solution directly resulting from the sulfuric acid leaching process, afterwards neutralized until pH 4.7. It contained the following interesting elements (in $g.L^{-1}$): Ca = 0.66, Cr = 0.09, Cu = 0.82, Na = 1.40, Ni = 16.0 and Zn = 1.45. The solution was dried at 105°C for 24 hours and the mass of crystals was subsequently analyzed by XRF. A sample of a commercial $NiSO_4$ salt was dried in parallel and analyzed by XRF, to be used as reference. Table 5.1 shows these results.

Table 5.1. Chemical composition (wt.%) of crystals obtained by evaporation of the tested solution and of a commercial $NiSO_4$ salt

	Crystals (wt.%)										
	Ca	Cr	Cu	Ni	Zn	SO_4	Na_2O	P	Cl	K_2O	Mn
Tested solution	1.49	0.36	1.42	24.09	3.81	63.32	1.45	0.18	0.03	0.08	0.19
Commercial $NiSO_4$	nd	nd	nd	40.31	nd	59.38	0.29	nd	nd	0.01	nd

nd – under the detection limit.

The crystals obtained contained a reasonably high Ni content, reaching about 60% of the amount of the pure-grade commercial sample. Zinc, Ca and Cu were the main contaminants.

This test has demonstrated that, to obtain a good purity grade for the sulfate salt, impurities should be removed from the feed solution, justifying then the use of a previous solvent extraction treatment.

5.2 Second Test

The starting solutions for this test resulted from the solvent extraction step (see sub-section 4.4), and are rich in Ni and Zn. They were submitted to evaporative crystallization by drying them inside an oven at 250°C for 24 hours, according to literature data [71]. As in the previous test, commercial $NiSO_4$ and now commercial $ZnSO_4$ were used for comparison purposes. After evaporation, the chemical composition of the crystals was estimated by XRF, as shown in Table 5.2.

Table 5.2. Chemical composition (wt.%) of crystals obtained by evaporation from Ni- and Zn sulfate rich solutions. For comparison, commercial sulfate salts were also tested

	Crystals (wt.%)										
	Ca	**Cr**	**Cu**	**Ni**	**Zn**	**SO_4**	**Na_2O**	**P**	**Cl**	**K_2O**	**Mn**
Ni-rich sample	0.03	nd	nd	34.74	0.12	64.76	0.17	nd	nd	nd	0.02
Commercial $NiSO_4$	nd	nd	nd	40.16	nd	59.23	0.59	nd	nd	0.01	nd
Zn-rich sample	0.91	nd	nd	0.08	10.45	79.01	6.0	nd	nd	0.14	0.02
Commercial $ZnSO_4$	0.06	nd	nd	0.01	32.03	59.55	nd	nd	0.13	0.15	0.18

nd – under the detection limit.

The chemical composition of the formed crystals is now much closer to that of the corresponding commercial sulfate salts. However, the relative amount of SO_4 appearing in the treated samples is over the stoichiometric ratio, probably due to the conditions of precipitation, that occurred at an acidic pH. This aspect can of course be corrected if the application envisaged for the sulfate salts would require it. Starting from this point of view, the re-use of these salts in plating or finishing activities would not demand pH adjustments before the crystallization step.

Crystals formed from the Ni-rich sample showed a slight Zn contamination. Those obtained from the Zn-rich sample presented higher impurity levels, Na, Ni and Ca being the main contaminants.

It is well known that it is feasible to separate two or more chemically similar metals by crystallization if their solubilities in aqueous solution are reasonably different [14]. However, specific evaporating crystallizers, placed in series, should be available, to assure that steam from one evaporator passes to the steam chest of the next. Gupta [44] states that forced circulating evaporators are in vogue to crystallize several salts, such as NaCl, $(NH_4)_2SO_4$, Na_2SO_4, $FeSO_4$ and $NiSO_4$.

Anyway, the results described in this sub-section, despite being obtained in incipient experimental conditions, demonstrate the potential viability of the actual process.

6. Flowsheet of the Integrate Hydrometallurgical Process to Treat the Plating Sludge

In this section, a complete flowsheet for the sludge treatment is suggested (see Figure 6.1). It integrates all the steps described along this chapter. A final test was carried out, in an attempt to simulate a pilot plant in which non-ferrous metals present in industrial sludges would be recovered. At the end of the process, valuable raw materials such as a copper cement, a chromium precipitate and both nickel and zinc sulfates are produced, together with a non- hazardous waste.

Data resulting from the final test were extrapolated considering 1 ton of dried plating sludge.

6.1 Final Test

In the sequence proposed in Figure 6.1, the leaching tank was fed by a sulfuric acid solution with a 100 $g.L^{-1}$ concentration and dried plating sludge < 1mm, with a liquid-solid ratio at 5:1. The sulfuric acid leaching was conducted under permanent stirring (700 r.p.m.) for 30 minutes. After filtration and washing of cake 1, it was re-leached at the above-mentioned conditions. The cake resulting from this re-leaching was classified as non-hazardous waste (NHW), according to European environmental regulations for waste disposal [72].

Afterwards, the leach liquor was neutralized by adding dried plating sludge (< 1mm), until pH 2. The neutralization operation was also conducted under permanent agitation (700 r.p.m.) for 1 hour. About 120 $g.L^{-1}$ of dried plating sludge were consumed.

Table 6.1 shows the average chemical composition of the plating sludge, used as feed for the leaching and neutralization steps. The cake and final resulting waste were also analyzed.

In this sequence, the neutralized liquor was sampled and analyzed by AAS to determine the Cu content. After then, the copper cementation tank was fed by the neutralized liquor and zinc dust, which was added in the same proportion of the Cu present in the liquor (plus 10 wt. % as margin of confidence). The Cu cementation operation was conducted under agitation (100 r.p.m.) for 15 minutes.

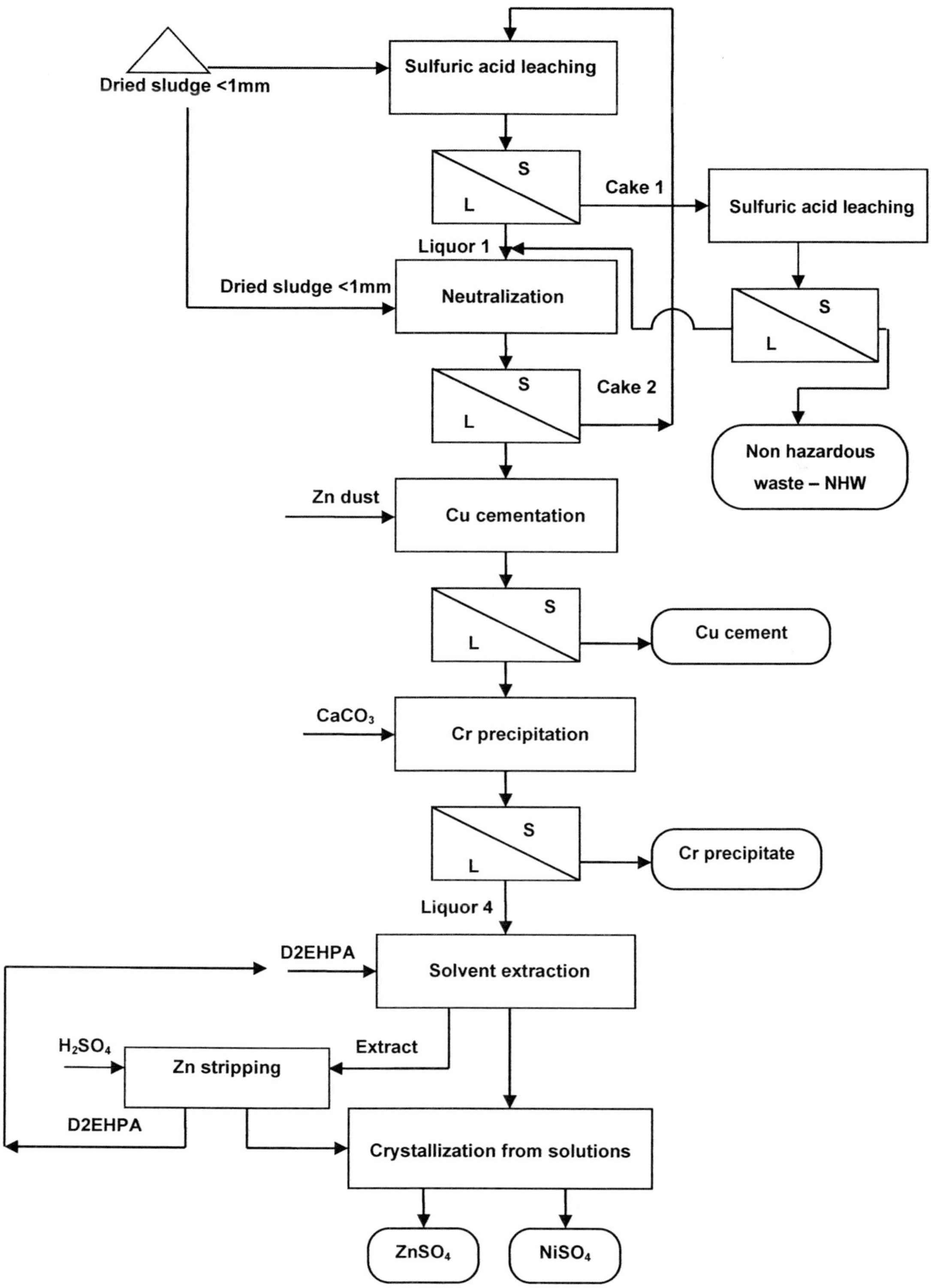

Figure 6.1. Flowsheet of the complete hydrometallurgical process to treat the plating sludge.

After filtration, the liquor was heated at 40 °C and then used as feed for the chromium precipitation tank. A gradual addition of a $CaCO_3$ 2 M solution (also pre-heated at 40°C) was made to the mixture and that was kept under agitation (at 100 r.p.m.) for about 1 hour until pH = 3.5.

Table 6.1. Chemical composition (wt.%) of the plating sludge and cakes generated from leaching operations

	Plating sludge	Cake 1	NHW*
Al_2O_3	2.55	2.33	1.84
CaO	0.35	0.11	0.88
Cl	0.22	0.2	0.65
Cr	7.02	0.97	0.71
Cu	4.19	1.9	0.62
Fe	1.53	2.23	1.29
MgO	0.11	0.09	0.09
Mn	0.37	0.02	0.05
Ni	9.46	3.51	0.93
P	4.08	0.58	0.99
Pb	0.36	1.91	2.21
S	5.83	6.1	4.11
SiO_2	59.86	76.61	80.19
Zn	2.96	0.44	0.19
Total	98.89	97.0	94.75
**	1.11	3.0	5.25

* NHW – Non Hazardous Waste

** Other elements or compounds such as Ba, K_2O, Sn, TiO_2, ZrO_2, Na_2O, etc.

Next, the copper and chromium free liquor was analyzed by AAS for determination of nickel and zinc contents, and was then treated by D2EHPA 2 M diluted in kerosene (equal volumes of organic and aqueous solutions - A/O = 1). Nickel and zinc separation, and zinc extraction, were then carried out under vigorous agitation (800 r.p.m.) for 30 minutes, followed by a resting period of 15 minutes.

The obtained Ni-rich aqueous solution was withdrawn and the H_2SO_4 2 M solution was added to the Zn-loaded organic phase for zinc stripping. This process was conducted under strong agitation (800 r.p.m.) for 30 minutes, followed again by a resting period of 15 minutes.

At the end, the Zn-rich aqueous solution was withdrawn and both nickel and zinc sulfate containing solutions were submitted to evaporative crystallization at 250°C for 8 hours.

It should be remembered that the liquor was filtered, sampled and analyzed by AAS for Cr, Cu, Ni and Zn determination after each step. The precipitate was washed, dried, sampled and analyzed by XRF.

6.2 Results

Table 6.2 shows the chemical composition of the resulting crystals while Table 6.3 shows the efficiency level reached in each hydrometallurgical individual step.

It must also be mentioned that after the sulfuric acid leaching step, significant volume losses of liquor occurred (due to water absorption upon neutralization, during filtration, etc.). Table 6.4 gives an estimation of these volume losses, assuming 1 ton of dried plating sludge as the starting amount.

Table 6.2. Chemical composition of the crystals resulting from evaporative crystallization of the raffinate (Zn-rich liquor) and of the extract (Ni-rich liquor)

	Crystals (wt. %)						
	Ca	Ni	Zn	SO_4	Cl	K_2O	Mn
Zn-rich sample	0.06	0.01	32.03	59.55	0.13	0.15	0.18
Ni-rich sample	0.03	36.63	3.77	59.43	nd	nd	nd

nd – under the detection limit.

Table 6.3. Efficiency level (%) of each individual hydrometallurgical step

	Sulfuric acid leaching	Liquor neutralization*	Copper cementation	Chromium Precipitation	Solvent Extraction	
					Raffinate	Extract
Cr	98	10	–	>99	–	–
Cu	88	54	97	–	–	–
Ni	98	140	–	–	~100	–
Zn	99	340	88	–	12	~100

* The values shown represent the percentual enrichment of the respective metal in the leach liquor, after neutralization by the plating sludge addition.

Table 6.4. Estimation of liquor volume losses after each sequential operation of the complete hydrometallurgical process

Volume (L)	Operation
5,000.0	Initial volume for 1 ton of dried plating sludge
4,750.0	Leaching (- 5% of the initial volume)
2,850.0	Neutralization (- 40% of the initial volume)
2,422.5	Cu cementation (- 15% of the prior volume)
969.0	Cr precipitation (- 60% of the prior volume)

Based on these estimations, a complete material balance was made, as follows:

Input

1. Dried plating sludge (< 1mm): 1 ton.
2. Water to leaching solution: L/S ratio = 5:1 or 200 g solid/1L solution, that is, 5,000 L water to 1 ton of plating sludge.
3. Sulfuric acid to leaching solution: 100 g/1L solution, that is, 500 kg H_2SO_4 to 5,000 L water.
4. Water for cake leaching: 60% by weight of the remained solid sludge, that is, 600 kg. Thus, 600 x 0,2 kg/L = 3,000 L.
5. Sulfuric acid for cake leaching: 0.1 kg/L x 3,000 L = 300 kg.
6. Plating sludge used for liquor neutralization: 100 g/L, that is, 500 kg.
7. Zinc dust for Cu cementation, assuming that 11.1 g/L Cu are transferred from the solution to the cement: 2,850 L solution = 31.6 kg Cu, but using 10% as margin of confidence = 34.8 kg of Zn dust.
8. Water to wash Cu cement: ~50 L to wash 38.1 kg of Cu cement.
9. Calcium carbonate for Cr precipitation: average of 70 g/L $CaCO_3$ 2 M, that is, 169.57 kg reagent for 2,422.5 L liquor.
10. Water to prepare $CaCO_3$ suspension: 1,000 L to 169.57 kg reagent;
11. Water added to warm (at 40 °C) the liquor and the $CaCO_3$ suspension: ~10,000 L.
12. Water to wash the Cr precipitate: ~150 L to 121.1 kg precipitate.
13. Zinc extractant (D2EHPA MM = 322.42 g/mol, d = 0.965 g/mL): 2 M corresponds to 645 g. However, the extractant can be recycled 20/30 times. Thus, considering a 25 times re-use: 645 g/25 x 969 L = 25 kg D2EHPA.
14. Sulfuric acid for Zn stripping: H_2SO_4 2 M, that is, 198 g x 969 L = 191.86 kg.

Output

1. Cake 1: 60 wt. % of the initial sludge is kept solid, that is, 600 kg;
2. NHW cake: only 8 wt. % of the initial cake is dissolved, that is, 600 kg x 0,92 = 550 kg;
3. Copper cement: considering a transfer of 11.1 g/L Cu from the solution (2,850 L) to the cement (85% purity grade) = 11.1 g/L x 2,850 L/0.85 = 37.22 kg;
4. Chromium precipitate: considering a transfer of 15 g/L Cr from the solution (2,422.5 L) to precipitate (containing 35% Cr) = 121.1 kg;
5. Zinc extracted in the SX step: considering 43.0 g/L of extracted Zn from 969 L liquor = 41.7 kg Zn, that is, 102.8 kg of zinc sulfate;
6. Nickel separated as raffinate: considering 44.4 g/L of separated Ni from 969 L liquor = 43.0 kg Ni, that is, 113.4 kg of nickel sulfate.

Once concluded this material balance, Table 6.5 shows the raw material production estimations.

Table 6.5. Estimation of the amounts of potential raw materials produced in the complete hydrometallurgical treatment

Product	Mass (kg)
Copper cement (90 wt.% Cu)	37.2
Chromium precipitate (35 wt.% Cr)	121.1
Zinc sulfate – technical grade	102.8
Nickel sulfate – technical grade	113.4

Finally, Table 6.6 compares two scenarios (the actual proposed treatment *vs.* landfilling as hazardous waste), from an economical point of view.

Table 6.6. Comparative economical evaluation of two different scenarios: landfilling as hazardous waste *vs.* the now proposed hydrometallurgical process

	Parameter	Cost (€/kg)	Total consumption (kg)	Income (€/kg)	Total Production (kg)	Total per parameter (€)
Landfilling	HW*	0.06	1,000	–	–	- 60.0
	Water	0.77/m^3	19.2	–	–	- 14.78
	Commercial H_2SO_4	0.3	991.86	–	–	-297.56
	Zn dust	1.52	34.80	–	–	-52.90
	Limestone ($CaCO_3$)	0.12	169.57	–	–	-20.35
Current proposal	Extractant (D2EHPA)	41.4	25.0	–	–	-1,035.0
	Kerosene	0.49	38.15	–	–	-18.69
	NHW**	0.033	550.0	–	–	-18.15
	Cu cement	–		2.58	37.2	95.97
	Cr precipitate	–		1.67	121.1	202.24
	Zinc sulfate	–		2.88	102.8	296.06
	Nickel sulfate	–		8.55	113.4	969.67
Total (€)						+ 106,51

*HW – hazardous waste; **NHW – non hazardous waste.

The estimatives for the cost of chemicals, raw materials and obtained products are based on the market prices provided by Portuguese suppliers. Cost values of landifilling were given by certified operators.

However, for copper cement and for chromium precipitate there are no reference prices. Thus, these were calculated taking into account the market price of the corresponding metal and the purity grade of the formed compounds.

From this rough economical estimation, a profit of about 106.51 € (~130 USD) per ton of treated sludge was obtained, while landfilling cost is 60.0 € (~73 USD) per ton However, several important costs were not considered in the present suggested method: the initial investment and equipment amortization, energy, labour, etc. All of those items will be included in a deeper future study that will firstly involve pilot plant tests.

7. Conclusion

This chapter details a complete flowsheet proved suitable to treat hazardous plating sludges and to recover existing valuable metals, namely Cr, Cu, Ni and Zn. The process comprises sequential application of simple, cheap and environmental-friendly hydrometallurgical steps that convert the plating sludge into interesting raw materials (Cu cement, Cr precipitate, and Ni and Zn sulfate salts), and a remaining non hazardous waste that might be landfilled at much lower costs.

References

[1] Crowson P., Minerals exploration: Too much or too little? In Astride Mining: Issues and Policies for the Minerals Industry, Mining Journal Books, London, UK, 2003. Accessed on November 2005 in http://www.curtin.edu.au/curtin/dept/wasm/postgraduate/courses/minecon/papers/Crowson%20Paper17203.pdf (2003).

[2] Pongrácz, E., Pohjola, V. (2004). Re-defining waste, the concept of ownership and the role of waste management. Resources Conservation and Recycling 40, 141-153.

[3] Magalhães, J.M., Silva, J.E., Castro, F.P., Labrincha, J.A. (2005). Physical and chemical characterisation of metal finishing industrial wastes. *Journal of Environmental Management* 75, 157-166.

[4] Dempsey, C.R., Oppelt. E.T. (1993). Air Waste 43, 25–73.

[5] U.S. Geological Survey (USGS), Washington, 2005. Accessed on February 2006 in http://minerals.usgs.gov/minerals/.

[6] London Metal Exchange. Accessed on February 2006 in http://www.lme.co.uk/dataprices_pricegraphs.asp.

[7] Castro, F. (1998). ECOCERAM I, Technical report, TecMinho – "Associação Universidade Empresa para o Desenvolvimento", Portugal.

[8] Castro, F. (1999). Ecoinert - inertization of metallurgical wastes by stabilization in ceramic products. Technical Report CRAFT Project no. BE-S2-2260.

[9] Castro, F. (2005). ECOCERAM II, Technical report, TecMinho – "Associação Universidade Empresa para o Desenvolvimento", Portugal.

[10] Magalhães, J.M., Silva, J.E., Castro, F.P., Labrincha, J.A. (2004). Effect of experimental variables on the inertization of galvanic sludges in clay-based ceramics. *Journal of Environmental Management* 106-B, 139-147.

[11] Magalhães, J.M., Silva, J.E., Castro, F., Labrincha, J.A. (2004). Role of the mixing procedure and composition of galvanic sludges on the inertization process in clay-based ceramics. *Journal of Hazardous Materials* 106B, 169–176.

[12] Magalhães, J.M., Silva, J.E., Castro, F.P., Labrincha, J.A. (2005). Kinetic study of the immobilization of galvanic sludge in clay-based matrix. *Journal of Hazardous Materials* B121, 69-78.

[13] Soil and waste pH. (1995). U.S. EPA method 9045 C, revision 3, January 1995.

[14] Gupta, C.K., Mukherjee, T.K. *Hydrometallurgy in extraction processes,* Vol I. CRC Press, Boca Raton, 1990, 225 pp.

[15] Bingöl, D., Canbazoglu, M., Aydogan, S. (2004). Dissolution kinetics of malachite in ammonia/ammonium carbonate leaching. *Hydrometallurgy* 76, 55-62.

[16] Chang, C.J., Liu, J.C. (1998). Feasibility of copper leaching industrial sludge using ammonia solutions. *Journal of Hazardous Materials* 58, 121-132.

[17] Gong, Y.J., Chen, J.Y. (1993). Kinetics of conversion of galena into lead carbonate in ammonium carbonate solution in the presence of cupric ion. *Hydrometallurgy* 33, 177-195.

[18] Jandová, J., Pedlík, M. (1994). Leaching behaviour of iron-nickel alloys in ammoniacal solution. *Hydrometallurgy* 35, 123-128.

[19] Jackson, E. (1986). *Hydrometallurgical extraction and reclamation.* Ellis Horwood Limited, Halsted Press, John Wiley & Sons (Ed.). 266 pp.

[20] Bell, S.L., Welch, G.D., Bennett, P.G. (1995). Development of ammoniacal lixiviants for the *in-situ* leaching of chalcopyrite. *Hydrometallurgy* 39, 11-23.

[21] Burkin, A.R. *The Chemistry of Hydrometallurgical Process.* E&F Spon, London, 1966, pp.157.

[22] Guan, Y.C., Han, K.N. (1997). The leaching kinetics of chalcopyrite (CuFeS2) in ammonium iodide solutions with iodine. *Metallurgical and Materials Transactions B* 28B, 979-1000.

[23] Richardson, H.W. Hydrometallurgical reclamation of copper from metal-bearing waste, in *Handbook of copper compounds and applications,* H. Wayne Richardson, New York, 1997, 339-342.

[24] Blanco, L.J.L., Zapata, V.F.M., Garcia, D.D.J. (1999). Statistical analysis of laboratory results of Zn wastes leaching. *Hydrometallurgy* 54, 41-48.

[25] Prosser, A.P. (1996). Review of uncertainty in the collection and interpretation of leaching data. *Hydrometallurgy* 41, 119-153.

[26] Flett D.S. (1999). Review — new reagents or new ways with old reagents. *J Chem Technol Biotechnol* 74, 99–105.

[27] Sole K.C., Solvent extraction and ion exchange, in: Short Course: Hydrometallurgy HYDRO 2003, 2003 *International Symposium on Hydrometallurgy,* 22–24 August 2003, Vancouver, Canada, 2003.

[28] Cox M., Solvent extraction in hydrometallurgy, in: J. Rydberg, C. Musikas, G.R. Choppin (Eds.), *Principles and Practices of Solvent Extraction,* Marcel Dekker Inc., New York, 1992, pp. 357–412.

[29] Cole P.M., Sole K.C., Solvent extraction in the primary and secondary processing of zinc, in: K.C. Sole, P.M. Cole, J.S. Preston, D.J. Robinson (Eds.), *Proceedings of the International Solvent Extraction Conference* ISEC 2002, vol. 2, 17–21 March 2002, Cape Town, South Africa, Chris van Rensburg Publications (Pty) Ltd., 2002, pp. 863–870.

[30] Kongolo K., Mwema M.D., Banza A.N., Gock E. (2003). *Cobalt and zinc recovery from copper sulphate solution by solvent extraction.* Miner Eng 16, 1371–1374.

[31] Cheng Y.C. (2000). Purification of synthetic laterite leach solution by solvent extraction using D2EHPA. *Hydrometallurgy* 56, 369–386.

[32] Reddy B.R., Sarma P.V.R.B. (2001). Transfer of nickel from sodium sulphate solutions to the spent electrolyte through solvent extraction and stripping. *Hydrometallurgy* 60, 123–128.

[33] Sarma P.V.R.B., Reddy B.R. (2002). Liquid–liquid extraction of nickel at macro-level concentration from sulphate/chloride solutions using phosphoric acid based extractants. *Miner Eng* 15, 461–464.

[34] Cerpa A., Alguacil F.J. (2004). Separation of cobalt and nickel from acidic sulfate solutions using mixtures of di(2-ethylhexyl) phosphoric acid (DP-8R) and hydroxyoxime (ACORGA M5640). *J Chem Technol Biotechnol* 79, 455–460.

[35] Alguacil F.J. (2002). Solvent Extraction with DP-8R/LIX 860 mixtures for the selective separation of cobalt and nickel. *Rev Metalurgia* 38, 205–209.

[36] Juan D. de, Meseguer V., Lozano L.J. (1998). Chromium extraction with organic solvents. Part 1: Extraction with DEHPA. *Rev Metalurgia* 34, 253–260.

[37] Tanaka M., Kobayashi M., Seki T., Recovery of nickel from spent electroless nickel plating baths by solvent extraction, in: K.C. Sole, P.M. Cole, J.S. Preston, D.J. Robinson (Eds.), Proceedings of the International Solvent Extraction Conference ISEC 2002, vol. 2, 17–21 March 2002, Cape Town, South Africa, Chris van Rensburg Publications (Pty) Ltd., 2002, pp. 787–792.

[38] Rickelton W.A., Nucciarone D., The treatment of cobalt/nickel solutions using CYANEX® extractants. Hydrometallurgy and refining of nickel and cobalt, in: W.C. Cooper, I. Mihaylov (Eds.), Proceedings of the Nickel-Cobalt 97 International Symposium, vol I, August 17–20, Sudbury, Ont., Canada, 1997, pp. 275–292.

[39] Lanagan M.D., Ibana D.C. (2003). The solvent extraction and stripping of chromium with Cyanex 272. *Miner Eng* 16, 237–245.

[40] Rickelton W.A., Flett D.S., West D.W. (1984). Cobalt–nickel separation by solvent extraction with bis(2,4,4-trimethylpentyl) phosphinic acid. *Solvent Extr Ion Exch* 2 (6), 815–838.

[41] Nucciarone D., Jakovljevic B., Fir Medeiros B.A., Hillhouse J., DePalo M., Extraction of selected metal cations with the alkylarylphosphinic acid: 2,4,4-trimethylpentylphenylphosphinic acid, in: K.C. Sole, P.M. Cole, J.S. Preston, D.J. Robinson (Eds.), *Proceedings of the International Solvent Extraction Conference ISEC* 2002, vol. 1, 17–21 March 2002, Cape Town, South Africa, Chris van Rensburg Publications (Pty) Ltd., 2002, pp. 402–407.

[42] Nucciarone D., Jakovljevic B., Fir Medeiros B.A., Hillhouse J., DePalo M., Extraction of selected metal cations with the alkylarylphosphinic acid: 2,4,4-trimethylpentylphenylphosphinic acid, in: K.C. Sole, P.M. Cole, J.S. Preston, D.J. Robinson (Eds.), *Proceedings of the International Solvent Extraction Conference ISEC* 2002, vol. 1, 17–21 March 2002, Cape Town, South Africa, Chris van Rensburg Publications (Pty) Ltd, 2002, pp. 402–407.

[43] Liranza E.G., Daudinot A.M., Martínez R.V., Bárzaga B.R., Separation of cobalt and nickel by solvent extraction from sulfate liquors obtained by acid leaching of a product from the Caron process, in: K.C. Sole, P.M. Cole, J.S. Preston, D.J. Robinson (Eds.), *Proceedings of the International Solvent Extraction Conference ISEC* 2002, vol 2, 17–21 March 2002, Cape Town, South Africa, Chris van Ransburg Publications (Pty) Ltd., 2002, pp. 952–957.

[44] Gupta, C.K., Mukherjee, T.K. *Hydrometallurgy in extraction processes,* Volume II. CRC Press, Boca Raton,1990, 262 pp.

[45] Sulka, G.D., Jaskuya, M. (2005). Influence of the sulphuric acid concentration on the kinetics and mechanism of silver ion cementation on copper. *Hydrometallurgy* 77, 131–137.

[46] Sedzimir, J.A. (2002). Precipitation of metals by metals (cementation) – kinetics, equilibria. *Hydrometallurgy* 64, 161-167.

[47] Moyes, A.J. The Intec Zinc Process (IZP). (2004). Accessed on April 2004 in http://www.intec.com.au/html/Technology/Lead_and_Zinc.shtm.

[48] Hansen, K.S. (1997). Model based analysis and control of a cementation process. *Computers Chem Eng* 21, S1099-S1103.

[49] Hsu, Y.J., Kim, M.J., Tran, T. (1999). Electrochemical study on copper cementation from cyanide liquors using zinc. *Electrochimica Acta* 44, 1617-1625.

[50] Karavasteva, M. (1998). The influence of copper on the effect of certain surfactants during the cementation of cadmium by suspended zinc particles. *Hydrometallurgy* 48, 361-366.

[51] Fabrice, G., Stéphane, B., Marc, A. (2004). Comparison of cementation in different reactor configurations: fixed bed or fluidised bed, assisted or not by electromagnetic field. *Proceedings from International Society of Electrochemistry,* 55th Annual Meeting, Thessaloniki, 19-24 September 2004.

[52] Karavasteva, M. (1996). The effect of certain surfactants on the cementation of copper by suspended zinc particles. *Hydrometallurgy* 43, 379-385.

[53] Reis, M.T.A., Carvalho, J.M.R. (1994). Recovery of heavy metals by a combination of two processes: cementation and liquid membrane permeation. *Minerals Engineering* 7, 1301-1311.

[54] Makhloufi, L., Saidani, B., Cachet, C., Wiart, R. (1998). Cementation of Ni ions from acidic sulfate solutions onto a rotating zinc disc. *Electrochimica Acta* 43, 3159-3164.

[55] Kuntyi, O.I., Znak, Z.O., Dyug, I.V. (2003). Contact deposition of copper powders onto zinc in H_2SO_4-$CuSO_4$ and H_2SO_4-$CuSO_4$-$ZnSO_4$ solutions and their morphology. *Russian Journal of Applied Chemistry* 76 (12), 1942-1945.

[56] Dvořák, P., Jandová, J. (2005). Hydrometallurgical recovery of zinc from hot dip galvanizing ash. *Hydrometallurgy* 77, 29-33.

[57] Espinosa, C.R.D., Tenório, J.A.S. (2000). Laboratory study of galvanic sudge's influence in the clinkerization process. Resources, *Conservation and Recycling* 31, 71-82.

[58] Segadães, A.M. Refractários. Fundação JJ Magalhães (Ed.), 1997, 249 pp.

[59] Aghababazadeh, R., Mirhabibi, A.R., Moztarzadeh, F., Salehpour, Z. (2003). Synthesis and characterisation of chromium oxide as a pigment for high temperature application. *Pigment & Resin Technology* 32 (3), 160-165.

[60] Ozel, E., Turan, S. (2003). Production and characterisation of iron-chromium pigments and their interactions with transparent glazes. *Journal of the European Ceramic Society* 23, 2097-2104.

[61] Flett, D.S. (2005). Solvent extraction in hydrometallurgy: the role of organophosphorus extractants. *J Org Chem* 690, 2426-2438.

[62] Sole, K.C. (2003). Solvent extraction and ion exchange. In: Dreisinger, D. (Organizer), *Proceedings of the International Symposium on Hydrometallurgy* HYDRO 2003, 22-24 August 2003, Vancouver, Canada, 28 pp.

[63] Motteram, G., Ryan, M., Berezowsky, R., Raudsepp, R., (1996). Murrin Murrin nickel/cobalt project – project development overview, in ALTA Nickel/Cobalt Pressure Leaching and Hydrometallurgy Forum. ALTA Metallurgical Services, Blackburn South, Australia.

[64] Burger, P.A., (1996). The Bulong, Western Australia, Ni/Co laterite deposits – a case history, in Nickel 96 Mineral to Market. Grimsey, E.J. & Neuss, I. (eds.). *The Australasian Institute of Mining and Metallurgy,* Melbourne, pp. 37-41.

[65] Chong, T.H., Sheikholeslami, R. (2001). Thermodynamics and kinetics for mixed calcium carbonate and calcium sulfate precipitation. *Chemical Engineering Science* 56, 5391-5400.

[66] Giuliette, M., Seckler, M.M., Derenzo, S., Ré, M.I., Cekinski, E. (2001). Industrial crystallization and precipitation from solutions: state of the technique. *Braz J Chem Eng* 18 (4), 423-440.

[67] Queneau, P.B. (2003). Recycling non-ferrous metals from U.S. industrial waste. In: Dreisinger, D. (Organizer), Proceedings of the International Symposium on Hydrometallurgy HYDRO 2003, Volume 2: *Electrometallurgy and Environmental Hydrometallurgy*, 22-24 August 2003, Vancouver, Canada, 1543-1553 pp.

[68] Zinc sulphate. Accessed on November 2005 in: http://www.cahe.nmsu.edu/employee/cca/docs/applications.pdf.

[69] Zinc sulphate. Accessed on November 2005 in: http://web.missouri.edu/~soilwww/webpub05/micro2_2004.htm.

[70] Zinc sulphate. Accessed on November 2005 in: http://www.cattletoday.com/archive/2005/April/CT391.shtml.

[71] Habashi, F. (1980). Principles of extractive metallurgy. Volume 1 – General Principles. New York: Gordon and Breach, 413 pp.

[72] Council Decision 2003/33/CE of 19 December. 2002. Official Journal of the European Communities L 11/27.

References of Reprinted Material

{I} Tables 3.1, 3.2, 3.3, 3.4 and Figures 3.1 and 3.2 – reprinted from Journal of Hazardous Materials, B121, Silva, J.E. et al., Leaching behaviour of a galvanic sludge in sulphuric acid and ammoniacal media, pp. 195-202, Copyright (2005), with permission from Elsevier.

{II} Figure 4.1 – reprinted from Solvent extraction and Ion Exchange, K.C. Sole, in: Short Course: Hydrometallurgy HYDRO 2003, 2003 International Symposium on Hydrometallurgy, 22–24 August 2003, Vancouver, Canada, 2003, with permission from the author.

{III} Figure 4.2 – reprinted from Journal of Minerals Engineering, 16, Lanagan, M.D., Ibana, D.C., The solvent extraction and stripping of chromium with Cyanex 272 , 237-245 pp., Copyright (2003), with permission from Elsevier.

{IV} Tables 4.24, 4.25, 4.26 – reprinted from Journal of Hazardous Materials, B120, Silva, J.E. et al., Solvent extraction applied to the recovery of heavy metals from galvanic sludge, 113-118 pp., Copyright (2005), with permission from Elsevier.

In: Trends in Hazardous Materials Research
Editor: Edward C. Booking, pp. 59-71
ISBN 1-60021-335-9

Chapter 2

DEVELOPMENT OF BENTONITE AND ZEOLITE-POLYACRYLAMIDOXIME COMPOSITES: THE USE IN URANIUM AND LEAD ADSORPTION

Selçuk Şimşek and Ulvi Ulusoy*
Cumhuriyet University, Department of Chemistry, Sivas, Turkey

ABSTRACT

Polyacrylonitryl (PAN) and bentonite (B)/zeolite (Z)-PAN composites were prepared by direct polimerization of pure AN and AN saturated suspensions of B and Z. PAN and the composites were subjected to amidoximation procedure to obtain PAO, B-PAO and Z-PAO. FT-IR and XRD were employed to characterize their structures. The sorption dependency of the materials on ion concentration and temperature were then investigated for Pb^{2+} and UO_2^{2+}. The reusability and storage ability of the adsorbents were also considered for Pb^{2+} and UO_2^{2+} adsorption.

All isotherms were L and H type of the Giles classification. For both ions, the adsorption capacities of B and Z-PAO composites were higher than that of pure PAO, when the PAO contents of composites were normalized to pure PAO. The introduction of B and Z in to PAO significantly increased the Langmuir equilibrium constants (L mol^{-1}), so as 353, 2180 and 149100 for Pb^{2+} 1980, 25900 and 284200 for UO_2^{2+} adsorption onto PAO, B-PAO and Z-PAO respectively. The adsorption was enthalpy controlled. For the use of 1 mol L^{-1} HCl as a regenerative effluent, the structure of PAO completely changed to be gel in the first use, whilst the composites were reusable for five sequential treatments without any change in their structures. Storage foregoing had also no effect on the structural stability.

The studied features of the composites suggest that these materials should be considered amongst the new adsorbents. It is envisaged that the use of B-PAO and Z-

---

* Corresponding author: Ulvi Ulusoy (MSc, PhD). Cumhuriyet University, Department of Chemistry, 58140-Sivas, Turkey. E-mail: ulusoy@cumhuriyet.edu.tr; Tel: +903462191010 Ext. 1623; Fax: +903462191186

PAO composites will provide practicality and effectiveness for separation and removal procedures involving di/trivalent cations.

Keywords: adsorption, PAN, composite, aluminosilicates, uranium, lead

1. Introduction

The adsorption processes are generally known to be one of the most effective techniques for removal and recovery of heavy metal ions because of the economic and environmental concerns. Adsorbents with strong affinity and high loading capacity for targeted metal ions have been developed by modifications of the surface of various substrates, such as polymers and clays with metal complexing groups [Celis et al., 2000 and Lagadic et al., 2001]. Whilst synthetic ion-exchange resins are expensive to use on a large scale, natural materials such as clay and zeolite are classified amongst the low-cost adsorbents [Ma et al., 2000, Babel et al.,2003 and Wang et al., 2003]. The mineralogical and physicochemical features of clay and zeolite are considered to be the basis for their immense importance in separation, ion exchange and catalysis.

The aggregation and coagulation of zeolite and clay particles under varying conditions of temperature and electrolytes lead variations in flow properties of these minerals. This is an undesired feature for their practical use as adsorbents [Luckham and Rossi, 1999, Starodoubtsev et al., 2002 and Godelitsas and Armbruster, 2003]. Having a composite of a mineral and a polymer, in which the mineral dispersed in the polymer network or *vice versa* may enable the use of mineral or polymer itself as an adsorbent confined in an isolated and practically usable medium in aquatic solutions. Beside this, the particles embedded in a network strengthen the structure and prevent its collapse in bad solvents. The enhancement in adsorptive features can also be possible with additional modifications.

Polyacrylamidoxime (PAO) is a well-known polymeric adsorbent prepared from modification of PAN and having considerable attention due its affinity particularly to uranium ions [Fisher and Lieser, 1993 and Dessouki et al., 1999]. Poor mechanical strength as depending on the applied procedures for polyacrylonitrile (PAN) modification to PAO and hydrolysis in acidic medium [Omichi et al., 1986 and Eldridge, 2000] are of limiting factors its use in adsorption and removal procedures.

In this study, the composites of bentonite (B) as of the clay of interest and a regional zeolite (Z) with PAN were introduced as new materials to provide an alternative solution for the above mentioned limitations. PAN was considered to be the primary adsorptive component after its modification to PAO, B and Z were attained as the supportive components for PAN besides their possible contribution to adsorptive features of the composites. The preparation and characterizations of PAO and B/Z-PAO composite, and their adsorptive features with reference to the dependency on concentrations and temperature for Pb^{2+} and UO_2^{2+} were investigated. The reason in choice of Pb^{2+} and UO_2^{2+} was due to their immense importance in view of environmental concerns. The adsorptive features were evaluated on the base of adsorption parameters derived from the compatibility of adsorption isotherms to the Langmuir and Freundlich models. Additional considerations were also given to the reusability and storage ability of the composites in Pb^{2+} and UO_2^{2+} adsorption.

2. Experimental

2.1 Reagents

Na-montmorillonite (bentonite) with a univalent CEC of 0.8 mol kg^{-1}, hydroxylamine hydrochloride ($NH_2OH.HCl$), acrylonitrile monomer (AN), N,N,N',N'-tetramethylethylenediamine (TEMED), ammonium peroxodisulfate, APS ($H_8N_2O_8S_2$), 4-(2-pyridylazo) resorcinol (PAR), $Pb(NO_3)_2$ and $UO_2(NO_3)_2.6H_2O$ were purchased from Sigma. All chemicals used were of analytical reagent grade.

Zeolite was obtained from Central North Anatolian occurrences associated with Eocene submarine volcanism. The certified chemical composition of zeolite is provided in Table 1. The mineral is composed of ~90% zeolite, as clinoptilolite $\{(Na,K)_6.[Al_6Si_{30}O_{72}].24H_2O\}$ and mordenite $\{Na_3KCa_2.[Al_8Si_{40}O_{96}].28H_2O\}$, 5% kuartz, 5% feldspar and smectite in trace level. The ratio of SiO_2/Al_2O_3 is 4.7, which suggests that the zeolite is clinoptilolite with reference to the classification of International Mineralogical Association [Godelitsas and Armbruster, 2003]. The zeolite rocks were crashed, ground and sieved to 100 mesh size. No pre-treatment was applied to the chemicals, clay and zeolite. All experiments were always performed in duplicates.

Table 1. Chemical composition of the studied zeolite, % (certified by IMO GmbH)

SiO_2	71.89	K_2O	0.57
Al_2O_3	15.16	TiO_2	0.36
CaO	6.51	BaO	0.13
Fe_2O_3	1.80	P_2O_5	0.08
MgO	1.80	MnO	0.05
Na_2O	1.06		
SrO	0.59	Total	100.00

2.2 Preparation of Composites and PAN and Amidoximation

B-PAN or Z-PAN was prepared by direct polymerization of acrylonitrile (AN) monomer adsorbed onto the clay or zeolite. A suspension of accurately weighed clay or zeolite in AN was mixed for 2 h on a magnetic stirrer, the excess of AN was then decanted. The AN adsorbed onto minerals was immediately polymerized by addition of 100 mg of APS in 1 mL water and 500 μL of TEMED. Bare PAN was also obtained by direct polymerization of AN monomer with the same procedure. The composites (B-PAN and Z-PAN) and PAN were subjected to several washes until the conductivity of distilled water was attained. The obtained materials were dried in ambient temperature. The final composition of the composites had 1/4 and 1/10 mass ratio of PAN to the clay and zeolite. The composites and PAN were ground and sieved to a particle size 50-100 mesh and stored in containers. One batch for each of B-PAN, Z-PAN and PAN were prepared at once to conduct the overall investigation.

The required amounts of the composites and PAN to conduct the adsorption investigations were subjected to amidoximation procedure modified from Şahiner et al., 1999. The materials were reacted with 0.05 mol L^{-1} of hydroxylamine hydrochloride solution (1:1 in $NH_2OH.HCl$ NaOH) in a rotary evaporator combined with a back coolant at 60°C for 3 h. Amidoximated samples were filtered to separate from solution, washed with several rinses of water and dried under ambient temperature.

FT-IR (Mattson 1000, UK) was employed to estimate chemical structure of bare B and Z, the composites, PAN and their amidoximated forms, while XRD (Rigaku Dmax 2200, using Cu K_α radiation, a proportional counter, 2θ/min scan rate) was used for powder diffractions of B, Z and their PAN composites. Swelling features and structural changes along with the study procedures were also considered.

2.3 Uranium and Lead Adsorption

Adsorptive features of the adsorbents were investigated for UO_2^{2+} and Pb^{2+}. 0.1g of the studied adsorbents (PAO, B-PAO and Z-PAO) in the studied solutions were equilibrated with 10 mL of UO_2^{2+} and Pb^{2+} at concentrations within the range from $2x10^{-4}$ to $10x10^{-3}$ $molL^{-1}$ (50-2500 ppm). The adsorbent-solution systems were equilibrated for 24 hours at 298 K in a thermostatic water bath and the suspensions were then centrifuged at 3000 rpm for 10 min. The initial and final pH values of the solutions were always within the range of 4-5, which were well below the precipitation pH level of the ions of interest, pH>6.5 [Celis et al., 2000].

The metal-dye detection procedure provided by Laussmann et al. [Laussmann et al., 1996] was modified for UO_2^{2+} and Pb^{2+} determination in the supernatants. A solution of $3.5x10^{-3}$ mol L^{-1} of PAR in 0.7 mol L^{-1} of Tris/HCl at pH 8-9 was prepared. A 50 μL fraction of supernatant was added onto 3 mL of the reagent and the absorbance of the formed metal complex was measured at 520 and 550 nm for UO_2^{2+} and Pb^{2+}.

To confirm accuracy of the results obtained from the metal-dye detection, UO_2^{2+} and Pb^{2+} in some selected samples were also determined by radioactivity measurements by means of a gamma spectrometer [NAI(Tl) detector combined with a EG&G ORTEC multi-channel analyzer and software, MAESTRO 32, MCA Emulator, USA] was employed. The activity measured at 186 keV of ^{235}U at natural level in U composition and that at 238.6 keV of ^{212}Pb at secular equilibrium with ^{232}Th extracted from florite rocks [Ulusoy and Şimşek, 2005].

2.4. Temperature Effect on Adsorption of UO_2^{2+} and Pb^{2+}

Temperature effect on adsorption for determination of thermodynamic parameters was studied for three temperatures; 283, 298 and 313 K. Duplicates (0.1 g each) of the materials were equilibrated with solutions at $2x10^{-3}$ mol L^{-1} (for B-PAO and Z-PAO) and $10x10^{-3}$ mol L^{-1} (for PAO) of UO_2^{2+} and Pb^{2+} at the chosen temperatures for 24 h. The samples were subjected to the same procedure described above and equilibrium concentrations were determined.

2.5 Reusability and Storage Ability

Hydrochloric acid at 0.5-1 mol L^{-1} was known to be an effective regenerant for resins with amidoxime but causes some hydrolysis of this active groups [Omichi et al., 1986 and Eldridge, 2000]. The effectiveness of the same effluent was also tested for the composites introduced in this study with a reusability study for UO_2^{2+} and Pb^{2+}.

To test of reusability, five duplicates of 0.1g PAO, B-PAO and Z-PAO samples in polypropylene tubes were equilibrated with 10 mL of $2x10^{-3}$ mol L^{-1} UO_2^{2+} and Pb^{2+} for 24 h and the adsorbed amounts were derived from the ion contents of supernatants. The contents of columns were eluted with 5 mL fractions of 10 mL of 1 molL^{-1} HCl and the columns were then washed with distilled water until the effluents had a neutral pH. The UO_2^{2+} and Pb^{2+} contents of the effluents (HCl) were determined. Each sample was subjected to the same procedure for five sequential times.

The storage ability of the composites was checked with reference to the repeatability of their adsorptive features for the ions of interest for one use for 2 months with a week interval.

2.6 Data Evaluation

The amounts of adsorption of UO_2^{2+} and Pb^{2+} (Q, mol kg^{-1}) were calculated from $Q=[(C_i-C_e)V/w]$, where C_i and C_e are the initial and equilibrium concentrations (mol L^{-1}), w is the mass of adsorbent (kg) and V is the solution volume (L). The Langmuir $[Q=(K_LX_LC_e)/(1+K_LC_e)]$ and Freundlich $(Q=aC_e^{\beta})$ models were fit to the isotherms experimentally obtained, where X_L is the monolayer sorption capacity (mol kg^{-1}), K_L is the adsorption equilibrium constant (L mol^{-1}) related to the adsorption energy. 'a' and 'β' are empirical Freundlich constants.

In obtaining thermodynamic parameters, the distribution coefficients (K_d) were derived from $K_d=Q/C_e$ for each temperature and 'ln K_d' was depicted against 1/T to provide adsorption enthalpy (ΔH^0, J mol^{-1}) and entropy (ΔS^0, J mol^{-1} K^{-1}) from the slopes ($\Delta H^0/R$) and intercepts ($\Delta S^0/R$) of the depictions with reference to '$\ln K_d = [(\Delta S^0/R)-(\Delta H^0/R)1/T]$', where R is ideal gas constant, 8.314 J mol^{-1} K^{-1} and T is the absolute temperature. Having had ΔH^0 and ΔS^0, ΔG^0 values were calculated from $\Delta G^0 = \Delta H^0-T\Delta S^0$ for 298K.

A t-test was applied to obtain the significance of regression coefficients (r) calculated from the compatibility of experimental data to the Langmuir and Freundlich models, and that to the linearity of the thermodynamic equation; $p<0.01$ was considered as the threshold for the significance [Miller and Miller, 1989].

3. Results and Discussion

3.1 Structural Evaluation

IR spectra of PAN and PAO; B, B-PAN and B-PAO; Z, Z-PAN and Z-PAO are compared in Figure 1-3.

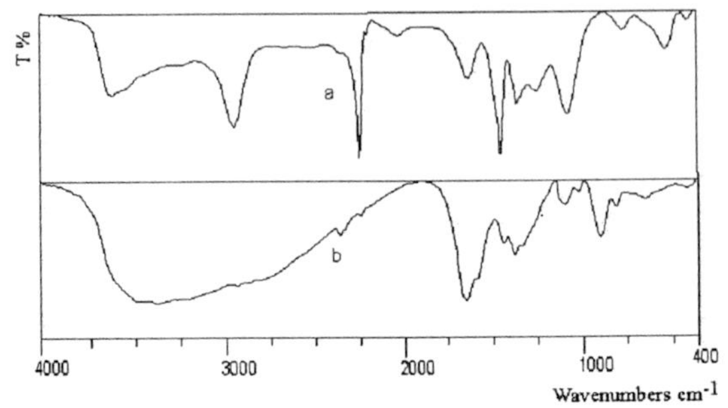

Figure 1. IR spectra of PAN (a) and PAO (b)

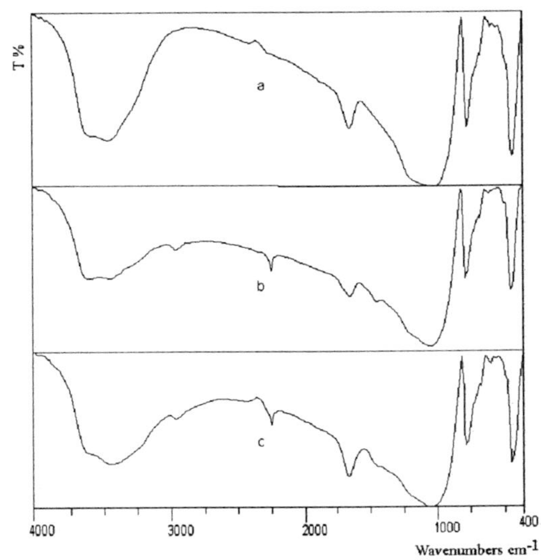

Figure 2. IR spectra of B (a), B-PAN (b) and B-PAO (c)

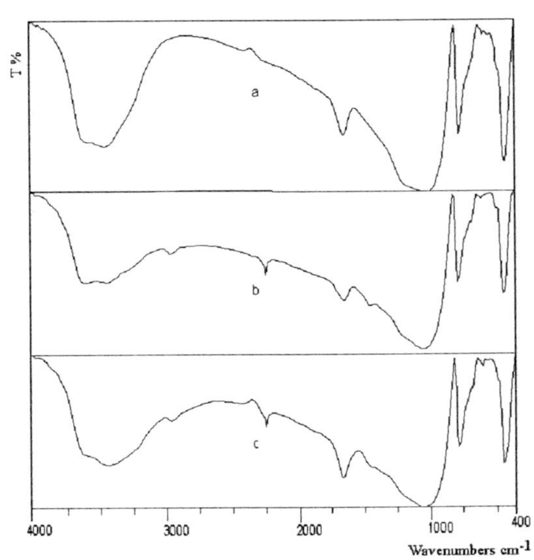

Figure 3. IR spectra of Z (a), Z-PAN (b) and Z-PAO (c)

The shrink in stretch around 2250 cm^{-1} and the growth in that around 1650 cm^{-1} was evidence for transformation of nitrile terminals to amidoxime in all figures associated with PAN and PAO. From comparison of the spectra b and c in Fig 2 and 3, it is seen that all nitrile groups were not transformed in to amidoxime. This was due to the time insufficiency for completion of amidoximation, which was not continued after 4 h reaction by considering its possible detrimental effect on the composites as reported earlier [Bryant et al., 2003]. The stretches at 400-1000 cm^{-1} of Si-O and Al-O of B and Z observed together with the nitrile and the C-H at 2960 cm^{-1} seen in spectra b and c of Fig 2 and 3 showed the inclusion of PAN in B and Z. The clear decrease in intensities at 425 and 1089 cm^{-1} of the Si-O, near 3700 cm^{-1} of the O-H stretching bands of the surface silanol, and 808 and 3642 cm^{-1} of Al-O-H of B and Z were observed. This might signify that there are interfacial interactions between the siloxane oxygen atoms or hydroxylated edges on the basal surface of the silicate layers of B with PAN [Shi et al., 1996]. This interaction is explained as the establishment of hydrogen bonds between the –Si–OH and =Al–OH surface groups and the neutral polyacrylonitrile, (-CH=CH-CN)$_n$ [Pefferkorn, 1999].

Besides the chemical bindings, there were also physical interaction between B and PAN, and Z and PAN. This effect appeared in XRD basal diffractions as plating in B and as amorphous change in Z (Fig 4 and 5). The clear increase in basal spacing from 12.4 Å of B to 13.6 Å of B-PAN should be an evidence for the presence of PAN in the interlayer of B (Fig 4). By considering these features, the both composites could be classified as the polymer-layered/filled crystal nanocomposites with reference to the classification by Alexandre and Dubois [Alexandre and Dubois, 2000].

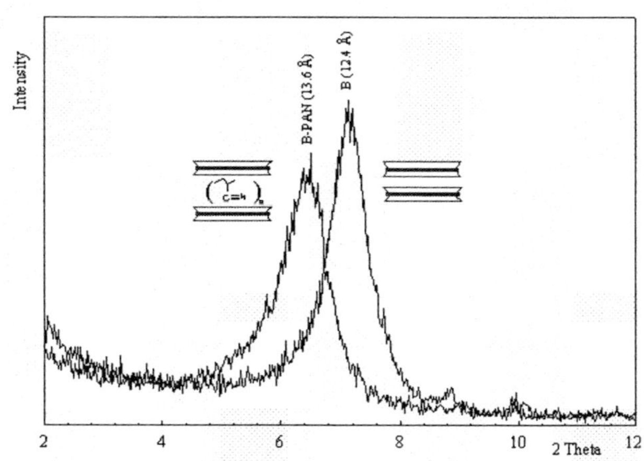

Figure 4. XRD basal diffraction of B and B-PAN, schemes represents PAN residence in B and its inter-layering effect

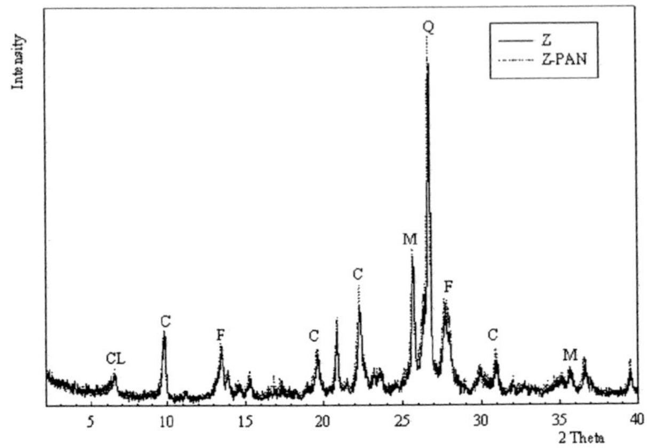

Figure 5. XRD spectra of Z (solid line) and Z-PAN [Clinoptilolite (C), Mordenite (M), Clay (CL), Kuartz (Q) and Feldspar (F) are the main composition of Z]

The adsorbed PAN onto B and Z can be considered as two parts: on the surface as a coating material and in the galleries and/or channels of B or Z. The incompleteness of transformation from PAN to PAO might be attributable to the latter as a result of possible strong physicochemical interactions between PAN and the minerals.

No swelling features were observed in both composites i.e. zeolite did not change its feature but the clay lost the water imbibing property.

3.2 The Adsorptive Features for UO_2^{2+} and Pb^{2+}

The adsorption isotherms of UO_2^{2+} and Pb^{2+} and their compatibility to the Langmuir and Freundlich models are compared for each adsorbent (PAO, B-PAO and Z-PAO) in Figure 6-8. Table 2 provides the Langmuir and Freundlich parameters derived from the fits.

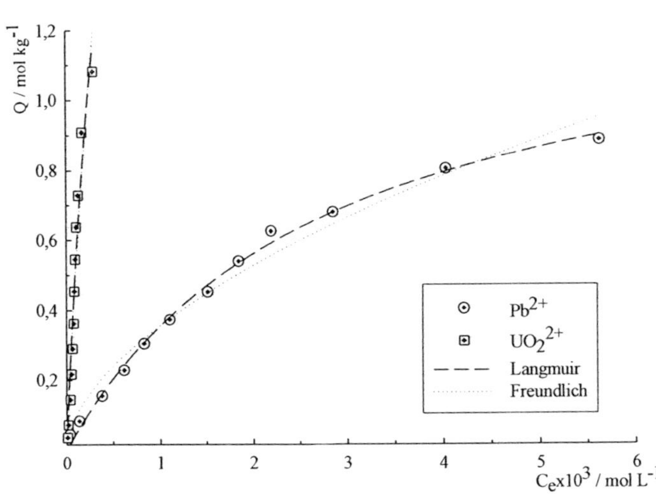

Figure 6. Isotherms with Langmuir and Freundlich fits for adsorption of Pb^{2+} and UO_2^{2+} onto PAO

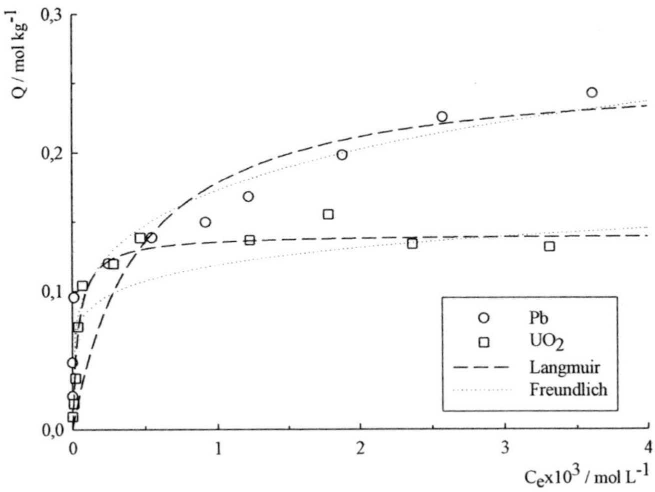

Figure 7. Isotherms with Langmuir and Freundlich fits for adsorption of Pb^{2+} and UO_2^{2+} onto B-PAO

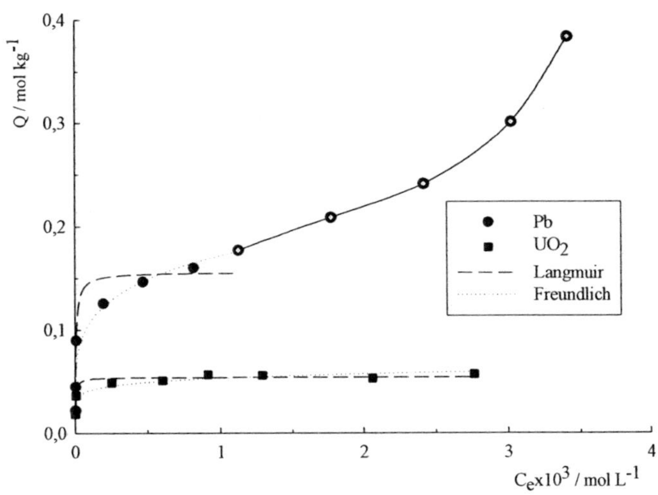

Figure 8. Isotherms with Langmuir and Freundlich fits for adsorption of Pb^{2+} and UO_2^{2+} onto Z-PAO (Continuation with solid line defines adsorption of Pb^{2+} as a second layer(s) after filling of the mono layer adsorption capacity)

The profiles of isotherms correspond to the high affinity class (L and H type) of the Giles classification, i.e. the adsorption has a chemical character for all. The adsorption hyperbolically increases with increasing concentration. The steep rise at the beginning lowers gradually and reaches a plateau defining the completion of filling the monolayer adsorption capacity (X_L). One exception was of Pb^{2+} adsorption onto Z-PAO for which the isotherm was evaluated in two parts; the first associated with monolayer coverage and the second considered being the second layer(s).

Table 2. Langmuir and Freundlich parameters obtained for Pb^{2+} and UO_2^{2+} adsorption onto PAO, B-PAO and Z-PAO

	Langmuir			*Freundlich*		
	X_l/ mol kg^{-1}	K_l/ L mol^{-1}	$^*R^2$	*a*	*β*	$^*R^2$
Pb^{2+}						
PAO	0.74	353	0.996	18.5	0.57	0.980
B-PAO	0.26 (1.04)**	2180	0.769	0.83	0.22	0.912
Z-PAO	0.16 (1.60)	149100	0.928	0.72	0.21	0.928
UO_2^{2+}						
PAO	0.31	1980	0.951	870	0.81	0.931
B-PAO	0.14 (0.56)	25900	0.963	0.32	0.15	0.791
Z-PAO	0.05 (0.50)	284200	0.958	0.11	0.11	0.872

* All coefficients of variations calculated for the Langmuir and Freundlich fits are significant at $p<0.01$

** The X_L values obtained for amounts normalized to the bare PAO contents of B-PAO (1/4) and Z-PAO (1/10).

The monolayer adsorption capacities (X_L) derived from the isotherms were in the order of PAO>B-PAO≥Z-PAO for both ions as being $Pb^{2+}>UO_2^{2+}$, but the sequence turned its reverse when the PAO contents of composites (1/4 for B-PAO and 1/10 for Z-PAO) normalized to the case if the composites was bare PAO, as whole (Table 2). In any case, the order obtained asserts that the adsorption capacity of PAO significantly increased with the inclusion of B or Z. However, the K_L values as the measure of spontaneity of adsorption also rose dramatically to provide a sequence of Z-PAO>B-PAO>PAO for both ions, as being in much favor of UO_2^{2+}. The reason of this can be explained by the fact that the additional electrostatic force contribution of negatively charged Si-O and Al-O of minerals to the readily exist in amidoxime terminals so that the electrostatic attractiveness for the cations increases significantly.

The preferable ion was UO_2^{2+} despite the fact that its radius is higher, ≈5 Å than that of Pb^{2+}, ≈1.2 Å. The preference for this ion is due to the high stability of complexes formed by UO_2^{2+} and the active sites of PAO and the minerals. However, it is known that UO_2^{2+} is a strong Lewis acid so the complexation with hard bases such as oxygen containing ligands is expected to be strong [Dai et al., 1997]. The required number of chelating amidoxime terminals for UO_2^{2+} [Kabay et al., 1998] are more than that of Pb^{2+} so that the monolayer adsorption capacity of the adsorbents for UO_2^{2+} was lower than that of Pb^{2+}.

Comparison of Freundlich and Langmuir parameters suggests that the presence of a systematic relation between the obtained for each ions with reference to the composites; the one with the higher X_L had the higher 'a' and that with the higher K_L had the lower 'β'. This relation might be interpreted as the 'a' is a measure of the adsorption intensity and 'β' is of the heterogeneity of adsorbents for each ion, as discussed in our previous communication [Ulusoy and Şimşek, 2005].

3.3 Thermodynamic Parameters for Adsorption of UO_2^{2+} and Pb^{2+}

Thermodynamic parameters derived from the depictions of 'ln Kd vs. 1/T' (Figure 9) and by using related equations are provided in Table 3.

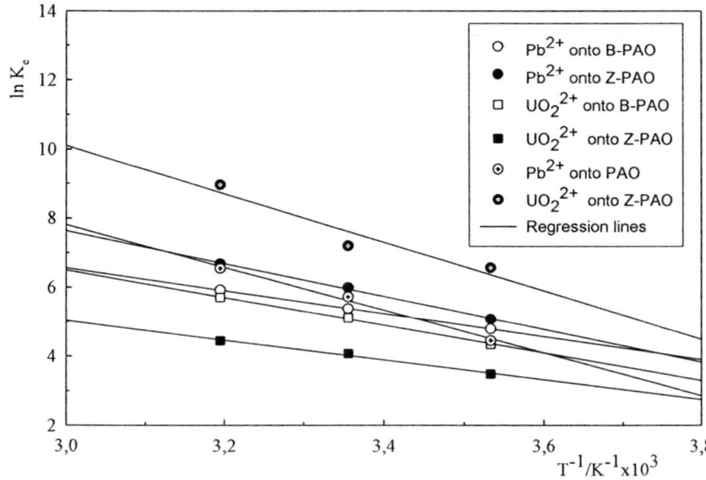

Figure 9. ln K_e vs 1/T plots in the sorption of Pb^{2+} and UO_2^{2+} onto PAO, B-PAO and Z-PAO to provide thermodynamic parameters

Table 3. Thermodynamic parameters for Pb^{2+} and UO_2^{2+} adsorption onto PAO, B-PAO and Z-PAO

	ΔH^0/ kJ mol^{-1}	ΔS^0/ J mol^{-1} K^{-1}	ΔG^0/ kJ mol^{-1}	$^*R^2$
Pb^{2+}				
PAO	39.6	180	-14.0	0.845
B-PAO	27.6	138	-13.5	0.999
Z-PAO	39.5	182	-14.7	0.997
UO_2^{2+}				
PAO	54.9	247	-18.7	0.899
B-PAO	33.1	153	-12.5	0.998
Z-PAO	23.8	113	-9.9	0.991

* Coefficients of variations for the linearity of ln K vs. 1/T depictions used in obtaining ΔH^0 and ΔS^0 are significant at $p<0.01$.

The ΔH^0, ΔS^0 and ΔG^0 values show that the adsorption has endothermic, increased entropy, and as expected spontaneity ($\Delta G^0<0$) features for all the studied adsorbents and ions. The adsorbent had a higher monolayer adsorption capacity had a higher adsorption enthalpy and spontaneity as well as that had a higher entropy change. This eventually asserts that the adsorption is an enthalpy-controlled process [Atun et al., 1998].

3.4 Reusability and Storage Ability

Reusability investigation showed that Pb^{2+} and UO_2^{2+} adsorbed onto PAO (0.15 mol kg^{-1}) completely recoverable with 20 mL of 1 mol L^{-1} HCl solution. After the first use, PAO turned to be gel as previously reported [Omichi et al., 1986 and Eldridge, 2000], which was not reusable at all. The adsorbed amounts of Pb^{2+} and UO_2^{2+} onto B-PAO were 0.12 and 0.09 mol kg^{-1} in the first and use so as the both ions were recoverable with the 20 mL of HCl. The adsorption lowered to 0.07 mol kg^{-1} for both ions (60% and 80% of the initial sorption) and remained recoverably constant for the other uses. For Z-PAO, the wholly recoverable adsorbed amounts also decreased from 0.13 mol kg^{-1} for Pb^{2+} and 0.06 mol kg^{-1} for UO_2^{2+} to constantly 0.05 and 0.04 mol kg^{-1} (40% and %70 of the initial sorption). The lowering capacity seen both composites was attributed to partial hydrolysis of PAO, which could not be hybridized with the texture of the minerals. Despite of the decline in these monolayer capacities, the values reached was still comparable with those of pure PAO in favor of the composites. The IR spectra obtained before and after reuses provided no evidence signifying any changes in the structures.

The storage ability tests for one use of the all adsorbents performed for the ions of interest for 4 weeks with one-week interval indicated that the fractional adsorptions remained unchanged, for which the unchanged IR spectra of the structures obtained with the intervals were also evidence.

4. CONCLUSION

This study introduced the composite of B and Z with PAN as alternative adsorbent materials. B or Z was the supportive components for PAN besides its adsorptive feature, whilst PAN was considered as the primary adsorptive component after its modification to PAO. Having synthesized the composites by a simply practical procedure, the adsorption dependency on concentration and temperature of PAO, B-PAO and Z-PAO were investigated for Pb^{2+} and UO_2^{2+}. The reusability and storage ability of the composites for the ions were also tested.

Langmuir parameters obtained for the adsorption of Pb^{2+} and UO_2^{2+} indicated that PAO immobilized on B and Z had substantially different adsorptive features from the pure PAO so that both adsorption capacity and its spontaneity of the adsorption for the studied ions increased. The adsorption was enthalpy controlled. For the use of 1 mol L^{-1} HCl as an regenerative effluent, the structure of PAO completely changed to be gel in the first use, whilst the composites were reusable for five sequential treatments without any change in structures.

The studied features of the composites suggest that these materials should be considered amongst the favorable adsorbents. It is envisaged that the use of B-PAO and Z-PAO composites will provide practicality and effectiveness for Pb^{2+} and UO_2^{2+} separation and removal procedures and they have potential research materials in studies of adsorption involving di/trivalent cations.

ACKNOWLEDGEMENT

This work was supported by The Research Fund of Cumhuriyet University (Project no: F-166) as the PhD study by S. Şimşek under U. Ulusoy's supervision, to which the authors are grateful. The authors acknowledge to Assoc. Prof. Filiz Ercan for her help in obtaining XRD spectra.

REFERENCES

Alexandre, M.; Dubois, P. *Mater Sci Eng.* 2000, 28, 1-63.

Atun, G.; Hisarlı, G.; Tunçay, M. *Coll Surf.* 1998, 143, 27-33

Babel, S.; Kurniawan, T.A. *J Hazard Mater.* 2003, 97, 219-243

Bryant, D.E.; Stewart, D.I.; Kee, T.P.; Barton, C.S. *Environ Sci Technol. 2003*, 37, 4011-4016

Celis, R.; Hermosin, M.C.; Cornjo, *J. Environ Sci Technol.* 2000, 34, 4593-4599.

Dai, S.; Shin, Y.S.; Barnes, C.E.; Toth, L.M. *Chem Mater.* 1997, 9, 2521-2525

Dessouki, A:M.; El-Tahawy, M.; El-Boohy, H.; El-Mongy, S.A.; Badawy, S.M. *Radiat Phys Chem.* 1999, 54, 627-635

Eldridge, R.J.; *Encyclopedia of Separation Science ;* 0-12-226770; Academic Press; 2000, 2271-2279

Fisher, H.J.; Lieser, K.H. *J Anal Chem.* 1993, 346, 934-942

Godelitsas, A.; Armbruster, T.; *Micropor Mesopor Mater.* 2003, 61, 3-24.

Kabay, N.; Demircioğlu, M.; Yaylı, S.; Günay, E.; Yüksel, M.; Sağlam, M.; Sreat, M. *Ind Eng Chem Res.* 1998, 37, 1983-1990

Lagadic, I.L., Mitchell, M.K.; Payne, B.D. *Environ Sci Technol.* 2001, 35, 984-990

Laussmann, T.; Eujen, R.; Weisshuhn, C.M.; Thiel, U.; Vogel, G. *Biochem J.* 1996, 315, 715-720.

Luckham, P.F.; Rossi, S. *Adv Coll Interf Sci.* 1999, 82, 43-92

Ma, Y.; Tong, W.; Zhou, H.; Suib, S.L. *Micropor Mesopor Mater.* 2000, 37 243-252

Miller, J.C.; Miller, J.N. Statistics for Analytical Chemistry; *Ellis Horwood series in Analytical Chemistry* 0-7458-0292-3; John Wiley & Sons; NY, 1989,108

Omichi, H.; Katakai, A.; Sugo, T.; Okamoto, J. *J Sep Sci Technol.* 1986, 21, 563

Pefferkorn, E. *J Colloid Interface Sci.* 1999, 216, 197-220

Şahiner, N.; Pekel, N.; Güven, O. *React Func Poly.* 1999, 39, 139-146

Shi, H.; Lan, T.; Pinnavaia, *J. Chem Mater. 1996*, 8, 1584-1587

Starodoubtsev, S.G.; Ryabova, A.A.; Dembo, A.T.; Dembo, K. A.; Aliev, I.I.; Wasserman, A.M.; Khokhlov, A.R. *Macromol.* 2002, 35, 6362-6369

Ulusoy, U.; Şimşek, S. *J Hazard Matter.* 2005, B127, 163-171

Wang, Y.H.; Lin, S.H.; Juang, R.S. *J Hazard Mater.* 2003, 102, 291-302

In: Trends in Hazardous Materials Research
Editor: Edward C. Booking, pp. 73-111

ISBN 1-60021-335-9

Chapter 3

STRATEGIES FOR REMOVAL OF TOXIC HEAVY METALS AND URANIUM FROM ENVIRONMENTAL SAMPLES

T. Prasada Rao**, Sobhi Daniel and K. Prasad
Inorganic Materials Group, Regional Research Laboratory (CSIR), Trivandrum, India

ABSTRACT

Certain inorganic species exert toxic effects when present in excess of threshold concentrations. Unlike toxic organics, inorganics are not biodegradable and so builds up in the environment. The adverse effects of heavy metals such as cadmium, lead, chromium, arsenic, mercury, etc at very low levels is well known and these are often regarded as highly toxic. In recent past, it is established that uranium and its compounds, like lead are highly toxic which cause progressive or irreversible renal injury and in acute cases, may lead to kidney failure and death. Hence, at this juncture, there is a need for exploring novel strategies for the removal of toxic heavy metals and uranium from hazardous environmental, biological and geological samples (particularly in view of the increasing realization of "Green Science and Technology").

The need for removal of toxic heavy metals and uranium from various hazardous materials is highlighted in the introductory part. Subsequently, the basic principles of different techniques employed for the removal of toxic inorganic species has been described. Finally, the removal strategies reported in literature by several researchers for toxic inorganic species viz. heavy metals, radio isotopes, uranium and binary/multivalent heavy metal mixtures from environmental, biological and geological samples have been discussed. The chapter ends with a concluding note highlighting the exciting and rapidly developing Solid Phase Extraction (SPE) technique for removal of micro and macro amounts of toxic inorganic species in a variety of hazardous materials.

* corresponding author:: tprasadaro@rediffmail.com

INTRODUCTION

Man and technology are inseparable parts of modern civilization, in order that technological advances make minimum impact on biosphere, it has become necessary to take adequate steps to cleanse the environment and check its further pollution through various control measures. Heavy metals such as lead, mercury, arsenic, chromium, nickel, cadmium, copper and zinc are usually associated with toxicity [1] which is aggravated by bioaccumulation. Major anthropogenic sources of heavy metals in the environment include metal extraction, metal fabrication and surface finishing, paints and pigments as well as manufacture of batteries. Guidelines for various hazardous materials have been put forward by different national governments for removal of toxic inorganics. Table 1 lists the potential toxic effects and maximum contaminant/permissible levels in drinking water, industrial effluent and sludge for some important toxic inorganics [2-4]. Researchers across the world have developed several removal strategies for treating the hazardous materials to bring down the concentration levels below the maximum permissible levels. The essential features of such removal strategies which render hazardous materials to harmless ones are discussed below in detail.

Table 1. Maximum permissible levels of toxic inorganics in drinking water, industrial effluent and sludge

Pollutant	Potential health effects	Maximum contaminant level (μg/ml)			Sewage sludge (mg/g) [4]	
		Drinking water (USEPA) [2]	Industrial effluents (Indian standards [3]			
			Inland surface waters	Public sewers	EU	USEPA
Arsenic	Skin damage & increased risk of cancer	0.01	0.2	0.2	-	41
Cadmium	Kidney damage	0.005	2	1	20-40	39
Chromium	Allergic dermatitis	0.1	0.1	2	-	-
Copper	Gastrointestinal distress, Liver or Kidney damage	1.3	3	3	1000-1750	1500
Lead	Kidney problems and high blood pressure	0.015	0.1	1	750-1200	300
Mercury	Kidney damage	0.002	0.01	0.01	16-25	17
Nickel	Decreased body weight, heart and liver damage & skin irritation	-	3	3	300-400	420
Uranium	Increased risk of cancer, kidney toxicity	0.03	-	-	-	-
Zinc		5	5	15	2500-4000	2800

TECHNIQUES FOR REMOVAL OF TOXIC INORGANIC SPECIES

The removal of toxic inorganic species from hazardous materials can be accomplished by several techniques. The techniques which find wide use include (a) liquid-liquid extraction, (b) ion exchange, (c) nano/ultra filtration, (d) biosorption, (e) coprecipitation, (f) electrodeposition and (g) solid phase extraction [5,6]. The basic principles of each of these techniques are described below.

Liquid-liquid Extraction

Liquid-liquid extraction (LLE) is based on the distribution of inorganic species of interest between two immiscible solvents. The formation of a charge saturated and/or coordination saturation of chemical species in aqueous phase by either chelation or ion-association is the prime requirement. The distribution ratio of particular inorganic species is usually defined as the ratio of its total concentration in organic phase to that in aqueous phase at equilibrium. While extracting the desired species, larger the distribution ratio of particular species and smaller that of coexisting species, higher the recovery of species of interest which will result in higher enrichment. The proper choice of extractant, pH or acidity of the aqueous phase, immiscible solvent, masking and salting out agents and modifiers are important. In some cases, the selective stripping of matrix species from the organic phase into a small volume of the aqueous phase can also improve the toxic species removal [7,8].

Ion Exchange

Ion exchange means the exchange of ions of like sign between an aqueous solution and a resin of highly insoluble body in contact with it. Ion exchangers consist of a framework carrying a positive or negative surplus charge which is compensated by mobile counter ionic species. To remove toxic inorganic species, it is passed through a column of an appropriate ion exchange resin; the exchangeable ionic species will be exchanged. Subsequently, the exchanged ionic species can be removed from the resin by eluting with a suitable eluent. The capacity of ion-exchangers is usually defined in terms of the number of exchangeable counter ions in the resin [9].

Nano/Ultrafiltration

Nanofiltration is a form of filtration that uses membranes to preferentially separate different fluids or ions. It is a pressure driven separation process with performance characteristics between reverse osmosis and ultra filtration. The filtration process takes place on a selective separation layer formed by an organic semi-permeable membrane. The driving force of the separation process is the pressure difference between the feed and the filtrate side at the separation layer of the membrane. The nanofiltration technique is mainly used for the removal of two valued ions and larger mono valued ions like heavy metals. Thus, nanofiltration can be viewed as a chemical free softening process as no regenerating chemicals are used [10].

Ultrafiltration, a form of filtration process, is not as fine as nanofiltration. Ultrafiltration also uses a membrane that is partially permeable to perform separation, but the membrane pores are typically much larger than the membrane pores that are used in nanofiltration [10].

Biosorption

Biosorption is a property of certain types of inactive, dead or live microbial biomass to bind and concentrate heavy metals from even very dilute aqueous solutions. Biomass exhibits this property, acting just as a chemical substance, as an ion exchanger of biological origin. It is particularly the cell wall structure of certain algae, fungi and bacteria which was found responsible for this phenomenon [11]. It can be defined as the ability of biological materials to accumulate heavy metals from wastewater through metabolically mediated or physicochemical pathways of uptake. The biosorption process involves a solid phase (sorbent or biosorbent; biological material) and a liquid phase (solvent, normally water) containing a dissolved species to be sorbed. Due to higher affinity of the sorbent for the sorbate species, the latter is attracted and gets bound by different mechanisms. The process continues till equilibrium is established between the amount of solid bound sorbate species and its portion remaining in the solution. The major advantages of biosorption over conventional methods include low cost, high efficiency, minimization of chemical or biological sludge, regeneration of biosorbent and possibility of metal recovery.

Coprecipitation

Coprecipitation is defined as the contamination of a precipitate by other toxic substances that remain in the solution under the conditions of the precipitation/surface sorption. Mixed crystal formation and occlusion mechanisms are considered to be responsible for coprecipitation [12]. The loss of toxic inorganic species can be minimized by following means (a) masking of the desired toxic species with a suitable complexing agent, (b) precipitation from homogenous solution, (c) prolonged digestion, (d) re-precipitation and (e) washing the precipitate with an appropriate solution.

Electrodeposition

The term electrodeposition has been used for the electrochemical reaction in which the product is insoluble in the solution phase and deposits on or in an electrode. The electrodeposition is usually carried out in an electrolysis cell, which essentially consists of two or three electrodes immersed in the electrolytic solution. In this cell, reduction takes place at one electrode which is called the cathode, and oxidation at the other which is called anode. The electrodeposition may occur at either electrode. Diffusion, convection and electrolytic migration are the three processes which regulate the movement of toxic inorganic species to the electrode [13].

Solid Phase Extraction

Solid phase extraction (SPE) is currently being used for the selective removal of toxic inorganic species. The basic principle of SPE is the transfer of toxic inorganic species from the aqueous phase to the active sites of the adjacent solid phase; it can also be termed as solid-liquid extraction. The transfer is simulated by the selection of appropriate optimal conditions in the system of 3 major components – water (liquid phase), target metal ion and sorbent [14]. The sorbed metal is then removed from sorbent with a suitable eluent (See Figure 1). The mechanism involved in SPE depends on the nature of the sorbent and toxic species to be removed. Thus, SPE is carried out by means of adsorption, ion exchange, chelation, ion pair formation etc.

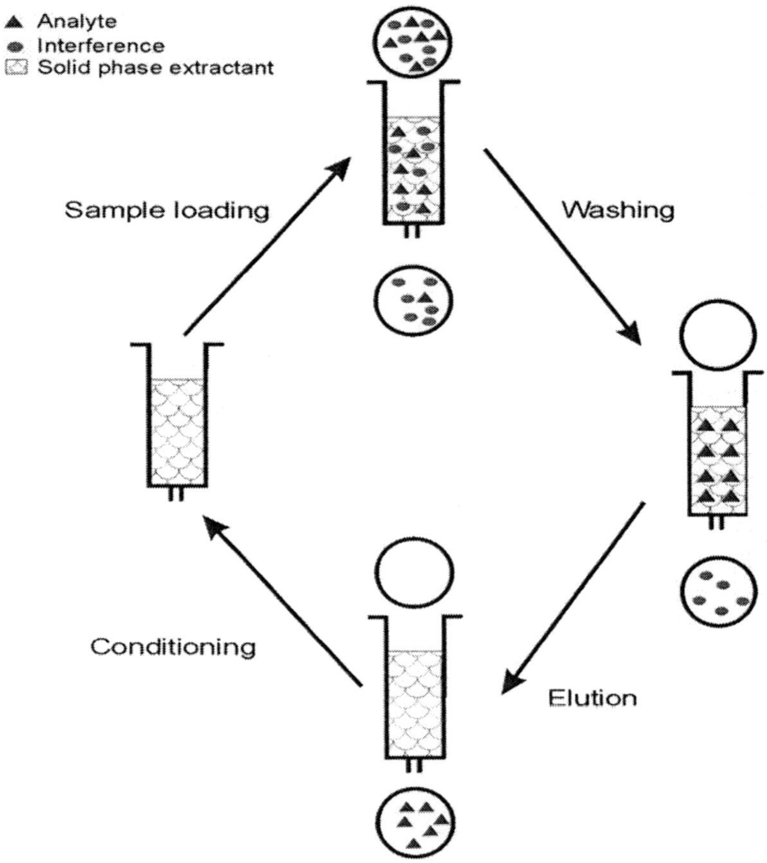

Figure 1. Priciple of Solidphase extraction

Hitherto, LLE was most often used among various toxic inorganic species removal strategies in view of its simplicity, rapidity, ready adaptability to scale up studies, easier removal of toxic metal species and recycling of extractant. SPE offers several advantages over LLE [15,16]. These include

a) flexibility
b) absence of emulsion
c) low cost because of lower consumption of reagents and
d) environmental friendly.

The sorbents employed in SPE can be broadly classified into inorganic based (inorganic oxides) and organic based (neutral and synthetic polymer) sorbent (See Figure 2). The most important inorganic based sorbents are silica gel, C_{18} bonded silica gel, modified silicas, alumina, florisil, diatomaceous earth, titania and zirconia. However, these inorganic based sorbents are beset with an additional problem due to the formation of suspensions. Hence, organic based sorbents are more popular and have been divided into polymeric and non-polymeric sorbents.

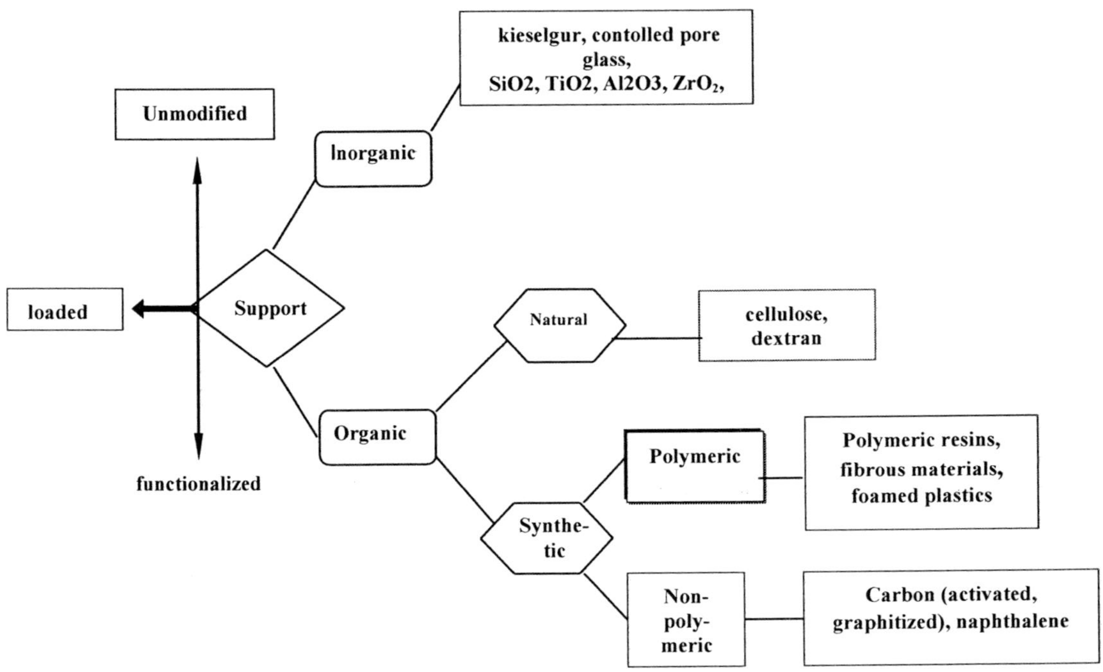

Figure 2. The various sorbents for solidphase extraction

Ion Imprinted Polymer – SPE

Ion imprinting is a versatile technique for creating macromolecular matrices that display selective ion recognition behaviour. This has shown remarkable promise as a technique for preparing materials with spatial recognition towards a target toxic metal ion by mimicking the "key" and "lock" principle. The concept of ion imprinting technology is described in Figure 3. Ion imprinted polymers are prepared by the copolymerization of functional and crosslinking monomers in presence of toxic target inorganic species (the imprint ion) which acts as an ionic template. The functional monomers initially form a complex with the imprint ion and following polymerization, the functional groups are held in position by the highly crosslinked polymeric structure. Subsequent leaching of imprint ion reveals binding sites that are complimentary in size and shape to the imprint ion [17,18].

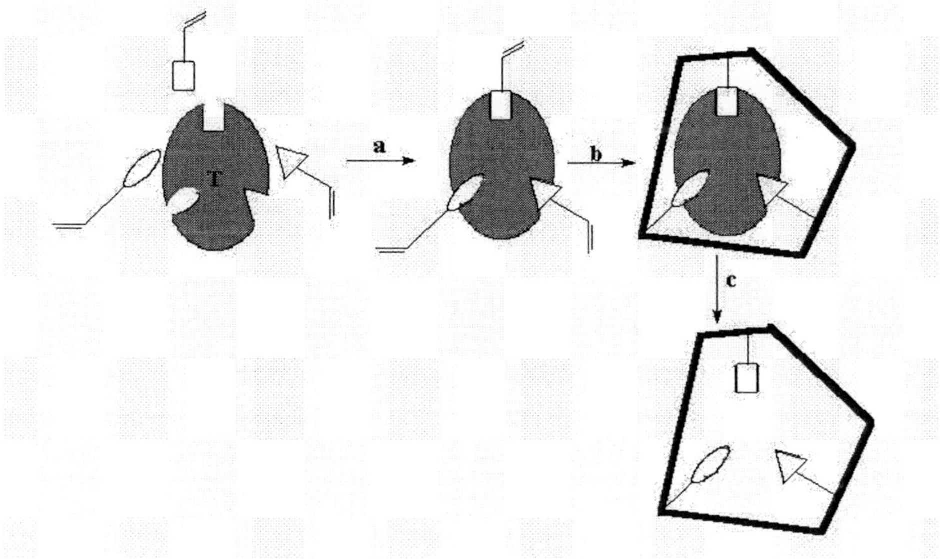

T-Template, a-Complexation, b-Polymerization, c-Leaching
Figure 3. Schematic representation of the imprinting process

Thus, an ionic memory is introduced into the polymer, which is now capable of rebinding the imprint ion selectively [19].

Removal Strategies for Toxic Inorganics

Chromium

Chromium is one of the contaminants, which exist in hexavalent and trivalent forms. Hexavalent form is more toxic than trivalent. Chromium compounds find their way into the environment mainly through tanning and electroplating industries. Strong exposure of chromiumVI) causes cancer in the digestive tract and lungs and may cause epigastric pain, nausea, vomiting, severe diarroea and hemorrage. It is therefore essential to remove chromium(VI) prior towaste water disposal. Various technologies are employed for removing toxic chromium from aqueous solutions. They are summarized as follows.

Ion Exchange

Batch adsorption experiments were carried out to evaluate the performance of IRN 77 and SKN 1 cation exchange resins in the removal of chromium from water and waste water solutions in the pH range 2.0-4.0 [20]. The percent removal of chromium was examined by varying experimental conditions viz. dosage of adsorbent, pH of solution and contact time. The adsorption capacity for chromium calculated from the Freundlich adsorption isotherm was found to be 35.38 and 46.34 mg/g of resins respectively. Solvent impregnated resins (SIR) have been prepared by a wet-impregnation technique using two different polymer matrices, Diaion HP 20 (hydrophobic styrene-divinyl benzene polymer) and HP-2MG (hydrophilic methacrylic based polymer) [21]. Aliquat 336 was employed as extractant and acetone as the solvent for impregnation. The optimal pH range and maximum adsorption capacity during removal of chromium(VI) were 3.2-8.1 and 10.32 mg/g of polymeric resin. The sorption of chromium(VI) from water and waste water on macroporous resins containing tertiary amine groups (Lewatit MP 62 and Lewatit M 610) was studied at varying chromium(VI) concentrations, adsorbent dose, pH, contact time and temperatures [22]. The optimal pH range and maximum adsorption capacity were found to be 2.0-6.0 and 148.4 mg/g respectively. Galan et al [23] have compared various ion exchange technologies for the removal of chromium(VI) from polluted ground waters. The experimental evidence points to a higher process efficiency of both anion exchange resins and emulsion pertraction despite the flexibility and ease of operation of the membrane assisted liquid-liquid extraction.

Nano/Ultrafiltration

Nanofiltration technology was adopted for the removal of chromium(VI) from aqueous solutions as an alternative to the conventional methods [24].Better removal was obtained at basic pH. Complexation – ultrafiltration process was used to remove chromium(VI) from waste waters [25]. The factors such as pH, concentration of polyethylene-imine, chloride and sulphate concentration, membrane pore size, applied pressure and extraction factor were investigated. Bohdziewicz [26] has adopted hybrid complexation-ultrafiltration process for removal of chromium(VI) from ground water. The most favourable pH of the filtered solution containing chromium(VI) was 6.0.

Table 2. Salient features of biosorption based removal processes for chromium(VI)

S.No.	Biomass	Sample	pH	Adsorption capacity (mg/g)	Ref.
1	*Eichhornia crassipes*	Aqueous solutions	1.0	36.6	[27]
2	Spirogara species (Green algae)	Aqueous solutions	2.0	14.7	[28]
3	*Dunaliella algae* (1&2) species	Saline waste waters	2.0	58.3/45.5	[29]
4	*Sargassum wightii* (brown sea weed)	Tannery waste water	3.5-3.8	82	[30]
5	Activated sludge	Synthetic waste water	-	-	[31]
6	Activated sludge	Synthetic waste water	-	-	[32]
7	Distillery sludge	-	2-3	5.7	[33]
8	Low cost biological wastes & Vermiculite	Tannery effluent	3.0	1.5	[34]
9	Tamarindus indica seeds	Aqueous solutins	2.0	90	[35]
10	Ocimum basilicum seeds	Aqueous solutions	1.5	205	[36]
11	Sugar beet pulp loaded with $Fe(OH)_3$	Aqueous solutions	4.4	81.5	[37]
12	Eucalyptus bark	Industrial waste	2.0	45.0	[38]
13	Fe modified steam exploded wheat straw	Aqueous solutions	1.0-3.0	-	[39]

Biosorption

Biosorption based processes employ either living or non-living biomass. The salient features of various biomass based sorbents for the removal of chromium(VI) are compiled in Table 2 [27-39].

Solid Phase Extraction

The important details of solid phase extraction procedures reported for chromium(VI) are compiled in Table 3 [40-53]. The sorbents employed in these procedures include acid/alkalic activated carbon or chemically immobilized or chelate sorbed inorganic/organic materials.

Table 3. Salient features of various SPE based procedures for removal of toxic chromium(VI)

S.No.	Sorbent	Chelating/activation agent	Sample	pH	Adsorption capacity (mg/g)	Ref.
1	Activated carbon (AC) (wood)	KOH/H_3PO_4	Aqueous solutions	3.0	315/186	[40]
2	AC (Agricultural waste)	Thermal (200-800°C)	Tannery waste water	2.0	9.53-22.29	[41]
3	AC (*Casurina equisetifolia* leaves)	H_2SO_4/Phosphate salt/$ZnCl_2$	Industrial waste water	2.5-3.0	12.0-20.0	[42]
4	AC (*Terminalia arjuna* nuts)	$ZnCl_2$	Aqueous solutions	1.0	-	[43]
5	Coals	-	Aqueous waste streams	1.0 & 3.0	10.4-31.2	[44]
6	Polyacrylamide saw dust	-	Aqueous solutions	3.0	12.5	[45]
7	Calcium alginate beads	With/without humic acid	Tannery effluent	8.2	-	[46]
8	Maghemite nanoparticles (Sol-gel)	-	Waste water	2.5	19.2	[47]
9	Ferrochromium slag	-	Industrial effluent	3.6-3.8	-	[48]
10	Ferrite process	-	Aqueous solutions	9.0	-	[49]
11	FeS (tetragonal)	-	Aqueous solutions	5.0/7.0	240/130	[50]
12	Bauxite	Thermal	Aqueous solutions	2.0	-	[51]
13	Akaganeite (Nanocrystalline)	-	Aqueous solutions	5.5	80	[52]
14	Alginate-goethite beads (composite)	-	Aqueous solutions	2.0	27.1	[53]

Miscellaneous

Hafez's group [54,55] have used reverse osmosis (RO) membrane for removal of chromium(VI) from spent tanning effluent. The low cost RO membrane units i.e. the medium and low pressure, could be used economically for separation and removal of chromium(VI) from waste water of small-size tanning shops. Rana et al [56] have developed an electrochemical based process for the removal of chromium(VI) from waste water. The metal ion removal was significantly increased at lower pH i.e. ~2.0. The effect of key variables such as pH, concentration and charge was determined. Chromium(VI) in contaminated soil has been removed by electrokinetic remediation [57]. A combined electro coagulation and electroflotation process was designed to reduce chromium(VI) to chromium(III) and then remove total chromium from waste water to a value below 0.5 mg/L [58]. Neto et al [59] have removed chromium(VI) from a tannery effluent by using a surfactant derived from a vegetable oil (coconut oil). Important variables such as agitation speed, solvent rate and total flow rate were investigated and optimized. The reuse of microemulsion after washing with HCl/H_2SO_4 is also demonstrated.

Nickel

Industrial wastewater containing nickel(II) is common because nickel(II) is used in a large number of industries such as electroplating, batteries manufacturing, mine, metal finishing and forging and so on. The nickel(II) concentration in wastewater from mine drainage, tableware plating, metal finishing and forging was up to 130 mg/L. Higher concentrations of nickel cause cancer of lungs, nose and bone. Dermatitis (Ni itch) is the most frequent effect of exposure to nickel, such as coins and jewellery. Acute poisoning of nickel(II) causes headache, dizziness, nausea and vomiting, chest pain, tightness of the chest, dry cough and shortness of breath, rapid respiration, cyanosis and extreme weakness. A number of methods are available for the removal of metal ions from aqueous solutions. The different strategies employed for the removal of nickel(II) are given below.

Ion exchange

Nickel present in waste water has been removed by cation exchange resin [60,61]. The effect of various parameters such as nickel ion concentration, degree of stirring, and temperature on the mass transfer coefficient [60] were studied. Tokuyama et al [61] have developed a novel moving bed with liquid pulse.

Electrodeposition

Nickel present in industrial waste water has been removed by electrochemical reduction [62,63]. A critical evaluation of different types of electrodes was also carried out.

Biosorption

Chlorella Sorokiniana [64] *Sargassum sp* [65] and strain of gram-negative *cocobacilli bacteria* [66] based biosorption have been used for removal of nickel present in aqueous solutions. The critical components of nickel removal viz. initial nickel concentration, pH and biomass concentration have been systematically optimized.

Solid Phase Extraction

Activated carbon prepared from hazelnut shell was used as an adsorbent for removal of nickel from aqueous solutions [67]. Batch mode adsorption studies were carried out by varying initial metal ion concentration, agitation speed, temperature and particle size. Saw dust [68] and clay based beds [69] have been employed for removal of nickel from aqueous solutions. The waste from tea factory has been used in batch mode [70] for removal of nickel from aqueous solutions. Madhan et al [71] and Vijayaraghavan et al [72] have removed nickel from aqueous solutions by using crab shell particles.

Copper

Copper is present in the wastewater of several industries. It is also a micronutrient in agriculture and can therefore accumulate in surface waters. The excessive intake of copper results in its accumulation in the liver and produces gastrointestinal problems. Consequently, it is essential that potable water be given some treatment to remove copper before domestic supply.

Liquid-liquid Extraction

The effect of formaldehyde was studied on removal of copper(II) via liquid-liquid extraction from an equimolar solution of EDTA and Aliquat-336 in kerosene from synthetically complexed solutions [73]. Variables such as initial concentration of copper(II), solution pH, HCHO and Aliquat 336 concentrations and temperature were optimized for efficient copper(II) removal.

Liquid Membranes

Liquid emulsion [74] and supported liquid membrane [75] processes are described for the removal of copper(II) from a waste mine water and aqueous solutions respectively.

Ion Exchange

Juang et al [76] have studied the effect of formaldehyde on the ion exchange based process for the removal of copper(II) from an equimolar EDTA solution with a strongly basic Amberlite IRA-400 resin. Experiments were conducted as function of initial copper(II) concentration, solution pH, HCHO concentration and temperature.

Ultra/Nanofiltration

Micellar enhanced ultrafiltration with surfactants [77,78] have been employed for the removal of copper(II) from water and synthetic waste water respectively. Carboxymethyl cellulose (CMC) as a metal binding polymer in combination with ultrafiltration was used in a hybrid approach for selective removal of copper(II) from water [79]. Ku et al [80] have demonstrated the feasibility of removal of copper(II) ions from aqueous solutions over a broad operational range.

Biosorption

The removal of copper(II) from aqueous solutions was demonstrated by using *Thuja Orientalis* [81], *Ulothrixzonata* [82], *Turbinaria ornata* [83] and *Ascophyllum nodosum* (immobilized in hydrophilic polyurethane foam) [84] biomass. Jalali et al [85] have investigated the role of sulphate reducing bacteria in copper removal from aqueous sulphate solutions. Dried sugar beet pulp, an agricultural solid waste byproduct, was used as a biosorbent for the removal of copper(II) from aqueous solutions [86]. The maximum uptake capacity of 28.5 mg/g at 25°C at initial pH of 4.0 was achieved by this biosorbent.

Solid phase extraction

The salient features of various SPE procedures described for copper(II) removal are described in Table 4 [87-112].

Miscellaneous

Various organic and inorganic ion exchangers with phosphate functional groups were studied as transport media for ion –exchange assisted electrodialysis of diluted solutions containing copper(II) ions [113]. The removal of copper metal was achieved by El-Batouti [114] using a rotating iron cylinder. Water clarifier sludge [115] and sand grains [116] packed in fluidized-bed reactor were employed for removal of copper(II) from aqueous solutions. A flotation based process for removal of copper(II) from effluents was described by Lazaridis [117].

Zinc

Zinc pollution is a very serious environmental problem. There are several sources of water pollution by zinc, such as petrochemicals, fertilizers and petroleum refining and steam generation power plants. The harmful effects of zinc are due to the diseases resulting from it, an irritant to skin itching and dermatitis, and kertinization of the hands and soles of the feet. Considering the harmful effect of this metal, it is necessary to remove from liquid wastes to a limit accepted by national and international regulatory agencies before their discharge to the environment.

Table 4. Salient features of various SPE based processes for removal of copper(II)

S.No.	Sorbent	Sample	pH	Adsorption capacity (mg/g)	Ref.
1	Activated carbon (Pecan shell)	Aqueous solutions	-	-	[87]
2	Activated carbon	Aqueous solutions	-	-	[88]
3	*Mesoporous aluminas* (aminated/protonated)	Aqueous solutions	-	-	[89]
4	TiO_2	Aqueous solutions	-	-	[90]
5	Anatase-type TiO_2	Aqueous solutions	> 5	-	[91]
6	Illuminated TiO_2	Water	-	-	[92]
7	Reclaimed wafers	Waters	-	-	[93]
8	Alumina/Chitosan composite	Polluted water	-	200	[94]
9	Chitosan/cellulose film	Aqueous solutions	-	121	[95]
10	Crosslinked chitosan beads	Aqueous solutions	-	80.71	[96]
11	Molecularly imprinted polymer microbeads	Aqueous solutions	-	48	[97]
12	Polyacrylic gel beads	Aqueous solutions	-	-	[98]
13	Poly(hydroxyethyl methacrylate- Co-methacrylamido histidine) beads	Aqueous solutions	7.0	122.7	[99]
14	Polyacrylamide hydrogels	Aqueous solutions	5.5	184.6	[100]
15	Poly-4-vinyl pyridine	Waste water	-	-	[101]
16	Tree fern	Aqueous solutions	-	11.7	[102]
17	Egg shell packed column	Aqueous solutions	-	-	[103]
18	Prawn shell	Aqueous solutions	-	-	[104]
19	Fly ash	Aqueous solutions	5.0	2.3-2.8	[105]
20	Sewage sludge ash	Waste water	-	3.2-4.1	[106]
21	Steel-making byproduct	Aqueous solutions	-	-	[107]
22	Saw dust	Aqueous solutions	-	-	[108]
23	Limestone	Water	-	-	[109]
24	Monmorillonite clays	Aqueous solutions	-	-	[110]
25	Phosphate rock	Aqueous solutions	-	10.7	[111]
26	Manganese-coated sand	Aqueous solutions	-	-	[112]

Biosorption

Zinc has been removed from electroplating effluents using *Azolla filiculoides* at an optimum pH of 6.0 [118] by both batch and column modes. The maximum uptake was found to be 45.2 mg/g. Complete desorption of zinc was accomplished with 120 ml of either 0.2 N H_2SO_4 or HCl. Sulphate reducing bacteria have been employed for removing zinc and sulphate from synthetic wastewater by using four parallel upflow-mode reactors [119].

Solid Phase Extraction

Zinc present in aqueous solutions was removed by sea nodule residue [120], waste green sands [121], dried animal bones [122] and limestone [123] as sorbents. The adsorption experiments were performed while varying sorbent particle size, sorbent concentration, initial zinc concentration, shaking time and temperature. Sodium titanate has been employed as solid phase extractant for removal of radioactive zinc(II) (^{65}Zn) present in micro/tracer levels [124]. The study comprises of the optimization of various physicochemical parameters viz concentration of extractant, temperature, pH and influence of extraneous inorganic species (cationic/anionic) on the adsorption process. The results show that the uptake of zinc(II) on a sodium titanate surface follows a first order rate law and that equilibrium data fit well with Freundlich adsorption isotherm. Zinc ions in aqueous solutions were removed by using hydrous zirconium oxide [125]. The adsorption behaviour of the material was studied as a function of concentration of zinc ions, temperature and pH. The removal was found to be increasing with pH of the adsorptive solution while it was suppressed in acid concentrations. The kinetics of adsorption follows first order rate law and agrees well with the classical Freundlich adsorption isotherm in the entire range of adsorptive concentration.

Miscellaneous

The effect of ultrasonication on the precipitation of zinc hydroxide as well as the separation of zinc hydroxide and gypsum precipitates by use of dissolved air flotation were described by Qi and Aldrich [126]. Both ultrasonic treatment and the addition of carboxymethyl cellulose (CMC) resulted in flocculation of the zinc hydroxide formed during precipitation. The best flotation results could be obtained using the CMC during precipitation of the hydroxides, followed by ultrasonic conditioning prior to flotation. Kelebek et al [127] have described acid leaching process for removal of zinc from oxygen furnace dust.

Arsenic

Arsenic is a common contaminant in natural water. The most dangerous problem in this context is that in many areas of the world, ground water is contaminated by arsenic. Due to its high toxicity, especially regarding arsenite, drinking water regulations only permit the presence of extremely low concentrations of arsenic. The presence of arsenic in water causes toxic and carcinogenic effects on human beings. Thus, different techniques have been developed and tried out for the removal of arsenic from water.

Liquid-liquid Extraction

Arsenic present in the form of anions such as arsenite and arsenate were removed from solid samples by supercritical CO_2 extraction coupled with ion pairing via tetrabutyl ammonium bromide [128].

Ion Exchange

Selective removal of arsenic(V) from drinking water was carried out with strong-base anion-exchange resin [129]. Recyling sodium chloride regenerant solution was carried out by precipitating the arsenic with $FeCl_3$.

Nano/ultrafiltration

The removal of arsenic from synthetic water and surface water [130] and drinking water [131] was possible by employing nanofiltration membranes. The use of ZW-1000 membranes remove arsenic from water sources [132]. Weng et al [133] have employed electroultrafiltration for the removal of arsenic. Arsenic present in drinking water [134] and ground water [135] was removed by flocculation and microfiltration process.

Biosorption

Fresh and immobilized *Garcinia cambogia* plant biomass [136], *oaringa oleifera* Lamarck seed powder [137], *Lessonia nigrescens* [138] and *P. chrysogenum* [139] biomass either in fresh or immobilized form have been used for removal of arsenic. Arsenic present in ground water was removed by waste tea fungal biomass [140] and bacteria [141].

Solid Phase Extraction

Activated carbon [AC) [142] / zirconium loaded carbon based sorbents [143] and AC prepared from solvent extracted olive pulp and olive stones [144] have been employed for removal of toxic arsenic in water. Molybdate impregnated chitosan [145], red mud [146], natural solids [147], and Portland cement [148] have been used as adsorbents for removal of arsenic. Granular TiO_2 [149], hydrous zirconium oxide [150], steel manufacturing byproducts [151] and zeolites [152,153] were found to remove toxic arsenic in water. DeMarco et al [154] have reported polymeric/inorganic hybrid sorbent for removal of arsenic. Polymer inclusion membrane (with a suitable carrier) [155] and zirconium(IV) loaded phosphoric chelate [156] sorbents have been reported for the removal of arsenic.

Detoxification of Arsenic with Fe(0) / Fe(II) / Fe(III)

The salient features of detoxification of arsenic based on use of Fe(0)/Fe(II)/Fe(III) are described in Table 5 [157-176].

Table 5. Significant features of detoxification processes for toxic arsenic

S.No.	Detoxification agent	Chemical specie	Sample	Ref.
1	Fe(0)	As(III)	Ground water	[157]
2	Fe(0)	Methylated arsenic	Ground water	[158]
3	Fe(0)	As(III)	Ground water	[159]
4	Fe(0)	As(III)	Aerated ground water	[160]
5	Fe(0)	As(V)	Water	[161]
6	Fe(0)	As(III)	Ground water	[162]
7	Fe(0)	As(III)	-	[163]
8	Fe(0) + Fenton's reagent	As(III)&(V)	Drinking water	[164]
9	Iron ores (natural)	As(V)	Contaminated water	[165]
10	Ferruginous manganese ore	As(III)&(V)	Ground water	[166]
11	Mixed valent iron oxides	As(V)	-	[167]
12	Granular AC based iron containing adsorbents	As(V)	Drinking water	[168]
13	Fe(II)	As(V)	Mining water	[169]
15	Silica containing Fe_2O_3	As	-	[170]
16	Iron oxide loaded slag	As(V)&(III)	Aqueous solutions	[171]
17	Bead cellulose loaded with iron oxyhydroxide	As(III)	Ground water	[172]
18	Polymer supported Fe(III) oxide nanoparticles	As(V)&(III)	-	[173]
19	Ferrihydrite crystallization control	As(V)	Aqueous solutions	[174]
20	Fe(III) & Coarse calcite	As(V)	Waters	[175]
21	Ce(IV) doped iron oxide	As(V)	Ground water	[176]

Miscellaneous

Combined use of photocatalyst and adsorbent [177] and solar light [178] have been employed for removal arsenic from waters. Electrocoagulation based processes [179,180] have been reported for removal of arsenic. Advanced oxidation process has been described by Zaw and Emett [181] for removal of arsenic. Ning [182] have investigated arsenic removal by a reverse osmosis process. Manganese green sand [183] has been used for removal of organic arsenic.

Technetium

Anion exchange resins were found to remove toxic ^{99}Tc (persistent and highly mobile) from waste streams and contaminated ground water [184]. These studies were performed on actual surrogate raw raffinates, filtrates and surrogate ground water samples.

Cadmium

The toxicity of cadmium given its persistence in the environment and relatively rapid uptake and accumulation by living organisms makes this element a serious environmental risk. Cadmium is a metal widely used in industries such as cadmium plating, alkaline batteries, copper alloys, paints an plastics. It is a non-essential and non-beneficial element to plants and animals. Diseases such as renal damage, anemia, and hypertension are associated with excess cadmium. Hence, it is important to eliminate trace amounts of cadmium from drinking water or to remove cadmium from wastewater before they are discharged into environment.

Liquid-Liquid Extraction

Cadmium in " wet process phosphoric acid (WPA)" was removed by solvent extraction procedure using amine extractant – notably Alamine 336 [185]. The 20 recyclings of organic phase to solvent extraction circuit with average removal of cadmium to an extent of 68.6% was reported.

Liquid Membranes

The viability of cadmium removal from wet phosphoric acid by means of three liquid membrane processes (emulsion (ELM) /supported (SLM) liquid membranes and non-dispersive solvent extraction (NDSX) using Cyanex 302 as organic extractant [186]. The rate of cadmium removal was found to be in the order: NDSX > ELM > SLM. Bulk liquid membrane using extractants such as Alamine 304 and Alamine 336 has been employed for removal of cadmium ions incorporated in calcium sulphate anhydrate [187].

Biosorption

The salient features of various biosorption procedures for removal of toxic cadmium are compiled in Table 6 [188-200].

Table 6. Salient features of biosorption based removal processes for cadmium

S. No.	Biomass	Sample	pH	Adsorptin capacity (mg/g)	Ref.
1	*Padina* sp.	Aqueous solutions	5.0	59.4	[188]
2	*Cladophora*	Synthetic waste water	-	-	[189]
3	*Tetraselmis suecica*	Aqueous solutions	-	-	[190]
4	*Trametes versicolor* (live/dead)	Waste water	-	102.3/120.6	[191]
5	*Phanrochaete chrysosporium* (immobilized/free)	Aqueous solutions	6.0	89/74	[192]
6	*Lentinus sajor* – caju (entrapped live/dead)	Aqueous solutions	-	104.8/123.5	[193]
7	Rice polish	Waste water	8.6	9.72	[194]
8	Wheat bran	Waste water	8.6	-	[195]
9	*Moringa oleifera* Lam seed powder	Aqueous solutions	6.5	-	[196]
10	Peat (NaOH activated)	Aqueous solutions	-	-	[197]
11	Acid sensitive biomaterials	Mid-gut gland of scalp	5.0	-	[198]
12	Elastin like polypeptide composed of a polyhistidine tail	Soil	-	-	[199]
13	Acid sensitive biomaterials	Dilute waste streams	-	-	[200]

Solid Phase Extraction

The important features of SPE procedures reported for cadmium are described in Table 7 [201-214].

Table 7. Salient features of various SPE based procedures for removal of toxic cadmium

S. No.	Sorbent	Chelating/activation agent	Sample	pH	Adsorption capacity (mg/g)	Ref.
1	Chitin	-	Aqueous solutions	-	-	[201]
2	Poly(2-hydroxyethyl (methacrylate) (p-HEMA)	Thionein	Human plasma	-	0.19	[202]
3	Polypropylene hallow fibre	PVA/cibacron	Aqueous solutions	5.0	22.6	[203]
4	2-Methacryloyl amido cysteine	-	Aqueous solutions	-	254	[204]
5	Junifer fibre	Base treated	Water	-	29.54	[205]
6	Carbonized jack fruit peel	-	Aqueous solutions	-	-	[206]
7	Synthetic hydroxyapatite	-	-	-	-	[207]
8	Rice husk	Carbonate treated	Waste water	-	45.9	[208]
9	TiO_2	Thiolactic acid	Water	-	-	[209]
10	Zirconium oxide	-	-	-	-	[210]
11	Hydrous manganese oxide	-	-	-	-	[211]
12	Bentonite	Quinoline-8-ol	Aqueous solutions	-	61.35	[212]
13	Kaolin/limestone/alumina/fly ash	-	Coal gasification flue gases	-	0.5-40	[213]
14	p-HEMA	N-methacryoloyl(L)-crysteine methyl ester	Human plasma	-	3.64	[214]

Miscellaneous

Jacobsen et al present for the first time electrodialytic removal of cadmium from wastewater sludge [215] and wood combustion fly ash [216]. Cadmium present in simulated wastewater was removed by electrodeposition on spiral wound steel electrode [217]. Dutra et al [218) removed cadmium from aqueous solutions by electrowinning in a flo-by cell. Cadmium present in phosphoric acid solutions was removed by precipitation with sodium sulphide [219].

Mercury

Mercury is generally considered to be one of the most toxic metals found in the environment. Also, it is well known that mercury is one of the most toxic pollutants with serious impact in the food chain due to potential bioaccumulation. Mercury, deriving from chlor-alkali, paper and pulp, oil refinery, paint, pharmaceutical and battery manufacturing industrial wastewater discharges, present an ongoing and serious threat to human health and natural water quality. The major effects of mercury poisoning manifest as neurological and renal disturbances as it can easily pass the blood-brain barrier and affect the foetal brain. High concentration of mercury(II) cause impairment of pulmonary function and kidney, chest pain and dyspnousea. Consequently, the removal of mercury in water and wastewater assumes importance. The necessity to reduce the amount of mercury ions in waste streams of industries and subsequent possible reuse of the metal ions, leads to the need for the development of better methodologies that can provide effective and total mercury removal from liquid and gas effluents.

Ion Exchange

Experimental and theoretical analysis of the adsorption process of mercury by Duolite GT-73 ion exchange resin (a macroreticular chelating resin containing thiol (S-H) functional groups) have been carried out [220]. The studied resin was found to have a very high adsorption efficiency, reaching 30-40% in weight and the efficiency decreases, at lower pH due to the competition between ions H^+ and Hg^{2+}. A simplified mathematical model has been identified, accounting for diffusional resistance inside solid particles, where the equilibrium relationship between mercury concentration in the liquid and solid is described by Freundlich isotherm, neglecting H^+ competition.

Liquid Membranes

Paez-Hernandez et al [221] for the first time have employed activated composite membrane (ACM) for the removal of mercury(II) using di(2-ethylhexyl) dithiophoric acid (DTPA) as ligand. The parameters that influence the above removal such as pH, concentration of mercury(II) and the ligand content were systematically studied. It was noticed that 93% of mercury(II) was removed with ACM prepared with 1.0 M DTPA casting solution and $2.49x10^{-4}$ mercury(II) feed solution containing 0.1M HCl.

Ultrafiltration

The removal of mercury(II) from aqueous solutions by complexation-ultrafiltration was investigated using polyethylenimine (PEI) as polymeric complexing agent [222]. The retention of mercury was found to dependent on pH and chloride ion concentration.

Biosorption

Tunable biopolymers based on elastin-like polypeptides (ELP), which undergo reversible thermal precipitation within a wide range of temperature enabled selective removal of mercury(II) even in presence of 100 fold amounts of cadmium, nickel and zinc [223]. The nanoscale biopolymers using metalloregulatory proteins employed in this study represent a "green" technology for environmentally benign mercury removal. Marine Microalga *crystoseira Baccata* based biosorption of mercury(II) was studied as a function of contact time, initial mercury concentration, solution pH, salinity and presence of several divalent cations [224]. The effect of addition of calcium, magnesium, zinc, cadmium, lead and copper has negligible effect on mercury(II) uptake. In addition, the authors report a simple competitive chemical model which explains the major characteristics of mercury(II) sorption data.

Electrolytic Deposition

The recycling of PEI and recovery of mercury from the concentrate of the first ultrafiltration was carried out by electrodeposition using a packed bed flow – through (PBFT) cathode [225]. Influence of operating parameters such as current intensity, pH, PEI and chloride concentrations was studied. A process based on coupled complexation – ultrafiltration and electrodeposition on PBFT cathode was proposed and tested.

Solid Phase Extraction

Solid phase extraction finds widespread use for removal of environmentally benign mercury by employing activated carbon, and several other inorganic and polymeric sorbents. The salient features of these reports are described in detail in Table 8 [226-243].

Miscellaneous

Mercury present in metallic form in waste incineration and soil thermal treatment off-gas, natural gas and glycol-overhead in a natural gas dryer was removed by oxidative membrane gas absorption [244]. Ku et al [245] studied the addition effect of the surfactant and other operating factors on the treatment of waste water containing mercury ions in aqueous solutions by cementation with sacrificing metal, zinc. The removal of mercury was found to depend on pH, temperature, amount of zinc, concentration of mercury ion, contact time and addition of several organic surfactants. Cementation of mercury was shown to be a feasible process to achieve a very high degree of mercury removal over a broad operational range within a fairly reasonable contact time. The presence of CTAB and Triton X-100 retarded the cementation of mercury by zinc.

Table 8. Salient features of various SPE based procedures for removal of toxic mercury(II)

S.No.	Sorbent	Chelating/activation agent	Sample	pH	Adsorption capacity (mg/g)	Ref.
1	Activated carbon (AC) (Bagasse pith)	Steam activation in presence of SO_2 & H_2S	Waste water	-	-	[226]
2	AC (biomass & coals)	Water vapor at 750°C	Aqueous solutions	5.0-10.0	-	[227]
3	AC (furfural)	Water vapour at 800°C	Aqueous solutions	4.0-8.0	174	[228]
4	Ac (virgin) & (sulphur impregnated)	Impregnation H_2SO_4, H_3PO_4	Gas phase sorption from flue gases	-	-	[229
5	ACs (Organic sewage sludge)	$ZnCl_2$	Water	5.0-12.0	44.8-130.8	[230]
6	Carbon aerogel	-	Aqueous solutions	5.0-7.0	-	[231]
7	Nanostructured sorbent (silica gel)	Cysteine	Gas phase sorption from flue gases	-	12	[232]
8	AC	-	Flue gases	-	0.89-1.39	[233]
9	Ac (chloride impregnated)	$ZnCl_2$	Flue gas	-	-	[234]
10	Fly ash	-	Aqueous solutions	-	3.2-5.0	[235]
11	Zeolite	-	Sewage & mixed industrial and domestic effluents	6.28/8.48	57.5	[236]
12	Chitosan	Animated	Aqueous solutions	7.0	4.6	[237]
13	Clay	2-mercaptobenzimidazole	Chlor-alkali industry waste water	4.0-8.0	35.73	[238]
14	Polyaniline-polystyrene	-	Aqueous solutins	-	160	[239]
15	Poly(2-hydroxyethyl methacrylate)	Poly(ethyleneimine)	Synthetic solutions	5.0	335	[240]
16	N-chlorosulphonamidated polystyrene resin	Acrylamide	Aqueous solutions	-	1380	[241]
17	Polyvinyl pyridine	Polycrylamide	Aqueous solutions	-	674	[242]
18	Keratin powder (Algerian sheep hooves)	-	Aqueous solutions	2-10	68	[243]

Lead

Lead is a recognized environmental pollutant that acts as a cumulative poison. The major environmental sources of metallic lead, its salts and oxides are pigments, battery industries, lead smelters, etc. Once mobile in the environment in ionic form, like many other toxic elements, it finds its way into the human body through drinking water, food and air. Lead has the ability to replace calcium in the bone to form sites for long-term release. So it must be removed from the wastewater before being discharged into the drainage system or drinking .

Ultrafiltration

A novel process combining an outside-in ultrafiltration membrane and suspended pellets in a stirred tank reactor for removing lead from aqueous solutions was investigated [246]. The incorporation of the membrane avoids breakthrough of the fine particulate matter, which can plague conventional fluidized pellet reactors. The lead removal efficiency increased with increasing pH of the solution and pellet surface area.

Biosorption

Bisorption based processes employ marine brown algae, leaves, coffee grounds or peat moss. The salient features of various biosorption procedures reported for removal of lead are compiled in Table 9 [247-252].

Table 9. Salient features of biosorption based removal processes for lead

S.No.	Biomass	Sample	pH	Adsorption capacity (mg/g)	Ref.
1	*Sargassum hystrix/sargassum Natans/ Padina pavonia*	Aqueous solutions	4-5	-	[247]
2	Cypress, cinchona and pine leaves	Aqueous solutions	-	-	[248]
3	Treefern	Water	4-7	38.1	[249]
4	Coffee grounds	Drinking water	-	0.7	[250]
5	*Hemidesmus indicus*	Aqueous solutions	5.0	100	[251]
6	Peat moss	Air pollution control residues of Municipal solid wastes	11.3-12.5	-	[252]

Solid Phase Extraction

The interesting features of various SPE based processes for removal of lead are described in Table 10 [253-268]

Table 10. Salient features of various SPE based processes for removal of toxic lead

S.No.	Sorbent	Chelating/activation agent	Sample	pH	Adsorption capacity (mg/g)	Ref.
1	CNTs	Chemical vapour deposition	Aqueous solutions	-	-	[253]
2	Activated carbon (AC) (coconut shell)	With/without sulphide treatment	Aqueous solutions	-	29.44	[254]
3	Polymetallic sea nodule	-	Contaminated water bodies	6.0	440	[255]
4	Zeolite and sepiolite	-	Aqueous solutions	-	-	[256]
5	Saw dust	-	Synthetic and real effluents	-	2.2	[257]
6	Granular blast furnace slag	-	Aqueous solutions	4.0	10.02	[258]
7	Carbon aerogel	-	Aqueous solutions	2.0-6.0	34.72	[259]
8	Polyacrylamide-bentonite and zeolite composites	Phytic acid	Aqueous solutions	4-5	-	[260]
9	Poly(hydroxyethyl methacrylate-N-methacroloyl-(L)-glutamic acid) copolymer beads	-	Aqueous solutions	5.0	348	[261]
10	Poly(amidoamine) dendrimers	-	Contaminated soils	5.0	-	[262]
11	Silica gel	Cyclam	Drinking water	-	-	[263]
12	Activated phosphate	HNO_3	Aqueous solutios	2-3	155.04	[264]
13	Insolubilized humic acid	-	Aqueous solutions	5.0	39.22	[265]
14	Natural clinoptilolite	-	Aqueous solutions	3.0-7.0	11.0	[266]
15	Comparison of adsorbents	-	Drinking water	-	-	[267]
16	Natural materials	-	Industrial waste water	-	-	[268]

Miscellaneous

Lead is removed from natural [269] and calcareous [270] soils by electrokinetic extraction. Mohamed [271] removed lead from silty clay polluted soil by a novel electro-dialysis method. Lead from aqueous solutions was removed by precipitation with sodium di(n-octyl)phosphinate [272] or by cementation on iron [273].

Uranium

Uranium can be found in the environment in very small amounts in rocks, soil, air and water. Man adds uranium metal and its compounds, because they are released during mining and milling processes and also through nuclear power reactor effluents. The uranium contamination causes progressive or irreversible renal injury and in acute cases may lead to kidney failure and death.

Solid phase extractants such as activated carbon (AC), AC modified with hydrophobic silica, elemental iron, terrestrial plants and ion imprinted polymers have been employed for removal of toxic uranium with various degrees. Nano- or ultra or rhizofiltration, electrolytic methods and sequential extraction approaches were also demonstrated for removal of toxic uranium(VI). The salient features of the above mentioned approaches for removal of toxic uranium from contaminated samples are summarized in Table 11 [274-281].

Table 11. Salient features of various uranium removal strategies

S.No.	Removal strategies	Sample	pH	Concentration of uranium (mg/L)	Sorbent extractant/ functional chelate	Ref.
1	SPE	Ground water	5.0-8.0	10	Polyacryloamidixime	[274]
2	SPE	Aqueous solutions	7.0	0.1	AC modified with hydrophobic aerogel	[275]
3	Rhizofiltration	Aqueous solutions	5.0	0.02-0.9	Terrestrial plants	[276]
4	Ultrafiltration	Contaminated waters	5.0-9.0	-	Polyethylenimine	[277]
5	Nanofiltration	Drinking water	5.9-8.3	-	Membranes	[278]
6	Electrolytic method	Ground water	8.0	19.04	-	[279]
7	Reductive precipitation	Ground water	-	18,000	Fe^{o}	[280]
8	Ion imprinted polymer (IIP) SPE	Synthetic nuclear power reactor effluent	3.5-5.0	300	Salicylaldoxime trapped IIP particles	[281]

Binary Toxic Species

Liquid Membranes

Silver and copper ions were removed from acidic thiourea solutions with a supported liquid membrane containing di-(2-ethylhexyl) phosphoric acid (D2EHPA) as carrier [282]. The fundamental parameters influencing transport of metal ions viz thiourea concentration in the feed phase, pH of strip solution, D2EHPA concentration in the membrane phase, silver and copper concentrations in the feed phase, pH of feed solution, temperature of the system, membrane support characteristics and the type of acid were studied and optimized.

Ion Exchange

Strong cation exchange resin, Amberlite IR-120, was used for the removal of toxic chromium and cadmium from waste waters [283]. Various parameters such as concentration, pH, stirring time and resin amount were varied and optimal conditions were established.

Ultrafiltration

Hybrid complexation – ultrafiltration [284] or simple ultrafiltration [285] based processes have been developed for removal of zinc and nickel and cadmium and lead respectively from aqueous solutions.

Biosorption

Microspora and Lemna minor [286], *Gloeothece magna* [287] and bacterial strain isolated from metal polluted soil [288] have been employed for removal of lead and nickel, cadmium and manganese and lead and copper respectively from aqueous solutions. Biomaterials such as chitosan, coffee, green tea, tea, yuzu, aloe and Japanece coarse tea were used for removal of cadmium and lead via adsorption in the pH range 4.0-6.7 [289]. *Pinus pinaster* bark [290] and grape stalk waste [291] have been used for removal of cadmium and mercury and copper and nickel respectively from neutral pH solutions.

Solid Phase Extraction

The salient features of various SPE procedures described for removal of binary mixtures of toxic metals are described in Table 12 [292-313].

Table 12. Salient features of SPE based processes for removal of binary mixtures of toxic heavy metals

S.No.	Toxic metals	Sorbent	Sample	pH	Adsorption capacity (mg/g)	Ref.
1	Cr(III) & Hg(II)	Sodium titanate	Aqueous solutions	3-10	-	[292]
2	Cr(VI) & Ni(II)	Raw bagasse	Aqueous solutions	6&8	-	[293]
3	Cr(VI) & Hg(II)	Moroccan stevensite	Aqueous solutions	2-5	10.9	[294]
4	Cr(VI) & Pb	Red mud	Aqeous solutions	2&4	-	[295]
5	Cr (VI)& Pb	Bagasse fly wash	Waste water	5&6	1.8 & 3.8	[296]
6	Cd & Ni	Bagasse fly ash	Waste water	6 & 6.5	-	[297]
7	Co & Ni	Sepiolite	Aqueous solutions	-	-	[298]
8	Pb & Cu	Animated polyacrylonitrile fibres	Diethylene triamine	Aqueous solutions	-	[299]
9	Cu & Zn	Bone char	Effluents	-	47.3 & 32.7	[300]
10	Cu & Zn	Bagasse fly ash	Waste water	-	-	[301]
11	Cu & Cd	Sea food processing waste sludge	Aqueous solutions	-	-	[302]
12	Cu & Cd	Activated caron (AC)	Aqueous solutions	6.0	20.8 & 19.5	[303]
13	Cu & Cd	Mn oxide coated AC	Aqueous solutions	-	-	[304]
14	Cu & Cd	Cellulose graft copolymers	Aqueous solutions	-	-	[305]
15	Zn & Cd	Zeolite	Waste solutions	-	-	[306]
16	Zn & Cd	Red mud	Aqueous solutions	4-5	-	[307]
17	As & Se	Calcined and uncalcined layered double hydroxides	Aqueous solutions	> 4	-	[308]
18	Ni & Cd	AC	Aqueous solutions	-	-	[309]
19	Cd & Pb	Zeolite (oil shale ash)	Waste water	-	-	[310]
20	Hg & Cs	Polymer-captopropylsilsesquioxane (PMPS) & copper(II) ferrocyanide complexed PMPS	Water	3.0/6.0	-	[311]
21	Hg & Pb	Hybrid macroporous materials	Waste water	-	283 & 257	[312]
22	U & Sr	Monosodium titanate	High level radioactive waste	-	-	[313]

Miscellaneous

A reverse osmosis process is described by Ipek [314] for removal of nickel and zinc from aqueous solutions. Paulson [315] removed zinc and cadmium from Coeur d'Alene river water (Idaho, USA) via biogeochemical process. Copper and nickel from dilute aqueous solutions was removed by a synthetic flocculant used in the water treatment field [316]. An innovative technique has been developed for removing mercury and lead via electrodeposition at a spinning graphite cathode [317]. Herck et al [318] reported a hydrometallurgical process for zinc and lead removal from blast furnace sludge.

Ternary/multicomponent Toxic Heavy Metal Mixtures

Liquid-liquid Extraction

Toxic heavy metals (Hg, Cu, Cd, Cr, Ni, Zn and Pb) present in waste water was removed by liquid-liquid extraction with complexing surfactants [319]. Chelation supercritical fluid extraction of toxic heavy metals and uranium was achieved by Lin et al [320] with sulphur containing organophosphorus reagents. Simultaneous and synergistic extraction of cationic and anionic heavy metallic species was reported using mixed sorbent extraction system and liquid membrane device [321]. Arous et al [322] removed Ag, Cu and Zn using supported liquid membrane containing cryptands as carriers.

Ion Exchange

Ion exchange resin based processes were developed for removal of heavy and transition metal ions from ground water [323] and plating factory waste water [324]. Economy et al [325] have reported the use of polymeric ion exchange fibres for removal of heavy metal contaminants. A multicomponent ion exchange method was reported by Mier et al [326] for removal of Pb, Cd and Cr(VI) using Mexican Clinoptilolite.

Microfiltration

Microfiltration membranes have been used for heavy metal capture [327] and in treatment of heavy metals [328]. A new hybrid flotation – membrane filtration process was described by Blocher et al [329] for removal of heavy metal ions from waste water. Back and Yang [330] have described humic substance enhanced ultrafiltration for removal of heavy metals.

Biosorption

Pantoea sp. TEM 18 [331], *Scenedesmus incrassatulus* [332] and *Plaerochaete chrysosporium* [333] have been used for removal of Cr(VI), Cd(II) & Cu(II), Cr(VI), Cd(II) & Cu(II) and Pb(II), Cu(II) and Zn(II) respectively. Blackgram husk [334] and Hazelnut shell [335] have been used for multimetal removal from toxic effluents. Chefetz et al [336] have adopted a new approach based on adsorption of Ag(I), Pb(II) and Ru(III) onto an aquatic plant, followed by the conversion of the adsorbed ions to the corresponding nanometallic particles by polyol reaction in a microwave oven.

Solid Phase Extraction

Salient features of the procedures developed for removal of ternary or multicomponent mixtures of toxic heavy metals are described in Table 13. [337-357].

Table 13. Salient features of SPE strategies for removal of ternary/multicomponent mixtures of toxic heavy metals

S. No.	Toxic metals	Sorbent	Chelating/activation agent	Sample	pH	Adsorption capacity (mg/g)	Ref.
1	Cr, Cu, Ni, Fe, Mn & Zn	Straw and straw xanthate	Alkali	Synthetic & actual industrial wastes	-	-	[337]
2	Cu, Zn, Cr(III), Ni & Hg(II)	Silica	Thiol/amino functionalization	Waste water	-	-	[338]
3	Cu, Zn & Cr	Activated carbon	Tetrabutyl ammonium iodide/sodium diethyl dithiocarbamate	Waste water	-	38, 9.9 & 6.84	[339]
4	Cr, Pb, Ni, Cd & Cu	Sand, silica, coal & alumina	-	Industrial effluents	-	-	[340]
5	Ni, Cr, Cd & Mn	Scolecite (Brazilian)	-	Waste water	6.0	-	[341]
6	Pb, Cr, Ni & Cu	Palygorskite clay	-	Aqueous solutions	7.0	62.1, 58.5, 33.4 & 30.7	[342]
7	Pb, Cd & Cr(III)	Polymer/organosmectite composites	Dithiocarbamate	Aqueous solutions	-	70.4, 31.8 & 20.3	[343]
8	Cu, Cd & Ni	Hydrogel	-	Waste water	2-4&8	-	[344]
9	Cd, Cu & As(III)	Glassy carbon	Cysteine methyl ester	Aqueous solutions	-	-	[345]
10	Co, Cu, Ni, Cd, Pb & Hg	Silica gel	Thiol adsorbent	Aqueous solutions	-	59-662	[346]
11	Zn, Cu & Pb	Zeolite	-	Aqueous solutions	-	-	[347]
12	Cu, Ni, Co & Mn	Kaolinite	-	Aqueous solutions	-	-	[348]
13	Ni, Cu & Zn	Clay sorbent	-	Waste water	5.0	-	[349]
14	Pb, Zn & Cu	Acidic polysaccharide gels	Amide derivative	-	-	-	[350]
15	Pb, Zn, Cd & Hg	Mesoporous silica	-	Waste water	-	-	[351]
16	Zn, Cd & Hg	Hydrous TiO_2	-	Aqueous solutions	3-10	-	[352]
17	Pb, Cu & Zn	Calcined phosphate	-	Water	5	85.6, 29.8 & 20.6	[353]
18	Hg, Cd & Pb	Polyurethane foam	Ammonium pyrollidine dithiocarbamate	Industrial effluents	-	-	[354]
19	Pb, Cd & Hg	Porous poly(MMA) beads	Methacryloylamidoglutamic acid	Waste water	5.0-6.0	65.2, 28.2 & 29.9	[355]
20	Cd, Pb & Hg	Poly(HEMA/Chitosan)	Procion Green H-4G	Aqueous solutions	7.5-8.5	43.6, 68.8 & 48.2	[356]
21	Cd, Pb & Hg	Poly(HEMA-methacryloyl amido hystidine)	-	Aqueous solutions	5.0	2.9, 14.8 & 9.4	[357]

Miscellaneous

Mohammadi et al [358] modelled metal ion removal from waste water by electrodialysis based on basic electrochemistry rules and copper ion separation experimental data. Subsequently,the same studies were extended to Zn, Pb and Cr ions.

CONCLUSION

As seen from the above account, it is clear that most of the strategies for removal of toxic heavy metals and uranium are for aqueous solutions or contaminated natural waters. Very few reports deal with real hazardous materials such as industrial effluents, mining/plating effluents, radioactive/nuclear wastes or contaminated soils/sediments and toxic biological fluids. We do hope this article will trigger the researchers worldwide to test their novel removal strategies (in particular currently emerging solid phase extraction) for toxic inorganics present in various hazardous materials.

REFERENCES

[1] Alve, M.M.; Ceca, C.G.G.; De, C.R.G.; Castanheira, J.M.; Periera, M.C.S. and Vascancelos, L.A.T. *Wat. Res.* 1993, 27, 1333-1338.

[2] www.epa.gov./safewater.

[3] Teri energy Data Directory Yearbook, 2000-2001.

[4] Sharpe, M.J. *Environ. Monit.* 2001, 3, 2N-6N.

[5] Rao, T.P.; Gladis, J.M. *Review in Anal. Chem.* 2001, 20, 145-159.

[6] Rao T.P.; Metilda, P.; Gladis, J.M. *Talanta,* 2006, 68,1047-1064.

[7] Morrison, G.H.; Freiser, H. (1957) *Solvent extraction in analytical chemistry,* Wiley; New York.

[8] Ritcey, G.M.; Ashbrook, A.W. (1984) *Solvent extraction: Principles and applications to process metallurgy* Part 1, Elsevier Science Publishers, London.

[9] Hellferich, F. (1962) *Ion exchange;* McGraw-Hill, New York.

[10] Schaefer, A. (1994) *Nanofiltration: Principles and applications,* Elsevier, London.

[11] Volesky, B. (2003) Sorption and Biosorption, B.V. *Sorbex Inc. St. Lambert* (Montreal) Quebec, Canada.

[12] Berg, E.W. (1963) *Physical and chemical methods of separation;* McGraw-Hill Book Company, New York.

[13] Vogel's text book of quantitative chemical analysis (1989) Longman Group, Uk Ltd.,; London.

[14] Junker-Bucheit. A. and Witzenbacher. M.J. *Chromatogr.* A 1996, 737, 67-74.

[15] Thuman. E.M.; Mills. M.S. (1998) *Solid phase extraction, in: Principles and practice;* Wiley; New York.

[16] Simpson. N.J.K. (Ed.) (2000) *Solid phase extraction: Principles, strategies and applications.* Marcel Dekker, New York.

[17] Sellergren. B. *Angew Chem. Int. Edn.* 2000, 39, 1031-1037.

[18] Yilmaz. E.; Haupt. K., and Mosbach. K. Angew. *Chem. Int. Edn.* 2000, 39, 2115-2118.

[19] Haupt. K. *Analyst* 2001, 126, 747-756.

[20] Rengaraj, S.; Yeon, Kyeong-Ho, and Moon, Seung-Hyeon *J. Hazard. Mater.* 2001, B87, 273-287.

[21] Kabay, N.; Arda, M.; Saha, B., and Streat, M. *React. Funct. Polym.* 2003, 54, 103-115.

[22] Gode, F., and Pehlivan, E.J. *Hazard. Mater.* 2005, B119, 175-182.

[23] Galan, B.; Castaneda,D., and Ortiz, I. *Wat. Res.* 2005, 39, 4317-4324.

[24] Hafiane, A.; Lemordant, D., and Dhahbi, M. *Desalination* 2000, 130, 305-312.
[25] Aliane, A.; Bounatiro, N.; Cherif, A.T., and Akretche, D.E. *Wat. Res.* 2001, 35, 2320-2326.
[26] Bohdziewicz, J. *Desalination* 2000, 129, 227-235.
[27] Mohanty, K; Jha, M; Meikap, B.C., and Biswas, M.N. *Chem. Engg. J.* 2006, 117, 71-77.
[28] Gupta, V.K.; Shrivastava, A.K., and Jain, N. *Wat. Res.* 2001, 35, 4079-4085.
[29] Donmez, G., and Aksu, Z. *Process Biochem.* 2002, 38, 751-762.
[30] Aravindhan, R.; Madhan, B.; Rao, J.R.; Nair, B.N., and Ramasami, T. *Environ. Sci. Technol.* 2004, 38, 300-306.
[31] Chen, Y., and Gu, G. *Bioresource Technol.* 2005, 96, 1713-1721.
[32] Chen, Y., and Gu, G. *Bioresource Technol.* 2005, 96, 1722-1729.
[33] Selvaraj, K.; Manonmani, S., and Pattabhi, S. *Bioresource Technol.* 2003, 89, 207-211.
[34] Sumathi, K.M.S.; Mahimairaja, S., and Naidu, R. *Bioresource Technol.* 2005, 96, 309-316.
[35] Agarwal, G.S.; Bhuptawat, H.K., and Chaudari, S. *Bioresource Technol.* 2006, 97, 949-956.
[36] Melo, J.S., and D'Souza, S.F. *Bioresource Technol.* 2004, 92, 151-155.
[37] Altundogan, H.S. *Process Biochem.* 2005, 40, 1443-1452.
[38] Sarin, V., and Pant, K.K. *Bioresource Technol.* 2006, 97, 15-20.
[39] Chun, L.; Hongzhang, Chen, and Zuohu, Li *Process Biochem.* 2204, 39, 541-545.
[40] Khezami, L., and Capart, R.J. *Hazard. Mater.* 2005, B123, 223-231.
[41] Mohan, D.; Singh, K.P., and Singh, V.K. *Ind. Eng. Chem. Res.* 2005, 44, 1027-1042.
[42] Ranganathan, K. *Bioresource Technol.* 2000, 73, 99-103.
[43] Mohanty, K.; Jha, M.; Meikap, B.C., and Biswas, M.N. *Chem. Engg. Sci.* 2005, 60, 3049-3059.
[44] Lakatos, J.; Brown, S.D., and Snape, C.E. *Fuel* 2002, 81, 691-698.
[45] Raji, C.; Anirudhan, T.S. *Wat. Res.* 1998, 32, 3772-3780.
[46] Pande, A.K.; Pandey, S.D.; Misra, V., and Srimal, A.K. *Chemosphere* 2003, 51, 329-333.
[47] Hu, J.; Chen, G., and Lo, I.M.C. *Wat. Res.* 2005, 39, 4528-4536.
[48] Erdem, M.; Altundogan, H.S.; Turan, M.D., and Tumen, F. *J. Hazard. Mater.* 2005, B126, 176-182.
[49] Erdem, M., and Tumen, F. *J. Hazard. Mater.* 2004, B109, 71-77.
[50] Mullet, M.; Boursiquot, S., and Ehrhardt, Jean-Jacques Colloids and Surfaces A: Physicochem. *Eng. Aspects* 2004, 244, 77-85.
[51] Erdem, M.; Altundogan, H.S., and Tumen, F. *Min. Engg.* 2004, 17, 1045-1052.
[52] Lazaridis, N.K.; Bakoyannakis, D.N., and Deliyanni, E.A. *Chemosphere* 2005, 58, 65-73.
[53] Lazaridis, N.K., and Charalanabous, Ch. *Wat. Res.* 2005, 39, 4385-4396.
[54] Hafez, A.I.; El-Manharawy, M.S., and Khedr, M.A. *Desalination* 2002, 144, 237-242.
[55] Hafez, A.I., and El-Manharawy, M.S. *Desalination* 2004, 165, 141-151.
[56] Rana, P.; Mohan, N., and Rajagopal, C. *Wat. Res.* 2004, 38, 2811-2820.
[57] Sawada, A.; Mori, Ko-ichi; Tanaka, S.; Fukushima, M., and Tatsumi, K. *Waste Management* 2004, 24, 483-490.
[58] Gao, P.; Chen, X.; Shen, F., and Chen, G. *Sep. Purif. Technol.* 2005, 43, 117-123.

[59] Neto, A.A.D.; De Castro Dantas, T.N., and Alencar Moura, A.J. *Hazard. Mater.* 2004, B114, 115-122.
[60] Elshazly, A.H., and Konosowa, A.H. *Desalination* 2003, 158, 189-193.
[61] Tokuyama, H.; Maeda, S., and Takahashi, K. *Sep. Purif. Technol.* 2004, 38, 139-147.
[62] Njau, K.N.; Woude, M.vd.; Visser, G.J., and Janssen, L.J.J. *Chem. Engg. J.* 2000, 79, 187-195.
[63] Koene, L., and Janssen, L.J.J. *Electrochim. Acta* 2001, 47, 695-703.
[64] Akhtar, N.; Iqbal, J., and Iqbal, M. J. *Hazard. Mater.* 2004, B108, 85-94.
[65] Kalyani, S.; Srinivasa Rao, P., and Krishnaiah, A. *Chemosphere* 2004, 57, 1225-1229.
[66] Ghafourian, H.; Rabbani, M.; Nazeri, Y., and Sadeghi, S. *Int. Congress Series* 2005, 1276, 270-271.
[67] Demirbas, E.; Kobya, M.; Oncel, S., and Sencan, S. *Bioresource Technol.* 2002, 84, 291-293.
[68] Shukla, S.S.; Yu, L.J.; Dorris, K.L., and Shukla, A.J. *Hazard. Mater.* 2005, B121, 243-246.
[69] Marquez, G.E.; Ribeiro, M.J.P.; Ventura, J.M., and Labrincha, J.A. *Ceramics Int.* 2004, 30, 111-119.
[70] Malkoc, E., and Nuhoglu, Y.J. *Hazard. Mater.* 2005, B127, 120-128.
[71] Pradhan, S.; Shukla, S.S., and Dorris, K.L. *J. Hazard. Mater.* 2005, B125, 201-204.
[72] Vijayaraghavan, K.; Jegan, J.; Palanivelu, K., and Velan, M. *J. Hazard. Mater.* 2004, B113, 223-230.
[73] Lin, S.; Kao, H.; Su, H., and Juang, R. *J. Hazard. Mater.* 2005, A120, 1-7.
[74] Valenzuela, F.; Fonseca, C.; Basualto, C.; Correa, O.; Tapia, C., and Sapag, *J. Min. Eng.* 2005, 18, 33-40.
[75] Molinari, R.; Argurio, P., and Pirillo, F. *J. Membr. Sci.* 2005, 256, 158-168.
[76] Juang, R.; Lin, S.; Kao, H., and Theng, M. *Chemosphere* 2005, 59, 1355-1360.
[77] Tung, Ch.; Yang, Y.; Chang, Ch., and Maa, J. *Waste Management* 2002, 22, 695-701.
[78] Liu, Ch.; Li, Ch., and Lin, Ch. *Chemosphere* 2004, 57, 629-634.
[79] Petrov, S., and Nenov, V. *Desalination* 2004, 162, 201-209.
[80] Ku, Y.; Chen, S., and Wang, W. *Sep. Purif. Technol.* 2005,43, 135-142.
[81] Nuhoglu, Y., and Oguz, E. *Process Biochem.* 2003, 38, 1627-1631.
[82] Nuhoglu, Y.; Malkoc, E.; Gurses, A., and Canpolat, N. *Bioresource Technol.* 2002, 85, 331-333.
[83] Vijayaraghavan, K.; Jegan, J.; Palanivelu, K., and Velan, M. *Chem. Engg. J.* 2005, 106, 177-184.
[84] Alhakawati, M.S., and Banks, C.J. *J. Environ. Management* 2004, 72, 195-204.
[85] Jalali, K., and Baldwin, S.A. *Wat. Res.* 2000, 34, 797-806.
[86] Aksu, Z., and Isoglu, I.A. *Process Biochem.* 2005, 40, 3033-3044.
[87] Dastgheib, S.A., and Rockstraw, D.A. *Carbon* 2001, 39, 1849-1855.
[88] Goyal, M.; Rattan, V.K.; Aggarwal, D., and Bansal, R.C. *Colloids and Surf. A. Pysicochem. and Eng. Aspects* 2001, 190, 229-238.
[89] Rengaraj, S.; Kim, Y.; Joo, Ch.K., and Yi, J. *JColloid. Interface Sci.* 2004, 273, 14-21.
[90] Yang, J., and Lee, S.J. *Colloid Interface Sci.* 2005, 282, 5-10.
[91] Kim, M.; Hong, K., and Chung, J.G. *Wat. Res.* 2003, 37, 3524-3529.
[92] Barakat, M.A.; Chen, Y.T., and Huang, C.P. *Applied Cat.* 2004, 53, 13-20.

[93] Abbadie, A.; Favier, S.; Giroud, C., and Billon, T. *Microelectronic Eng.* 2004, 71, 310-320.
[94] Steenkamp, G.C.; Keizer, K.; Neomagus, H.W.J.P., and Krieg, H.M.J. *Member Sci.* 2002, 197, 147-156.
[95] Lima, I.S.; Lazarin, A.M., and Airoldi, C. *Int. J. Biol. Macromol.* 2005, 36, 79-83.
[96] Wan Ngah, W.S.; Endud, C.S., and Mayanar, R. *React. Funct. Polym.* 2002, 50, 181-190.
[97] Say, R.; Birlik, E.; Ersoz, A.; Yilmaz, F.; Gedickbey, T., and Denizli, A. *Anal. Chim. Acta* 2003, 480, 251-258.
[98] Laguecir, A.; Frere, Y.; Danichev, L.; Loureiro, J.M.; Ernst, B., and Burgard, M. *Desalination* 2002, 146, 311-317.
[99] Say, R.; Garipcan, B.; Emir, S.; Patir, S., and Denizli, A. Colloids and Surf. A. *Physicochem, and Eng. Aspects* 2002, 196, 199-207.
[100] Kasgoz, H.; Ozgumus, S., and Orbay, M. *Polym.* 2001, 42, 7497-7502.
[101] Viel, P.; Palacin, S.; Descours, F.; Baureau, C.; Derf, F.Le, Lyskawa, J., and Salle, M. *Appl. Surf. Sci.* 2003, 212-213, 792-796.
[102] Ho, Y. *Wat. Res.* 2003, 37, 2323-2330.
[103] Vijayaraghavan, K.; Jegan, J.; Palanivelu, K., and Velan, M. *Min Eng.* 2005, 18, 545-547.
[104] Chu, K.H. J. *Hazard. Mater.* 2002, B 90, 77-95.
[105] Lin, Ch., and Chang, J. *Chemosphere* 2001, 44, 1185-1192.
[106] Pan, S.; Lin, Ch., and Tseng, D. *Resources, Conservation and Recycling* 2003, 39, 79-90.
[107] Lopez, F.A.; Martin, M.I.; Perez, C.; Lopez-Delgado, A., and Alguacil, F.J. *Wat. Res.* 2003, 37, 3883-3890.
[108] Yu, B.; Shukla, A.; Shukla, S.S., and Dorris, K.L. *J. Hazard. Mater.* 2000, B 80, 33-42.
[109] Aziz, H.A.; Othman, N.; Yusuff, M.S.; Basri, D.R.H.; Ashaari, F.A.H., Adlan, M.N.; Othman, F.; Johari, M., and Perwira, M. *Environ. Int.* 2001, 26, 395-399.
[110] Cox, M.; Rus-Romero, J.R., and Sheriff, T.S. *React. Funct. Polym.* 2004, 60, 215-222.
[111] Sarioglu, M.; Atay, U.A., and Cebeci, Y. *Desalination* 2005, 181, 303-311.
[112] Lee, Ch.; Yang, W., and Hsieh, Ch. *J. Hazard. Mater.* 2004, B114, 45-51.
[113] Vasilyuk, S.L.; Maltseva, T.V., and Belyakov, V.N. *Desalination* 2004, 162, 249-254.
[114] El-Batouti, M. J. *Colloid and Interface Sci.* 2005, 283, 123-129.
[115] Lee, Ch.; Yang, W., and Chiou, Ch. *J. Hazard. Mater.* 2006, B129, 58-63.
[116] Lee, Ch.; Yang, W., and Hsieh, Ch. *Chemosphere* 2004, 57, 1173-1180.
[117] Lazaridis, N.K.; Peleka, E.N.; Karapantsios, Th.D., and Matis, K.A. *Hydrometallurgy* 2004, 74, 149-156.
[118] Zhao, M.; Duncan, J.R., and Van Hille, R.P. *Wat. Res.* 1999, 33, 1516-1522.
[119] Tuppurainen, K.O.; Vaisanen, A.O., and Rintala, J.A. *Min. Eng.* 2002, 15, 847-852.
[120] Agrawal, A.; Sahu, K.K., and Pandey, B.D. Colloids. and Surf. A. *Physicochem.Eng. Aspects* 2004, 237, 133-140.
[121] Lee, T.; Park, J., and Lee, J. *Chemosphere* 2004, 56, 571-581.
[122] Banat, F.; Al-Asheh, S., and Mohai, F. *Sep. Purif. Technol.* 2000, 21, 155-164.
[123] Nuttall, C.A., and Younger, P.L. *Wat. Res.* 2000, 34, 1262-1268.
[124] Mishra, S.P.; Singh, V.K., and Tiwari, D. *Radiochim. Acta* 1997, 79, 11-16.

[125] Mishra, S.P.; Singh, V.K., and Tiwari, D. *J. Radioanal. Nucl. Chem.* 1996, 210, 207-217.

[126] Qi, B.C., and Aldrich, C. *Min. Engg.* 2002, 15, 1105-1111.

[127] Kelebek, S.; Yoruk, S., and Davis, B. *Min Eng.* 2004, 17, 285-291.

[128] Wang, T., and Guan, Y. *Sep. Purif. Tech.* 2003, 31, 225-230.

[129] Korngold, E.; Belayev, N., and Aronov, L. *Desalination* 2001, 141, 81-84.

[130] Saitua, H.; Campderros, M.; Cerutti, S., and Padilla, A.P. *Desalination* 2005, 172, 173-180.

[131] Kosutic, K.; Furac, L.; Sipos, L., and Kunst, B. *Sep. Purif. Technol.* 2005, 42, 137-144.

[132] Floch, J., and Hideg, M. *Desalination* 2004, 162, 75-83.

[133] Weng, Y.; Chaung-Hsich, L.H.; Lee, H.; Li, K., and Huang, C.P. *J. Hazard. Mater.* 2005, B122, 171-176.

[134] Han, B.; Runnells, T.; Zimbron, J., and Wikramasinghe, R. *Desalination* 2002, 145, 293-298.

[135] Wikramasinghe, S.R.; Han, B.; Zimbron, J.; Shen, Z., and Karim, M.N. *Desalination* 2004, 169, 231-244.

[136] Kamala, C.T.; Chu, K.H.; Chary, N.S.; Pandey, P.K.; Ramesh, S.L.; Sastry, A.R.K., and Chandrasekhar, K. *Wat. Res.* 2005, 39, 2815-2826.

[137] Kumari, P.; Sharma, P.; Srivastava, S., and Srivastava, M.M. Int. *J. Miner. Process* 2006, 78, 131-139.

[138] Hansen, H.K.; Ribeiro, A., and Mateus, E. *Miner. Eng.* 2006, 19, 486-490.

[139] Loukidou, M.X.; Matis K.A.; Zouboulis, A.I., and Liakopoulou-Kyriakidou *Wat. Res.* 2003, 37, 4544-4552.

[140] Marugesan, G.S.; Sathishkumar, M., and Swaminathan, K. *Bioresource Technol.* 2006, 97, 483-487.

[141] Katsoyiannis, I.A., and Zouboulis, A.I. *Wat. Res.* 2004, 38, 17-26.

[142] Daus, B.; Wennrich, R., and Weiss, H. *Wat. Res.* 2004, 38, 2948-2954.

[143] Pattanayak, J.; Mondal, K.; Mathew, S., and Lalvani, S.B. *Carbon* 2000, 38, 589-596.

[144] Budinova, T.; Petrov, N.; Razvigorova, M.; Parra, J., and Galiatsatou, P. *Ind. Eng. Chem. Res.* 2006, 45, 1896-1901.

[145] Dambies, L.; Guibal, E., and Roze, A. Colloids and Surf. A. *Physicochem. and Eng. Aspects* 2000, 170, 19-31.

[146] Altundogan, H.S.; Altundogan,S.; Tumen, F., and Bildik, M. *Waste Management* 2000, 20, 761-767.

[147] Elizalde-Gonzalez, M.P.; Mattusch, J.; Einicke, W.D., and Wennrich, R. *Chem. Eng. J.* 2001, 81, 187-195.

[148] Kundu, S.; Kavalakatt, S.S.; Pal, S.; Ghosh, S.K.; Mandal, M., and Pal, T. *Wat. Res.* 2004, 38, 3780-3790.

[149] Bang, S.; Patel, M.; Lippincott, L., and Meng, X. *Chemosphere*, 2005, 60, 389-397.

[150] Suzuki, T.M.; Bomani, J.O.; Matsunaga, H., and Yokoyama, T. *Reactive Funct. Polym.* 2000, 43, 165-172.

[151] Ahn, J.S.; Chon, Ch.; Moon, H., and Kim, K. *Wat. Res.* 2003, 37, 2478-2488.

[152] Xu, Y.; Nakajima, T., and Ohki, A.J. *Hazard. Maer.* 2002, B92, 275-287.

[153] Shevade, S., and Ford, R.G. *Wat. Res.* 2004, 38, 3197-3204.

[154] DeMacro, M.J.; Sengupta, A.K., and Greenleaf, J.E. *Wat. Res.* 2003, 37, 164-176.

[155] Ballinas, M.; De SanMiguel, E.R.; Rodriguez, Ma T.; Silva, O.; Munoz, M., and De Gyves *J. Environ. Sci. Technol.* 2004, 38, 886-891.

[156] Seko, N.; Basuki, F.; Tamada, M., and Yoshii, F. *React. Funct. Polym.* 2004, 59, 235-241.

[157] Kober, R.; Welter, E.; Ebert, M., and Dhamke, A. *Environ. Sci. Technol.* 2005, 39, 8038-8044.

[158] Cheng, Z.; Van Geen, A.; Louis, R.; Nicolaidis, N., and Bailey, R. *Environ. Sci. Technol.* 2005, 39, 7662-7666.

[159] Kanel, S.R.; Manning, B.; Charlet, L., and Choi, H. *Environ. Sci. Technol.* 2005, 39, 1291-1298.

[160] Leupin, O.X., and Hug, S.J. *Wat. Res.* 2005, 39, 1729-1740.

[161] Bang, S.; Korfiatis, G.P., and Meng, X.J. *Hazard. Mater.* 2005, 121, 61-67.

[162] Lung Lein, H., and Wilkin, R.T. Chemosphere .2005, 59, 377-386.

[163] Nikolaidis, N.P.; Dobbs, G.M., and Lackovic, J.A. *Wat. Res.* 2003, 37, 1417-1425.

[164] Balaramakrishna, M.V.; Chandrasekharan, K.; Karunasagar, D., and Aurnachalam, J.J. *Hazard. Mat.* 2001, B84, 229-240.

[165] Zhang, W.; Singh, P.; Paling, E., and Delides, S. *Miner. Eng.* 2004, 17, 517-524.

[166] Chakravarty, S.; Dureja, V.; Bhattacharyya, G.; Maity, S., and Bhattacharjee, S. *Wat. Res.* 2002, 36, 625-632.

[167] Mishra, D., and Farrell, J. *Environ. Sci. Technol.* 2005, 39, 9689-9694.

[168] Gu, Z.; Fang, J., and Deng, B. *Environ. Sci. Technol.* 2005, 39, 3833-3843.

[169] Dousova, B.; Kolousek, D.; Kovanda, F.; Machovic, V., and Novotna, M. *Appl. Clay Sci.* 2005, 28, 31-42.

[170] Zeng, L., *Wat. Res.* 2003, 37, 4351-4358.

[171] Zhang, F., and Itoh, H. *Chemosphere* 2005, 60, 319-325.

[172] Guo, F.X., and Chen, F. *Environ. Sci. Technol.* 2005, 39, 6808-6818.

[173] Cumbal, L., and Sengupta, A.K. *Environ. Sci. Technol.* 2005, 39, 6508-6515.

[174] Richmond, W.R.; Loan, M.; Morton, J., and Parkinson, G.M. *Environ. Sci. Technol.* 2004, 38, 2368-2372.

[175] Song, S.; Lopez-Valdivieso, A.; Hernandez-Campos, D.J., Peng, C., and Monroy-Fernandez, M.G. *Wat. Res.* 2006, 40, 364-372.

[176] Zhang, Y.; Yang, M., and Huang, X. *Chemosphere* 2003, 51, 945-952.

[177] Nakajima, T.; Xu, Y.; Mori, Y.; Kishita, M.; Takanashi, H.; Maeda, S., and Ohki, A. *J. Hazard. Mat.* 2005, B120, 75-80.

[178] Garcia, M.G.; dHiriart, J.; Giullitti, J.; Lin, H.; Custo, G.; Hidalgo, M.; Litter, M.I., and Blesa, M.A. *Solar Energy* 2004, 77, 601-613.

[179] Kumar, P.R.; Chaudhari, S.; Khilar, K.C., and Mahajan, S.P. *Chemosphere* 2004, 55, 1245-1252.

[180] Parga, J.R.; Cocke, D.L.; Valenzuela, J.L.; Gomes, J.A.; Kesmez, M.; Irwin, G.; Moreno, H., and Weir, M. *J. Hazard. Mater.* 2005, B124, 247-254.

[181] Zaw, M., and Emett, M.T. *Toxicol. Lett.* 2002, 133, 113-118.

[182] Ning, R.Y. *Desalination* 2002, 143, 237-241.

[183] Thirunavukkarasu, O.S.; Viraraghavan, T.; Subramanian, K.S., and Tanjore, S. *Urban Wat.* 2002, 4, 415-421.

[184] Delcul, G.D.; Bostick, W.D.; Trotter, D.R., and Osborne, P.E. *Sep. Sci. Technol.* 1993, 28, 551-564.

[185] Nazari, K.; Ghadiri, A., and Babaie, H. *Miner. Eng.* 2005, 18, 1233-1238.

[186] Urtiaga, A.M.; Alonso, A.; Ortiz, I.; Daoud, J.A.; El-Reefy, S.A.; Perez de ortiz, S., and Gallego, T. J. *Membr. Sci.* 2000, 164, 229-240.

[187] Kontrec, J.; Kralj, D., and Brecovic, L. Colloids and Surfaces A. Physicochem. *Eng. Aspects* 2003, 223, 239-249.

[188] Kaewsarn, P., and Yu, Q. *Environ. Pollution* 2001, 112, 209-213.

[189] Sternberg, S.P.K., and Dorn, R.W. *Bioresource Technol.* 2002, 81, 249-255.

[190] Perez-Rama, M.; Alonso, J.A.; Lopez, C.H., and Vaamonde, E.T. *Bioresource Technol.* 2002, 84, 265-270.

[191] Arica, M.Y.; Kacar, Y., and Genc, O. *Bioresource Technol.* 2001, 80, 121-129.

[192] Iqbal, M., and Edyvean, R.G. J. *Chemosphere* 2005, 61, 510-518.

[193] Bayramoglu, G.; Denizli, A.; Bektas, S., and Arica, M.Y. *Microchem. J.* 2002, 72, 63-76.

[194] Singh, K.K.; Rastogi, R., and Hasan, S.H. *J. Hazard. Mater.* 2005, A121, 51-58.

[195] Singh, K.K.; Singh, A.K. , and Hasan, S.H. *Bioresource Technol.* 2006, 97, 994-1001.

[196] Sharma, P.; Kumari, P.; Srivastava, M.M., and Srivastava, S. *Bioresource Technol.* 2006, 97, 299-305.

[197] Fine, P.; Scagnowwi, A.; Chen, Y., and Mingelgrin, U. *Environ. Pollution* 2005, 138, 358-367.

[198] Seki, H., and Suzuki, A. J. *Colloid Interface Sci.* 1997, 190, 206-211.

[199] Prabhukumar, G.; Matsumoto, M.; Mulchandani, A., and Chen, W. *Environ. Sci. Technol.* 2004, 38, 3148-3152.

[200] Kostal, J.; Mulchandani, A., and Chen, W. *Macromol.* 2001, 34, 2257-2261.

[201] Benguella, B., and Benaissa, H. *Wat. Res.* 2002, 36, 2463-2474.

[202] Denizli, F.; Arica, Y., and Denizli, A. *React. and Funct. Polym.* 2000, 44, 207-217.

[203] Buyuktuncel, E.; Bektas, S.; Genc, O., and Denizli, A. *React. Funct. Polym.* 2001, 47, 1-10.

[204] Denizli, A.; Garipcan, B.; Karabakan, A.; Say, R.; Emir, S., and Patir, S. *Sep. Purif. Technol.* 2003, 30, 3-10.

[205] Min, S.H.; Han, J.S.; Shin, E.W., and Park, J.K. *Wat. Res.* 2004, 38, 1289-1295.

[206] Inbaraj, B.S., and Sulochana, N. *Bioresource Technol.* 2004, 94, 49-52.

[207] Lusvardi, G.; Malavasi, G.; Menabue, L., and Saladini, M. *Waste Management* 2002, 22, 853-857.

[208] Kumar, U., and Bandyopadhyay, M. *J. Hazard. Mat.* 2006, B129, 253-259.

[209] Skubal, L.R.; Meshkov, N.K.; Rajh, T., and Thurnauer, M. J. *Photochem. Photobiol. A. Chemistry* 2002, 148, 393-397.

[210] Mishra, S.P.; Singh,V.K., and Tiwari, D. *Radiochim. Acta* 1997, 76, 97-101.

[211] Mishra, S.P., and Tiwari, D. *Radiochim. Acta* 1998, 80, 213-217.

[212] Bentouami, A., and Ouali, M.S. *J. Colloid Interface Sci.* 2006, 293, 270-277.

[213] Diaz-Somoano, M., and Martinez-Tarazona, M.R. *Fuel* 2005, 84, 717-721.

[214] Anac, M.; Say, R., and Denizli, A. J. *Chromatogr.* B. 2004, 811, 119-126.

[215] Jacobsen, M.R.; Fritt-Rasmussen, J.; Nielsen, S., and Ottosen, L.M. *J. Hazard. Mat.* 2004, 106B, 127-132.

[216] Pedersen,A. J. *Biomass and Bioenergy,* 2003, 25, 447-458.

[217] Elsherief, A.E. *Electrochim. Acta* 2003, 48, 2667-2673.
[218] Dutra, A.J.B.; Espinola, A., and Borges, P.P. *Miner. Eng.* 2000, 13, 1139-1148.
[219] Ennaassia, Et.; El-Kaeemi, K.; Kossir, A., and Cote, G. *Hydrometallurgy* 2002, 64, 101-109.
[220] Chirale, S.; Ratto, M., and Rovatti, M. *Wat. Res.* 2000, 34, 2971-2978.
[221] Paes-Hernandez, M.E.; Aguilar-Artega, K.; Galan-Vidal, C.A.; Romero-Romo, M., and Ramirez-Silva, M.T. *Environ. Sci. Tech* 2005, 39, 7667-7670.
[222] Barron-Zambrano, J.; Laborie, S.; Viers, Ph.; Rakib, M., and Durand, G. *Desalination* 2002, 144, 201-206.
[223] Kostal, J.; Mulchandani, A.; Gropp, K.E., and Chen, W. *Environ. Sci. Technol.* 2003, 37, 4457-4462.
[224] Herrero, R.; Lodeiro, P.; Rey-Castro, C.; Vilarino, T., and Sastre deVicente, M.E. *Wat. Res.* 2005, 39, 3199-3210.
[225] Barron-Zambrano, J.; Laborie, S.; Viers, Ph.; Rakib, M., and Durand, G. J. *Membr. Sci.* 2004, 229, 179-186.
[226] Anoopkrishnan, K., and Anirudhan, T.S. J. *Hazard. Mater.* 2002, B92, 161-183.
[227] Ekinci, E.; Budinova, T.; Yardim, M.F.; Petrov, N.; Razvigorova, M., and Minkova, V. *Fuel Processing Technol.* 2002, 77-78, 437-443.
[228] Yardim, M.F.; Budinova, T.; Ekinci, E.; Petrov., N.; Razvigorova, M., and Mincova, V. *Chemosphere* 2003, 52, 835-841.
[229] Lee, Si-Hyun, and Park, *Young-ok Fuel Processing Technol.* 2003, 84, 197-206.
[230] Zhan, Fu-Shen; Nriagu, J.O., and Itoh, H. *Wat. Res.* 2005, 39, 389-395.
[231] Goel, J.; Kadirvelu, K.; Rajagopal, C., and Garg, V.K. *Carbon* 2005, 43, 197-200.
[232] Abu-Daabes, M.A., and Pinto, N. *Chem. Engg. Sci.* 2005, 60, 1901-1910.
[233] O'Dowd, W.J.; Hargis, R.A.; Granite, E.J., and Pennline, H.W. *Fuel Processing Technol.* 2004, 85, 533-548.
[234] Zeng, H.; Jin, F., and Guo, J. *Fuel* 2004, 83, 143-146.
[235] Rio, S., and Delebane, A. *Fuel* 2003, 82, 153-159.
[236] Chojnaki, A.; Chojnaki, K.; Hoffmann, J., and Gorecki, H. *Miner. Eng.* 2004, 17, 933-937.
[237] Jeon, Ch., and Holl, W.H. *Wat. Res.* 2003, 37, 4770-4780.
[238] Manohar, D.M.; Anoopkrishnan, K., and Anirudhan, T.S. *Wat. Res.* 2002, 36, 1609-1619.
[239] Gupta, R.K.; Singh, R.A., and Dubey, S.S. *Sep. Purif. Technol.* 2004, 38, 225-232.
[240] Denizli, A.; Senel, S.; Alsancak, G.; Tuzmen, N., and Say, R. *React. and Funct. Polym.* 2003, 55, 121-130.
[241] Bulbul Sonmez, H.; Senkal, B.F.; Sherrington, D.C., and Bicak, N. *React. Funct. Polym.* 2003, 55, 1-8.
[242] Yavuj, E.; Senkal, B.F., and Bicak, N. *React. and Funct. Polym.* 2005, 65, 121-125.
[243] Touaibia, D., and Benayada, B. *Desalination* 2005, 186, 75-80.
[244] Vander Vaart, R.; Akkerhuis, J.; Feron, P., and Jansen, B. J. *Membr. Sci.* 2001, 187, 151-157.
[245] Ku, Y.; Wu, Ming-Huan, and Shen, Yung-Shuen, *Waste Management* 2002, 22, 721-726.
[246] Li, Ch.; Liang, Y., and Chen, Y. *Sep. Purif. Technol.* 2005, 45, 213-219.

[247] Jalali, R.; Ghafourian, H.; Asef, Y.; Davarpanah, S.J., and Sepehr, S. *J. Hazard. Mater.* 2002, B92, 253-262.

[248] Al-Subu, M.M. *Adv. Environ. Res.* 2002, 6, 569-576.

[249] Ho, Y. *Bioresource Technol.* 2005, 96, 1292-1296.

[250] Tokimoto, T.; Kawasaki, N.; Nakamura, T.; Akutagawa, J., and Tanada, S. *J. of Colloid and Interface Sci.* 2005, 281, 56-61.

[251] Chandrasekhar, K.; Kamala, C.T.; Chary, N.S.; Sastry, A.R.K.; Nageswara Rao, T., and Vairamani, M. *J. Hazard. Mater.* 2004, B108, 111-117.

[252] Hammy, F.; Mercier, J., and Blais, *J. Hydrometallurgy* 2005, 80, 232-240.

[253] Li, Y.; Zhu, Y.; Zhao, Y.; Wu, D., and Luan, Z. *Diamond and Related Mater.* 2006, 15, 90-94.

[254] Goel, J.; Kadirvelu, K.; Rajagopal, Ch., and Garg, V.K. *J. Hazard. Mater.* 2005, B 125, 211-220.

[255] Bhattacharjee, S.; Chakrabarty, S.; Maity, S.; Kar, S.; Thakur, P., and Bhattacharyya, G. *Wat. Res.* 2003, 37, 3954-3966.

[256] Turan, M.; Mart, U.; Yuksel, B., and Celik, M.S. *Chemosphere* 2005, 60, 1487-1492.

[257] Taty-Costodes, V.C.; Fauduet, H.; Porte, C., and Ho, Y. J. *Hazard. Mater.* 2005, B123, 135-144.

[258] Dimitrova, S. *Wat. Res.* 2002, 36, 4001-4008.

[259] Goel, J.; Kadirvelu, K.; Rajagopal, C., and Garg, V.K. *Ind. Eng. Chem. Res.* 2005, 44, 1987-1994.

[260] Ulusoy, U., and Simsek, S. *J. Hazard. Mat.* 2005, B127, 163-171.

[261] Denizli, A.; Garipcan, B.; Karabakan, A., and Senoz, H. *Mat. Sci. Eng.* 2005, C25, 448-454.

[262] Xu, Y., and Zhao, D. *Ind. Eng. Chem. Res.* 2006, 45, 1758-1765.

[263] Cuenot, F.; Meyer, M.; Bucaille, A., and Guilard, R. *J. Mol. Liquids.* 2005, 118, 89-99.

[264] Mouflih, M.; Aklil, A., and Sebti, S. *J. Hazard. Mat.* 2005, B119, 183-188.

[265] Baker, H., and Khalili, F. *Anal. Chim. Acta* 2004, 516, 179-186.

[266] Bektas, N., and Kara, S. *Sep. and Purif. Technol.* 2004, 39, 189-200.

[267] Sublet, R.; Simonnot, M.; Boireau, A., and Sardin, M. *Wat. Res.* 2003, 37, 4904-4912.

[268] Abdel-Halim, S.H.; Shehata, A.M.A., and El-Shahat, M.F. *Wat. Res.* 2003, 37, 1678-1683.

[269] Altin,A., and Degirmenci, M. *Sci. Total Environ.* 2005, 337, 1-10.

[270] Amrate, S.; Akretche, D.E.; Innocent, C., and Seta, P. *Sci. Total Environ.* 2005, 349, 56-66.

[271] Mohamed, A.M.O. *J. Hazard. Mat.* 2002, B90, 297-310.

[272] Esalah, J.O.; Weber, M.E., and Vera, J.H. *Sep. and Purif. Technol.* 2000, 18, 25-36.

[273] Makhloufi, L.; Saidani, B., and Hammache, H. *Wat. Res.* 2000, 34, 2517-2524.

[274] Bryant, D.E.; Stewart, D.I.; Kee, T.P., and Barton, C.S. *Environ. Sci. Technol.* 2003, 37, 4011-4016.

[275] Coleman, S.J.; Coronado, P.R.; Maxwell, R.S., and Reynolds, J.G. *Environ. Sci. Technol.* 2003, 37, 2286-2290.

[276] Dushenkov, S.; Vasudev, D.; Kapulnik, Y.; Gleba, D.; Fleisher, D.; Ting, K.C., and Ensley, B. *Environ. Sci. Technol.* 1997, 31, 3468-3474.

[277] Kryvoruchko, A.P.; Yurlova, L.Yu.; Atamanenko, I.D., and Kornilovich, B.Yu. *Desalination* 1999, 122, 147-150.

[278] Raff, O., and Wilken, R. *Desalination* 1999,122, 147-150.
[279] Gregory, K.B., and Lovely, D.R. *Environ. Sci. Technol.* 2005, 39, 8943-8947.
[280] Gu, B.; Liang, L.; Dickey, M.J.; Yin, X., and Dai, S. *Environ. Sci. Technol.* 1998, 32, 3366-3373.
[281] Preetha, C.R.; Gladis, J.M.; Rao, T.P., and Venkateswaran,G. *Environ. Sci. Technol.* 2006, 40, 3070-3074.
[282] Gherrou, A.; Kerdjoudj, H.; Molinari, R., and Drioli, E. *Sep. Purif. Technol.* 2002, 28, 235-244.
[283] Kocaoba, S., and Akcin, G. *Desalination* 2005, 180, 151-156.
[284] Korus, I.; Bodzek, M., and Loska, K. *Sep. Purif. Technol.* 1999, 17, 111-116.
[285] Canizares, P.; Perez, A.; Camarillo, R., and Linares, J.J. *J. Membr. Sci.* 2004, 240, 197-209.
[286] Axtell, N.R.; Sternberg, S.P.K., and Claussen, K. Bioresource Technol. 2003, 89, 41-48.
[287] Mohamed, Z.A. *Wat. Res.* 2001, 35, 4405-4409.
[288] Tunali, S.; Cabuk, A., and Akar, T. *Chem. Eng. J.* 2006, 115, 203-211.
[289] Minamisawa, M.; Minamisawa, H.; Yoshida, S., and Takai, N. J. *Agric. Food Chem.* 2004, 52, 5606-5611.
[290] Vazquez, G.; Ganzalez-Alvarez, J.; Freire, S.; Lopez-Lorenzo, M., and Antorrena, G. *Bioresource Technol.* 2002, 82, 247-251.
[291] Villaescusa, I.; Fiol, N.; Martinez, M.; Miralles, N.; Poch, J., and Serarols, *J. Wat. Res.* 2004, 38, 992-1002.
[292] Mishra, S.P.; Singh, V.K., and Tiwari, D. *Radiochim. Acta* 1998, 81, 227-233.
[293] Rao, M.; Parwate, A.V., and Bhole, A.G. *Waste Management* 2002, 22, 821-830.
[294] Benhammou, A.; Yaacoubi, A.; Nibou, L., and Tanouti, B. *J. Hazard. Mater.* 2005, B117, 243-249.
[295] Gupta, V.K.; Gupta, M., and Sharma, S. *Wat. Res.* 2001, 35, 1125-1134.
[296] Gupta, V.K., and Ali, I. J. Colloid. *Interface Sci.* 2004, 271, 321-328.
[297] Gupta, V.K.; Jain, C.K.; Ali, I.; Sharma, M., and Saini, V.K. *Wat. Res.* 2003, 37, 4038-4044.
[298] Kocaoba, S., and Akyuz, T. *Desalination* 2005, 181, 313-318.
[299] Deng, S.; Bai, R., and Chen, J.P. *Langmuir* 2003, 19, 5058-5064.
[300] Cheung, C.W.; Porter, J.F., and McKay, G. *Sep. Purif, Technol.* 2000, 19, 55-64.
[301] Gupta, V.K., and Ali, I. *Sep. Purif. Technol.* 2000, 18, 131-140.
[302] Lee, S.M., and Davis, A.P. *Wat. Res.* 2001, 35, 534-540.
[303] Rao, M.M.; Ramesh, A.; Rao, G.P., and Seshaiah, K. *J. Hazard. Mater.* 2006, B129, 123-129.
[304] Fan, H., and Anderson, P.R. *Sep. Purif. Technol.* 2005, 45, 61-67.
[305] Okieimen,F.E.; Sogbaike, C.E., and Ebhoaye, J.E. *Sep. Purif. Technol.* 2005, 44, 85-89.
[306] El-Kamash, A.M.; Zaki, A.A., and Abed El Geleel, M. *J. Hazard. Mater.* 2005, B127, 211—220.
[307] Gupta, V.K., and Sharma, S. *Environ. Sci. Technol.* 2002, 36, 3612-3617.
[308] Yang, L.; Shahrivari, Z.; Liu, P.K.T.; Sahimi, M., and Tsotsis, T.T. *Ind. Eng. Chem. Res.* 2005, 44, 6804-6815.
[309] Basso, M.C.; Cerrella, E.G., and Cukierman, A.L. *Ind. Eng. Chem. Res.* 2002, 41, 180-189.
[310] Shawabkeh, R.; Al-Harahsheh, A.; Hami, M., and Khlaifat, A. *Fuel* 2004, 83, 981-985.

[311] Liu, Ch.; Huang, Y.; Naismith, N., and Ecomony, J. *Environ. Sci. Technol.* 2003, 37, 4261-4268.

[312] Schroden, R.C.; Al-Daous, M.; Sokolov, S.; Melde, B.J. Lytle, J.C.; Stein, A.; Carbajo, M.C.; Fenandez, J.T., and Rodriguez, E.E.J. *Mater. Chem.* 2002, 12, 3261-3267.

[313] Duff, M.C.; Hunter, D.B.; Hobbs,D.T.; Fink, S.D.; Dai, Z., and Bradley, J.P. *Environ. Sci. Technol.* 2004, 38, 5201-5207.

[314] Ipek, U. *Desalination* 2005, 174, 161-169.

[315] Paulson, A.J. *The Sci. Total Environ.* 2001, 278, 31-44.

[316] Morlay, C.; Cromer, M., and Vittori, O. *Wat. Res.* 2000, 34, 455-462.

[317] Lin, S.W., and Navarro, R.M.F. *J. Electrochem. Soc.* 2001, 148, C284-C288.

[318] Van Herck, P.; Vandecasteele, C.; Swennen, R., and Mortier, R. *Environ. Sci. Technol.* 2000, 34, 3802-3808.

[319] Schwuger, M.J.; Subklew, G., and Woller, N. Colloids. Surf. Phisicochem. *Eng. Aspects* 2001, 186, 229-242.

[320] Lin, Y.; Liu, Ch.; Wu, H.; Yak, H.K., and Wai, Ch. M. *Ind. Eng. Chem. Res.* 2003, 42, 1400-1405.

[321] Fang, Z.; Guha, A.K., and Sirkar, K.K. *Ind. Eng. Chem. Res.* 1996, 35, 4214-4220.

[322] Arous, O.; Gherrou, A., and Kerdjoudj, H. *Desalination* 2004, 161, 295-303.

[323] Vilensky, M.Y.; Berkowitz, B., and Warshawsky, A. *Environ. Sci. Technol.* 2002, 36, 1851-1855.

[324] Brower, J.B.; Ryan, R.L., and Pazirandeh, M. *Environ. Sci. Technol.* 1997, 31, 2910-2914.

[325] Economy, J.; Dominguez, L., and Mangun, Ch. L. *Ind. Eng. Chem. Res.* 2002, 41, 6436-6442.

[326] Mier, M.V.; Callejas, R.L.; Gehr, R.; Cisneros, B.E.J. and Alvarez, P.J.J. *Wat. Res.* 2001, 35, 373-378.

[327] Ritchie, S.M.C.; Kissick, K.E.; Bachas, L.G.; Sikdar, S.K.; Parikh, C., and Bhattacharyya, D. *Environ. Sci. Technol.* 2001, 35, 3252-3258.

[328] Broom, G.P.; Squires, R.C.; Simpson, M.P.J., and Martin, I. *J. Membr. Sci.* 1994, 87, 219-230.

[329] Blocher, C.; Dorda, J.; Mavrov, V.; Chmiel, H.; Lazaridis, N.K., and Matis, K.A. *Wat. Res.* 2003, 37, 4018-4026.

[330] Back, K., and Yang, J.W. *Sep. Sci. Technol.* 2005, 40, 699-708.

[331] Ozdemir, G.; Ceyhan, N.; Ozturk, T.; Akirmak, F., and Cosar, T. *Chem. Eng.* J. 2004, 102, 249-253.

[332] Pena-Castro, J.M.; Martinez-Jeronimo, F.; Esparza-Garcia, F., and Canizares-Villanueva, R.O. *Bioresource Technol.* 2004, 94, 219-222.

[333] Iqbal, M., and Eyvean, R.G.J. *Miner. Eng.* 2004, 17, 217-223.

[334] Saeed, A.; Iqbal, M., and Akhtar, M.W. J. *Hazard. Mater.* 2005, B117, 65-73.

[335] Cimino, G.; Passerini, A., and Toscano, G. *Wat. Res.* 2000, 34, 2955-2962.

[336] Chefetz, B.; Sominski, L.; Pinchas, M.; Ginsburg, T.; Elmachliy, S.; Tel-or, E., and Gedanken, A. *J. Phys. Chem.* B 2005, 109, 15179-15181.

[337] Kumar, A.; Rao, N.N., and Kaul, S.N. *Bioresource Technol.* 2000, 71, 133-142.

[338] Liu,A.M.; Hidajat, K.; Kawi, S., and Zhao, D.Y. *Chem. Commun.* 2000, 1145-1146.

[339] Monser, L., and Adhoum, N. *Sep. Purif. Technol.* 2002,, 26, 137-146.

[340] Yabe, M.J.S., and de Oliveira, E. *Adv. Environ. Res.* 2003, 7, 263-272.

[341] Bosco, S.M.D.; Jimenez, R.S., and Carvalho, W.A. *J. Colloid. Interface Sci.* 2005, 281, 424-431.

[342] Potgeiter, J.H.; Potgeiter-Vermaak, S.S., and Kalibantonga, P.D. *Miner. Eng.* 2006, 19, 463-470.

[343] Say, R.; Birlik, E.; Denizli, A., and Ersoz, A. *Appl. Clay Sci.* 2006, 31, 298-305.

[344] Essawy, H.A., and Ibrahim, H.S. *React Funct. Polym.* 2004, 61, 421-432.

[345] Wildgoose, G.G.; Leventis, H.C.; Simm, A.O.; Jones, J.H., and Compton, R.G. *Chem. Commun.* 2005, 3694-3696.

[346] Arakaki, L.N.H.; Filha, V.L.S.A.; Espinola, J.G.P.; da Fonseca, M.G.; de Oliveira, S.F.; Arakaki, T., and Airoldi, C.J. *Environ. Monit.* 2003, 5, 366-370.

[347] Peric, J.; Trgo, M., and Medvidovic, N.V. *Wat. Res.* 2004, 38, 1893-1899.

[348] Yavuj, O.; Altunkaynak, Y., and Guzel, F. *Wat. Res.* 2003, 37, 948-952.

[349] Vengris, T.; Binkiene, R., and Sveikauskaite, A. *Appl. Clay Sci.* 2001, 18, 183-190.

[350] Dhakal, R.P.; Ghimire, K.N.; Inoul, K.; Yano, M., and Makino, K. *Sep. Purif. Technol.* 2005, 42, 219-225.

[351] Sayari, A.; Hamoudi, S., and Yang, Y. *Chem. Mater.* 2005, 17, 212-216.

[352] Mishra, S.P., and Singh,V.K. J. Radioanal. *Nucl. Chem.* 1999, 241, 341-346.

[353] Aklil, A.; Mouflih, M., and Sebti, S. J. *Hazard. Mater.* 2004, A112, 183-190.

[354] Anjaneyulu, Y., and Rao, R.V.S. Can. J. *Chem. Eng.* 2001, 79, 71-79.

[355] Denizli, A.; Sanli, N.; Garipcan, B.; Patir, S., and Alsancak, G. *Ind. Eng. Chem. Res.* 2004, 43, 6095-6101.

[356] Genc, O.; Soysal, L.; Bayramoglu, G.; Arica, M.Y., and Bektas, S. *J. Hazard. Mater.* 2003, B97, 111-125.

[357] Denizli, A.; Bektas, S.; Arica, Y., and Genc, O.J. *Appl. Polym. Sci.* 2005, 97, 1213-1219.

[358] Mohammadi, T.; Moheb, A.; Sadrzadeh, M., and Razmi, A. *Sep. Purif. Technol.* 2005, 41, 73-82.

In: Trends in Hazardous Materials Research
Editor: Edward C. Booking, pp. 113-136
ISBN 1-60021-335-9

Chapter 4

CONTROL OF COLLOIDAL PARTICLES FOR THE SOIL AND GROUNDWATER REMEDIATION

Ilwon Ko[1]*, So-Young Kang[1], Kyoung-Woong Kim[1] and Cheol-Hyo Lee[2]

[1]Arsenic Geoenvironment Laboratory (NRL), Department of Environmental Science and Engineering, Gwangju Institute of Science and Technology, Oryong-dong, Buk-gu, Gwangju, Korea

[2]Environmental Research Center, OIKOS Co. Ltd, Geumcheon-gu, Seoul, Korea

ABSTRACT

Colloidal particles have a few micro-scale or nano-scale diameters and are suspended as charged particles in aqueous solution. The geochemical environment and engineering process generate colloidal particles and their suspension exposed to aqueous and terrestrial environments. These colloidal particles are very reactive due to large surface area and then highly contaminated with inorganic or organic contaminants. The colloidal stability in the aqueous solution is significant for its deposition condition. Due to the long suspension time of the particles it is difficult to precipitate them. Evidently, the colloidal phenomenon is applicable for the remediation process of soil and groundwater environment. The practical design in many remedial processes was not suitable for the application of nano-scale particles. This study presents the principle of colloidal control and their case studies of lab-scale and pilot-scale the remedial design using fine particles. The reactive barrier using nano-scale particles is effective as colloid flushing or colloid barrier formation for the removal through adsorption and degradation. Particularly, the deposition of nano-scale particles was highlighted for the removal of various contaminants. Soil washing with surfactant removes inorganic and organic contaminants through solubilization, mobilization, detachment and emulsion by formation of colloidal micelle above CMC (critical micelle concentration).

*corresponding author: iwkoh@hanmail.net

1. Introduction

Recent development of analytical technologies leads to the microscopic observation and interpretation of natural and engineering phenomena. The formation of stable colloidal dispersions in water-based system is an issue in applied colloid chemistry and numerous industrial applications, like food processing, the production of paints or printer ink, the development of nano-scale materials, cosmetic products, wastewater treatment and so on. Particularly, organic and inorganic colloids are present in various concentrations in natural soil and groundwater (Degueldre et al., 2000). In terms of the geoenvironment remediation, our understanding of the factors governing the behavior of colloidal particles is important for *in-situ* and *ex-situ* environment remediation (Yao et al., 1971; Ryan and Elimelech, 1996). The colloid-associated contamination in excavated soil or subsurface porous media has been investigated in terms of colloidal fine release, mobilization, entrapment, plugging and migration process (Roy and Dzombak, 1997; Sen et al., 2002; Sen et al., 2006). Despite many recent studies of contaminants removal using nano-scale materials (Liang et al., 1997; Kanel et al., 2005), the detail control studies of colloid particles in soil and groundwater environment have been little performed except for some studies, like the feasibility of colloid barrier formation (Thomas and Ward, 189; Cantrell et al., 1995; Tufenkji et al., 2002). Therefore, this study provides a significant applicability to control the fine particles like colloidal particles and silt- and clay-sized particles. Physical and chemical control of colloidal particles will be main factors in the engineering application for the geoenvironment remediation.

The aims of this book chapter are to present the case studies of the environmental remediation using nano-scale particles, actually colloidal fine particles, and to address the scheme of the remedial design.

2. Colloidal Behavior

2.1 Importance of Colloid

Colloid is a kind of forgotten dimensions or larger than molecule but too small to be seen in an optical microscope. The difference between particle and dissolved molecule is characteristics of transition zone of chemical and physical properties. This regard is distinguished by the characteristic size, actually, about 3×10^{-4} μm as a water molecule diameter (Probstein, 2003). Colloidal particles about a micro-scale or nano-scale in concerned size settle down very sluggish under gravity that may often be considered as suspended in the flow. In the concerned geochemical environment, colloidal particles includes such fine particles as silt and clay size (<75 μm, in a textural viewpoint), mineral (clay, silicates, oxyhydroxides, sulfides etc.) fragment (< 1,000 μm), organic molecule (< 10 μm), microbes (a few μm) and so on. μm

Many organic and inorganic colloids are reported in the geochemical environment (Ryan and Elimelech, 1996). Particularly, colloids in soil and groundwater environment can transport together with the contaminants, and be deposited on the ecosystem and actually human being as a final receptor. Recent experimental observations have reported colloid-facilitated transport of contaminants as a significant mechanism through vadose zone and subsurface zone. An understanding of colloid behavior in natural systems is essential for predicting the colloidal coagulation, deposition, and transport based on the colloidal stability in model systems designed to describe their theories in natural systems. Particularly, the colloidal stability is important in describing its dispersions in liquid media related with both chemical and physical perturbation (Ryan and Elimelech, 1996).

Depending the stability of the colloid, stable or unstable colloid becomes disperse or aggregated, respectively. One controlling factor is the aqueous chemical condition. The surface charge of colloid is affected by the charge of the aqueous species or colloids in the specific aqueous composition. The colloidal surface is neutralized by the opposite charge and then coagulated in the aggregates. However, in some cases, the colloid becomes stably suspended in aqueous solution. The other factor is the hydrodynamic interaction. The hydrodynamic interaction is related with the physical perturbation in flow, like rapid flow through fractures, rapid infiltration of rainfall, fast pumping and so on (Probstein, 2003).

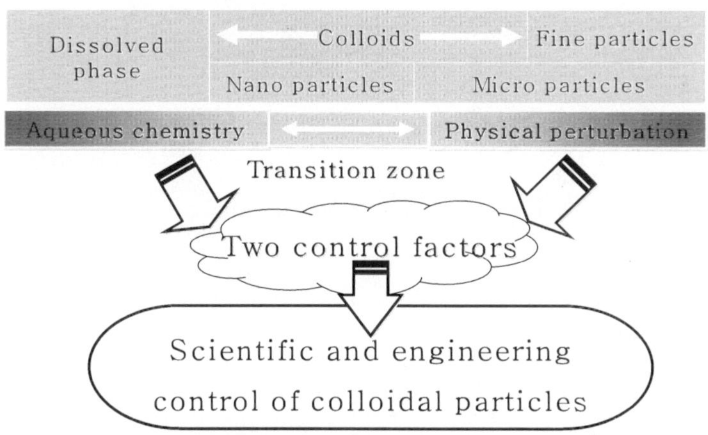

Figure 1. Schematics in the physico-chemical behavior and control of colloidal particles

Therefore, the colloid properties can be controlled by both the aqueous chemistry and physical perturbation (Figure 1). Before some decades, only soluble chemical species were dispersive contaminants in the subsurface flow condition. However, the colloid-facilitated transport became a significant mechanism in the dispersion of contaminants, and then their engineering control were a current issue as the development of micro- and nano-scale analysis technology. In specific, the engineering control is our focusing topic as well as future development of environmental technology controlling the colloid particle in the environment. Therefore, the colloidal stability should be controlled by the chemical and physical factors for the environmental remediation as the final application.

2.2 DLVO Theory

The understanding of colloid stability has been traditionally treated in terms of DLVO theory. The fundamental DLVO mean field theory of colloidal interactions, due to Derjaguin, Landau, Verwey and Overbeek, has been established since the 1940s and shown to apply successfully to a wide range of colloidal systems. The theory describes the force between surfaces interacting through a liquid medium. It combines the effects of the van der Waals attraction and the repulsion.

In colloidal surface, the electrical double layer consists of stern layer and diffuse layer. Coulombic force is electrostatic repulsion interaction to be dependent on the overlap of diffuse double layer, and potential energy of electrostatic repulsion, V_R to be obtained by integration of repulsion force with respect to distance. In the similar way, London-van der Waals force is potential energy of attraction, V_A to be obtained by integration of attraction force with respect to distance. DLVO theory tells that attraction decays as the inverse power of distance of separation, and repulsion decays as an exponential of distance. At both short and long distances of separation, the attraction is larger than repulsion force. In this regard, the critical coagulation concentration is a concentration at which the primary maximum of potential energy becomes zero (Figure 2).

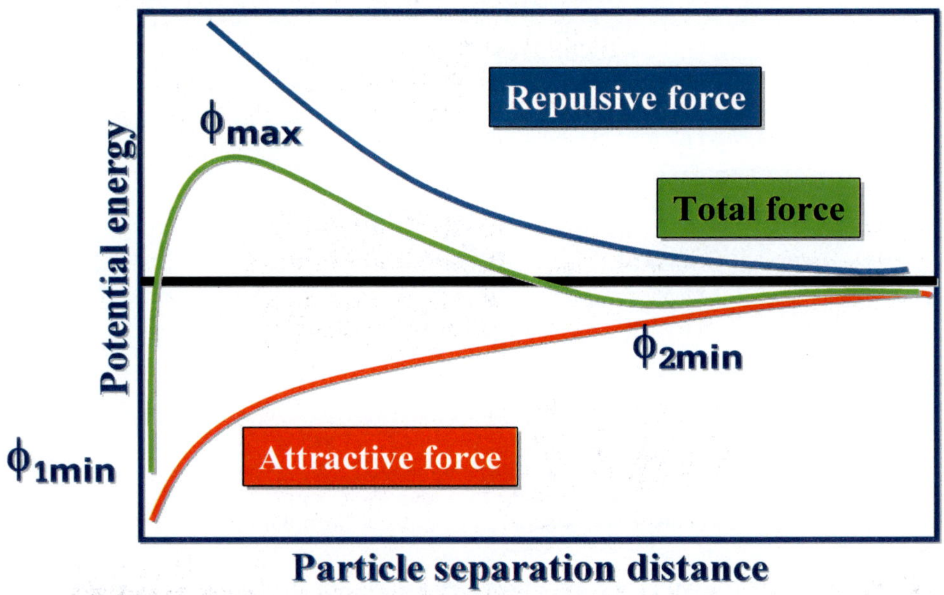

Figure 2. DLVO feature by summing the attractive and repulsive forces

Derjaguin and Landau(1941) and Verwey and Overbeek(1948) suggested the total potential energy of attraction and repulsion (Eqn. 1).

$$V_T = V_R + V_A \qquad \text{Equation (1)}$$

Giving the potential energy of electrostatic repulsion (Eqn. 2) and potential energy of the van der Waals attraction (Eqn. 3), respectively.

$$V_R = 3.469 \times 10^{19} \varepsilon (kT)^2 a \gamma^2 \exp(-\kappa h) / v^2 \qquad \text{Equation (2)}$$

$$V_A = -\frac{A}{12} \frac{1}{x^2+2x} + \frac{1}{x^2+2x+1} + 2\ln\left(\frac{x^2+2x}{x^2+2x+1}\right) \qquad \text{Equation (3)}$$

2.3 Colloidal Stability and Coagulation

A system is said to be colloidally unstable if collisions lead to the formation of aggregates; such a process is called coagulation or flocculation. On the contrary, there is no change in the number of suspended particles with time in a suspension medium, and then it is called as high stability of colloid. Principally, interparticle forces such as vander Waals attraction and electrolytic repulsion are concerned with colloidal stability. Further hydration repulsive forces, hydrophobic attraction, polymer bridging or steric repulsions also contribute to the overall stability of suspended particles. Clay mineralogy, ionic strength, and pH are known to affect colloid dispersibility. The surface charge of various mineral particles is a fundamental factor determining the intersurface forces between particles or between opposite chemical species. In particular, natural organic matter and increasing pH can stabilize the colloid. A decreasing ionic strength expands the diffuse layers and increases the repulsive interactions.

The stabilizing ways to prevent colloids from forming aggregates during their collision is summarized by three kinds of the characteristic interaction of colloidal surface (Figure 3).

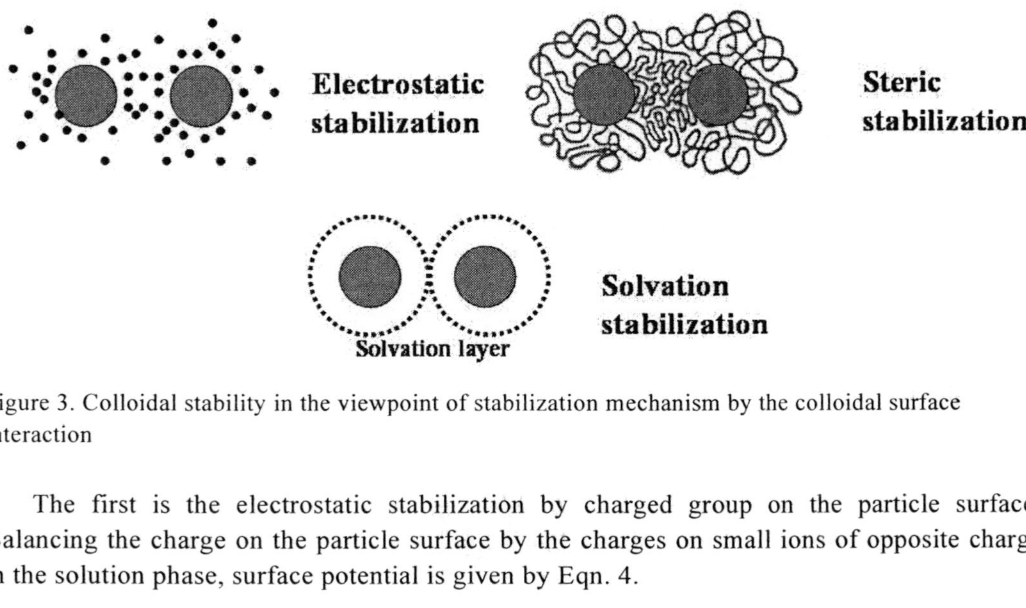

Figure 3. Colloidal stability in the viewpoint of stabilization mechanism by the colloidal surface interaction

The first is the electrostatic stabilization by charged group on the particle surface. Balancing the charge on the particle surface by the charges on small ions of opposite charge in the solution phase, surface potential is given by Eqn. 4.

$$\psi_r = \psi_s (a/r)\exp[(-x(r-a)] \quad \text{Equation (4)}$$

where

1/x : Thickness of electrical double layer

$x = (8\pi e^e n_A / 1000\varepsilon kT)I$ ε : dielectric constant of the medium

I : Ionic strength

Providing tolerance against high level of electrolyte, high stability of colloid can be kept in the low ionic strength.

The second is the steric stabilization by an adsorbed layer of some substance. Polymer, like natural organic matter, must be strongly adsorbed on particle surface but not so strong as to be collapsed. Polymer chains should extend as far as possible into solution and then be enhanced by solution in which polymer is readily soluble (Figure 4). Most effective polymers are block copolymers with partly strong acids and partly weak acids.

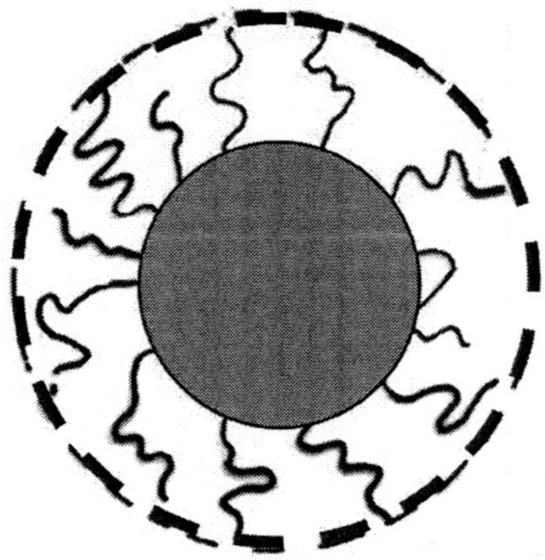

Figure 4. Schematic illustration of a particle with adsorbed nonionic polymer chain giving the range of influence of steric forces.

Electrosteric stabilization as a similar mechanism is combination of both electrostatic and steric stabilization by grafting to a core particle and polyelectrolyte chains.

The third is the solvation stabilization. This is true that colloids with adsorbed layers strongly approach with one another (Figure 3).

Considering a natural environment, steric and electrostatic stabilization are two main mechanisms for colloid stabilization. Electrostatic stabilization is based on the mutual repulsion of like electrical charges. Different phases generally have different charge affinities, so that a charge double-layer forms at any interface. In a stable colloid, its kinetic energy is too small to overcome the electrostatic repulsion between charged layers of the dispersing phase. The dispersed particles under an electric field migrate to the same electrode, and then must have the same sign charge. Unstable colloids form flocs as the aggregate due to interparticle attractive interaction. Removal of the electrostatic barrier prevents aggregation of the particles. This can be accomplished by adding ionic salt into a suspension or changing the pH of a suspension to neutralize their surface charge. It indicates that the decreasing repulsive forces that keep colloids separate, allowing for coagulation due to van der Waals forces. In addition, a charged polymer, like natural organic matter, can bridge each colloid by attractive electrostatic interactions. Clusters of colloids fall to the bottom of the suspension only if they are sufficiently small volume or size for the Brownian forces that work to keep the particles in suspension to be overcome by gravitational forces.

2.4 Colloidal Transport and Deposition

Transport of colloid in the porous media has been considered in the system of single particle of filter media, that is, a collector (Yao et al., 1971). The main flow direction comes from gravitational force. A colloid, following the streamline of the flow, may come in contact with the collector; this is interception (Figure 5). The density of suspended colloid is greater than that of water, and then the colloid is influenced by the buoyant weight of the particle and fluid drag on the particle under the gravitational force field. This is called as sedimentation. The third transport mechanism is diffusion from the Brownian motion of colloid. Suspended colloid is subject to random movement by molecules of the suspending medium.

The following general equation can describe the temporal and spatial variation of colloid in a porous media (Eqn. 5).

$$\frac{\partial C}{\partial t} + v\frac{\partial C}{\partial x} = D\frac{\partial^2 C}{\partial x^2} + (1 - \frac{\rho_w}{\rho_p})\frac{mg}{3\pi\mu d_p}\frac{\partial C}{\partial x} \qquad \text{Equation (5)}$$

where C is the local concentration of colloid,
v is the velocity of the water, t is time, D is the diffusion coefficient
ρ_w and ρ_p are the density of water and particle, respectively.
m is the water viscosity; m and d_p are the mass and diameter of colloid, respectively.
and g is the gravitational acceleration.

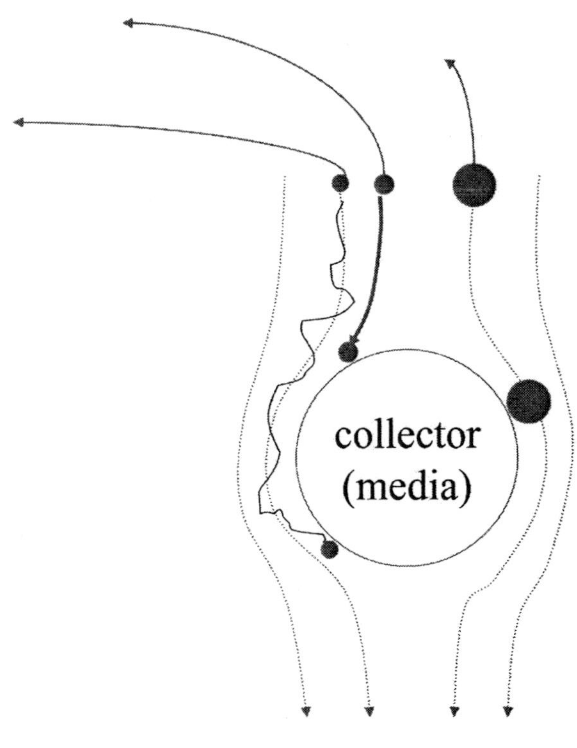

Figure 5. Mechanism of the colloidal deposition

Colloid is attached to collector grains under attractive DLVO conditions, by changing pH, ionic strength, and some aqueous species. Particularly, divalent ions, like Ca, Mg, destabilize colloids, and limit their transport through charge neutralization of colloidal particles.

3. Colloidal Phenomena in the Environment

3.1 Colloid in Soil and Sediment

The source of colloid comes from the soil and sediment, as geological materials. With the development of urbanization, due to the removal of the original vegetation cover and the exposure of the soil, soil particles are washed away by rainfall-runoff, and then settle out of the water in a lake, river or bay onto aquatic plants, rock and the bottom. Organic and inorganic contaminants also adhere to colloidal particles and are transported with them. Particularly, clay and humus, which are main constituents in soil, have a large surface area and negatively charged or amphoteric surface with high cation exchange capacity.

In a steady and dry weather flow condition, as a hydrodynamic condition, residence time condition is from days to weeks or longer. The colloidal particles predominate in the water column since the sediment fractions have separated from the water column at much shorter residence time and sediment-transport capacity is not sufficient to entrain these coarser fractions (Alexander et al. 1989). However, in flowing natural waters such as stream and river, the coarser sediment material can be transported along the channel bed as deposition load (FISRWG 1998). Much research have focused on both the transport of colloidal and suspended material in the porous media and water column from an environmental perspective as well as the sediment material as bed load from a sediment transport and hydraulic perspective (Michelbach and Wohrle 1993; Alber 2000; Roth et al. 2001). They have reported that long residence time and equilibrium conditions have allowed a steady-state coagulation and flocculation.

3.2 Colloid in Groundwater

The unsaturated zone as a colloidal source is typically viewed as a buffer that minimizes groundwater contamination from chemicals applied to the ground surface. However, elevated contaminant concentrations in groundwater evidently occur. In fact, the unsaturated zone may be a source of groundwater contamination, as some research has shown that contaminants may sorb to and be mobilized by colloids. Organic and inorganic colloids are present in various concentrations in natural groundwaters (Degueldre et al., 2000). From field studies it is known that colloids can have a strong impact on the transport of contaminants, since they migrate at velocities similar to the groundwater velocity and act as a carrier by adsorbing a significant amount of contaminants (McCarthy et al., 1998). This process can be of importance for the long-term safety, as a significant barrier for contaminants, due to high sorption capacities of the sediments. Rather high concentrations of humic colloids have been observed in the natural environment (Buckau et al., 2000). According to measurements of concentrations and ages of humics in real aquifer systems there is evidence for the stability and transport of humic colloids over large distances. Several column experiments with humic-associated aqueous system have investigated the impact of humic colloids on the contaminant transport. The results indicate that kinetically controlled sorption processes play an essential role in the contaminant transport, influencing the flow-velocity, column length and chemical composition of colloid.

4. Application to the Environmental Remediation

4.1 Colloid Barrier Formation / Colloid Flushing

The groundwater treatment using a subsurface treatment zone has been discussed and alternative reactive treatment materials and emplacement methods are being developed (USEPA, 1996). Emplacement of permeable reactive barrier has been limited to relatively shallow depth and width (USEPA, 1998). Chemical reactive barrier is alternative technique to immobilize contaminants, and colloid barrier or colloid flushing is to inject colloids to desired subsurface zone perpendicular to groundwater flow for target contaminants (Cantrell, et al., 1997). Durmusoglu et al. (2000) also formed a horizontal barrier by injecting gelling liquids to isolate the downward migration of NAPLs in the subsurface. The in-situ chemical reactive barrier is an inexpensive and effective means than the conventional trench-and-fill barrier due to no excavation and refilling solid materials. The most generally proposed a material for chemical reactive barriers is zero-valent iron (Fe^0). Many barrier materials have been tested for the conventional trench-and-fill barrier like granular activated carbon, pyrite, magnetite, a mixture of compost and saw dust, peat soils, zeolite and bimetallic compounds consisting of iron and palladium (USEPA, 1996; 1998).

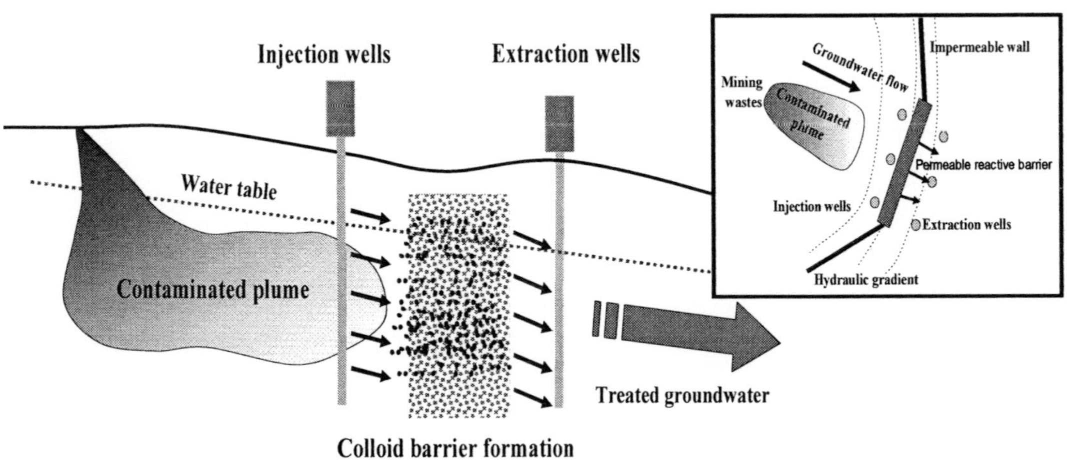

Figure 6. Schematics of the colloid barrier formation using nano-scale iron-oxide particle (Ko et al., 2003)

This study forms a chemical reactive barrier to inject nanoscale hematite colloids into the subsurface environment (Figure 6). Injection well is installed before zones apart from contaminated plume, and extraction well is installed to regulate the colloidal movement in the desired direction. This study used hematite colloids with about 750 nm diameter as an excellent chemical material to adsorb strongly arsenic and heavy metals (Ko et al., 2004). Zero valent iron lowers the pH of groundwater, removes inorganic oxyanions by reductive precipitation (Cantrell et al., 1995), and arsenic by surface precipitation or adsorption in the limited pH and Eh (Lackovic, 2000). Whereas hematite changes little pH of groundwater, and adsorb arsenic and heavy metals kinetically faster than zero valent iron (Jain et al., 1999).

In order to investigate the effect of pH and ionic strength on the deposition of hematite onto sand, various pH (3 to 10) and ionic strength (10^{-4} to 10 M) was controlled. Hematite colloids have the positively charged surfaces below PZC due to PZC of 9.2, but the sand particles consisting mainly of quartz have the negatively charged surfaces above PZC due to PZC of 2 of quartz (Ko et al., 2005). Therefore, the favorable deposition condition was assumed because positively charged hematite colloids attached dominantly onto negatively charged sand. The pH-and ionic strength- dependence of the coating reaction of hematite on sand is shown in Figure 7. A pronounced effect of ionic strength on the amount of adsorbed hematite is also shown in Figure 7a at constant pH 6.8. At the intermediate ionic strength, the adsorbed amount of hematite attained the maximum. This indicates that the surface charge can be balanced by electrolytes and then, the repulsion between colloidal particles is reduced (Kuhnen et al., 2000). The adsorbed amount of hematite increases with increasing pH (Figure 7b). The increase becomes very steep near the PZC of pure hematite. For pH values higher than PZC, a sharp decrease of the adsorbed amount of hematite can be observed. The hematite colloid adsorbed onto feldspar and quartz as dominant sand minerals (Figure 7c), which have low PZC (Stumm, 1992). It also indicates that the electrostatic attraction act as main mechanism. Therefore, hematite colloid will deposit the minimization of repulsive force on the surface of sand particles by the surface charge neutralization.

In the column of 1.5 meter height, hematite colloids were injected into column for various inlet concentrations (0.1, 1 and 10 mg/L) and various flow rates (611, 1,100, 1450 and 2,100 cm/day). Three experiments were conducted to determine the distribution of retained colloids within the sand columns by injecting the inlet concentrations (Figure 8). The concentration of retained colloids increased in these columns as influent colloid concentrations increased (Figure 8a). In addition, the concentration of retained colloids was evenly distributed throughout the 1.5 m-length of the column.

Collector efficiency based on the filtration theory was compared with collector efficiency determined by the experiment (Figure 8b). Although several assumptions can compute the collision efficiency, it may explain the qualitative and quantitative trend of colloidal deposition in the similar way suggested by Cantrell, and Kaplan (1997). Clean-bed removal model were used to illustrate the deposition mechanism of hematite colloids (Cantrell, et al., 1997). The overall single collector efficiency is the sum of three transport mechanisms (Yao et al., 1971):

$$\eta_0 = \eta_D + \eta_I + \eta_G \qquad \text{Equation (6)}$$

where η_D is the transport by interception, and η_I is the transport by interception, and η_G is the transport by gravity. These transport efficiencies for a spherical collector in a porous media are defined as following equations (Eqn. 7-10).

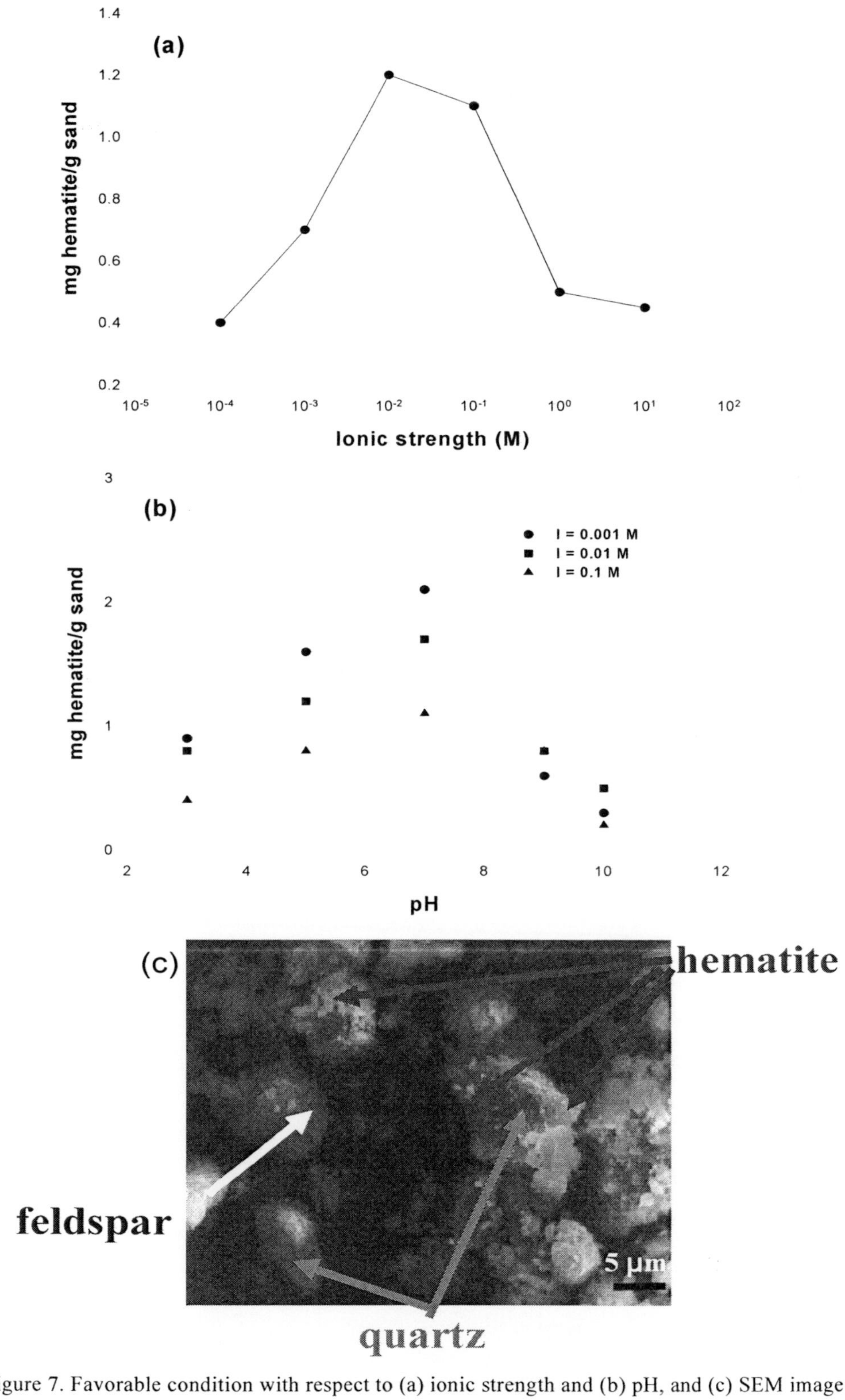

Figure 7. Favorable condition with respect to (a) ionic strength and (b) pH, and (c) SEM image of sand particles coated with hematite colloids.

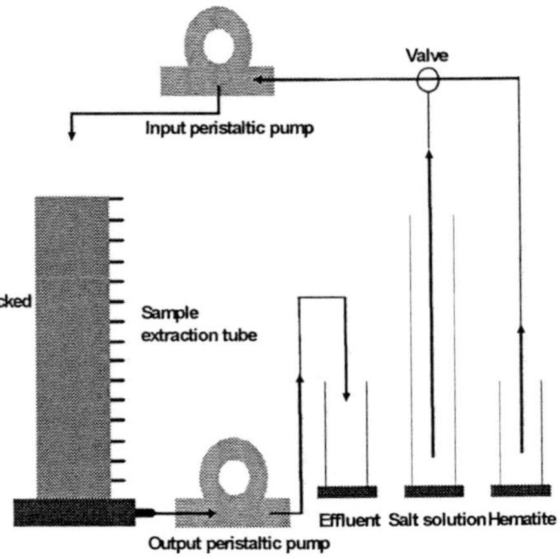

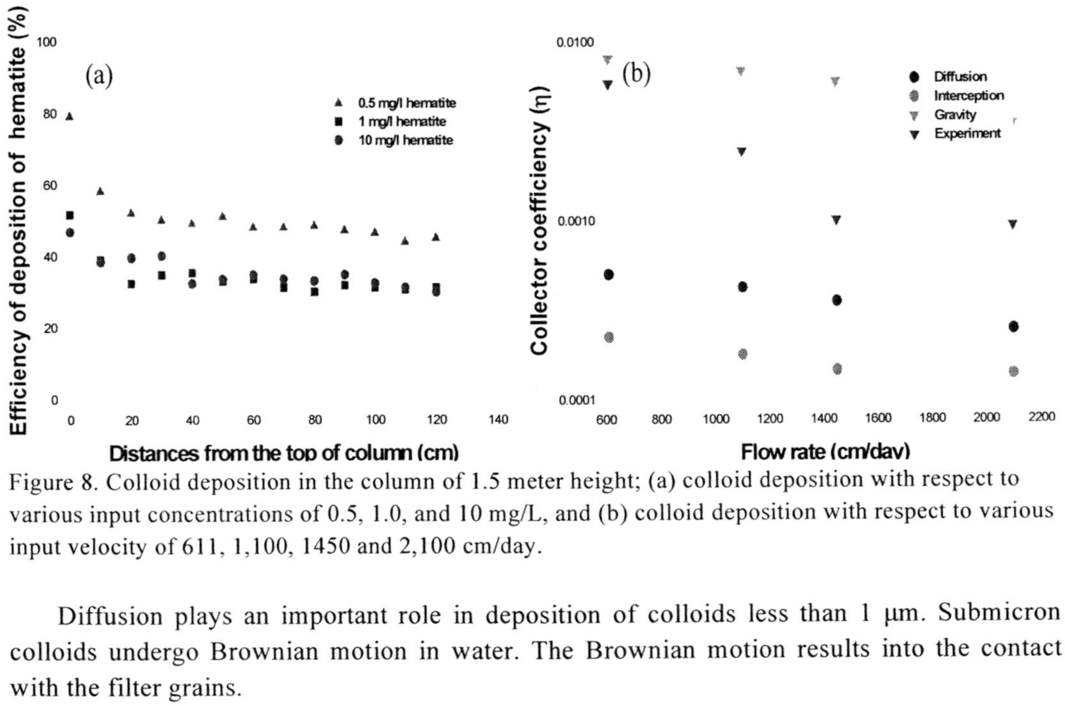

Figure 8. Colloid deposition in the column of 1.5 meter height; (a) colloid deposition with respect to various input concentrations of 0.5, 1.0, and 10 mg/L, and (b) colloid deposition with respect to various input velocity of 611, 1,100, 1450 and 2,100 cm/day.

Diffusion plays an important role in deposition of colloids less than 1 μm. Submicron colloids undergo Brownian motion in water. The Brownian motion results into the contact with the filter grains.

$$\eta_D = 4A_S^{1/3}\left(\frac{D_\infty}{Ud_c}\right)^{2/3} \qquad \text{Equation (7)}$$

where D_∞ is diffusion coefficient of particles and U is pore flow rate.

If sand grains were spherical, Happel's sphere-in-cell model is applied to the flow parameter of A_S. The porosity-dependent flow parameter is defined as:

$$A_S = 2(1-p^5)/(2-3p+3p^5-2p^6) \qquad \text{Equation (8)}$$

where $p = (1-f)^{1/3}$.

Transport of particles by interception occurs when the particle moving with the streamline comes into contact with the filter grain because of the finite volume.

$$\eta_I = \frac{2}{3} A_S (\frac{d_p}{d_c})^2 \qquad \text{Equation (9)}$$

where d_p and d_c are diameter of particle and collector, respectively.

Large colloidal particles, gravitationally, deposited onto filter grains.

$$\eta_G = \frac{(\rho_p - \rho)}{18\mu U} g d_p^2 \qquad \text{Equation (10)}$$

where ρ_p and ρ is density of particle and water, respectively.

Based on filtration theory, the interception of particles played a relatively small role in colloidal deposition. The collision efficiency by interception was less than the collision efficiency by gravitational settling and experiment because of the very low ratio of colloid to grain. The experimental collision efficiency followed the pattern of the diffusion and gravity. This finding is consistent that the collision efficiency by diffusion is significant in the size less than 1 μm (Yao et al., 1971; Cantrell et al., 1997). Further, gravitational settling may be attributed to the high density of colloids relative to that of the suspending liquid. Therefore, a model computation describes the controlling factors of gravitational settling and diffusion in the colloid barrier formation using nano-scale particles.

Further other controlling factors of colloidal particles are also found in in the chemical and physical deposition condition (Elimelech, 1991). As the grain size increases, the deposition of particles onto the collector decreases, and as a result, the barrier length of the particles will increase. Because the size of hematite colloids was less than 1 μm, colloids deposited onto sand by Brownian motion. In addition, deposition of hematite by diffusion and interception results in the decreases as the grain size increases in terms of depositional mechanism (Elimelech, 1991). The permeability of soil medium is mainly determined by pore size due to the size fractionation of soil particles. Therefore, the size of soil particles plays a significant role in the deposition of the nano-scale hematite. Also, according to DLVO theory, at high ionic strength, the double layer of surface shrinks, and the high ionic strength will enhance the coagulation because of the dominant attraction force among particles. Therefore, larger size of particles and enhanced deposition onto sand due to the increased attraction result in the decrease of colloid barrier length. Particularly, divalent cations like Ca^{2+} strongly can increase the coagulation of the hematite colloids due to the large charge balance, and then reduce the barrier length. As a physical parameter, the flow rate also affect that the particles were more mobile at higher flow rate. So, at higher flow rate, deposition rate of particles is smaller, and then results in greater barrier length. Although flow rate is higher than the average flow rate of typical aquifer, the high flow rate can be induced by pumping between injection and extraction well. Therefore, the chemical and physical parameters can be controlled for the formation of colloid barrier in the subsurface environment.

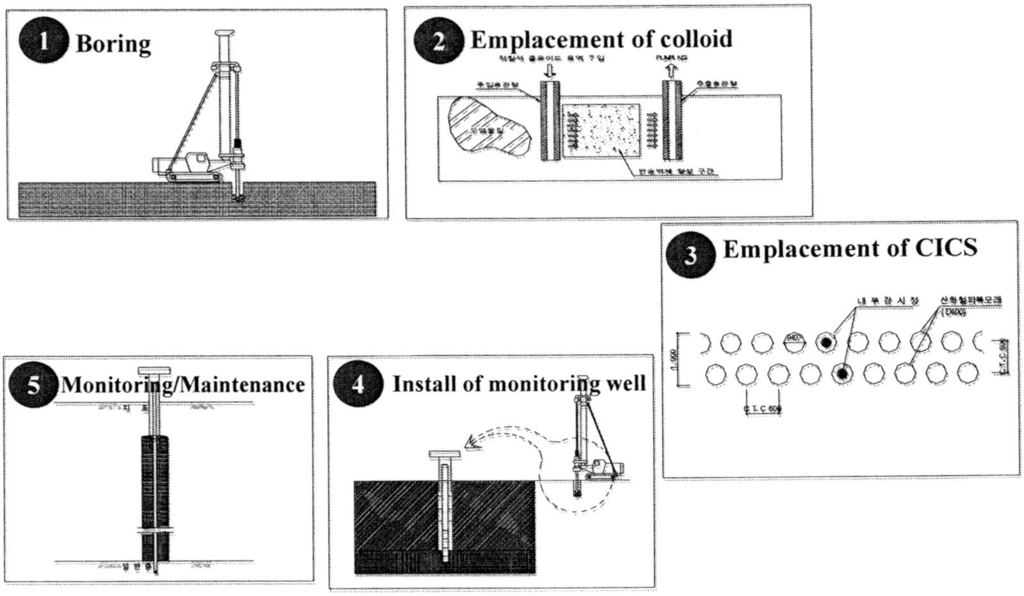

Figure 9. Install of chemical colloid barrier formation (Ko et al., 2003, Patent No. 10-0476064)

Conventional trench-and-filling method has been applied to permeable reactive barrier with excavation. The technique with excavation has the limitation of cost and barrier depth in the large and deep contaminated area. However, the colloid emplacement by the well nozzle can permit no excavation in the permeable chemical barrier. For the colloid barrier formation, the injection and extraction well are required. The injection wells are placed near the front of plume, and the extraction wells are located in the area apart from the injection wells in order to induce the groundwater flow (Figure 9). The groundwater flow reaches the proper flow rate, and then the colloidal particles introduced into the subsurface through the injection wells. As the colloidal particles are deposited between injection and extraction well, the colloid barrier is set up with finite depth and width depending on the colloid deposition rate.

4.2 Physical Separation of Fine Particles

Generally, many cases of contaminated waste, soil and sediment are related with particulate contaminants, like mine tailings, fine particles in soil and sediment, and sludge. Fine particles, like silt and clay, have reactive surfaces, and then are severely contaminated in the high concentration level. In this section, physical separation of fine particles will be applied for the removal of highly contaminated particulate form. In specific, the soil remediation technology is applied for the river deposit contaminated with arsenic, in the Kunwi county, Korea, from an abandoned zinc mine at the upper of river (Lee et al., 2005; Ko, 2006). Sand and gravel, with less than about 5 % silt and clay, are dominantly distributed in this river deposit from a textural analysis (Lee et al., 2003; 2004). Arsenic is incorporated in the fine particles in the river deposit, and then high concentration of arsenic is concerned with fine particles, actually < 149 and < 75 μm (Figure 10). In the deep and wide river environment, the fine particles are generally sedimented as the river deposit. The pathway of arsenic contamination is deduced from its correlation (Figure 11). In this regard, arsenic is released form mining site, and then it migrates to the river deposit. In particular, surface and subsurface water flow repeatedly occur during the cycled wet and dry seasons. The fine particles, enriched with arsenic, migrate through the suspended flow or the river deposit.

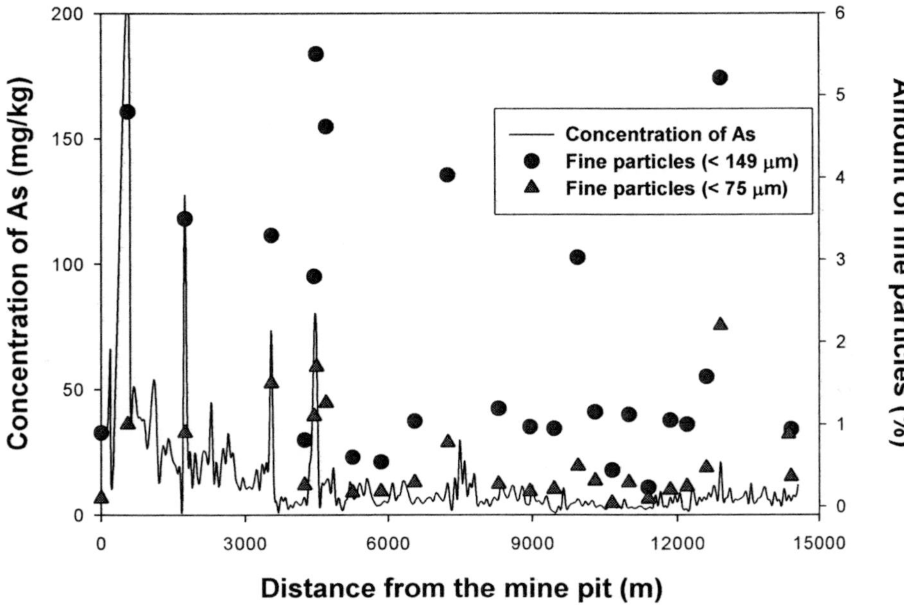

Figure 10. Distance vs. concentration of arsenic in the fine particles (Ko et al., 2006)

Therefore, the predicted pathway of arsenic contamination can be involved with arsenic leaching and loss of contaminated particulate from mining waste for several decades or storm accident periods (Figure 11). Further the pathway of contamination suggests the feasibility of the decontamination of river deposit to physically remove the fine particles.

Classical soil washing technology is based on both the chemical extraction and physical separation (Ko et al., 2005). The chemical extraction is suitable for the complete removal of contaminants in soil and sediment. However, the delayed residence time need in the washing scrubber, and then the efficiency of soil washing process is related directly with long time and high cost. Furthermore, chemical species of arsenic is soluble in an intermediate to alkaline aqueous condition (Smedley and Kinniburgh, 2002; Ko et al., 2004). Then, addition of chemical is questionable for the enhancement of arsenic decontamination together with the following heavy wastewater treatment. Then this study uses the water-based separation technology only with physical size separation of fine particles. The characterization of contamination with respect to particle size indicates the enrichment of contaminants in fine particles rather than mining soil, considered as recalcitrant contamination in both fine and large particles.

Figure 12 shows the effect of physical separation of fine particles on the removal efficiency. The overall removal efficiencies can be calculated from the reductions in the amounts of contaminants that are removed by physical separation. The bulk concentrations of arsenic were obtained from each mass ratio of particle size fraction using the following equation (Eqn. 11).

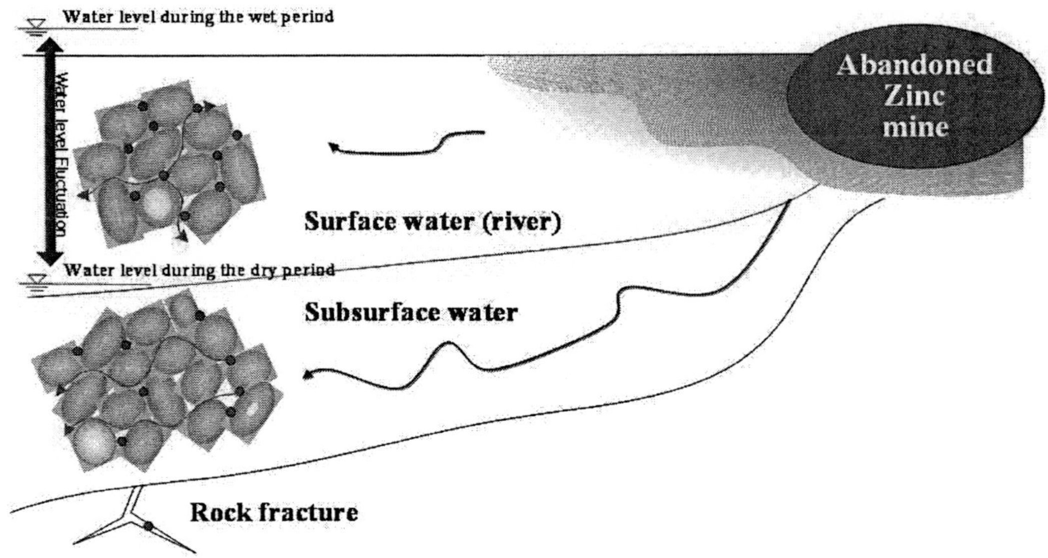

Figure 11. Transport pathway of soluble and colloidal particles from the zinc mining site, Korea (Ko et al., 2006)

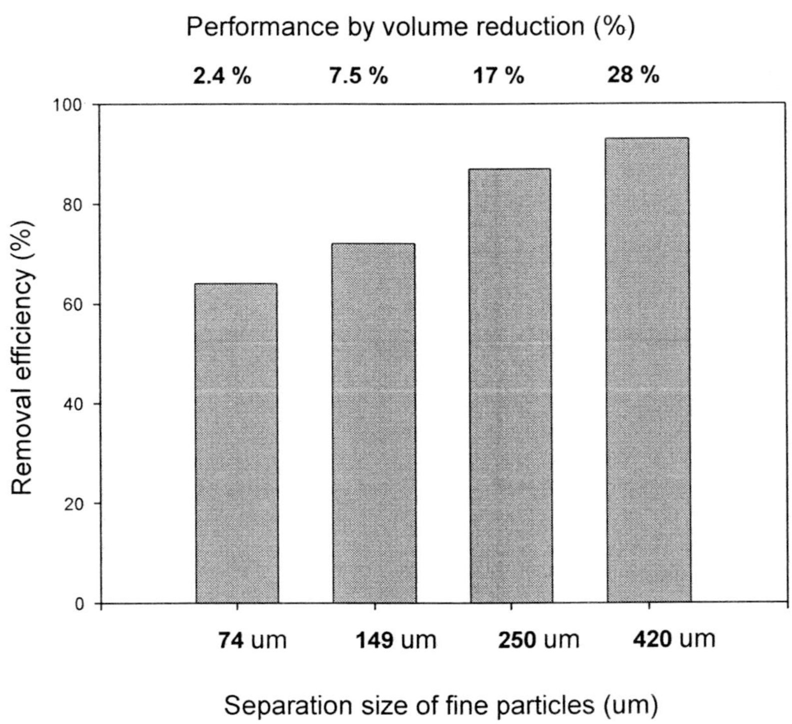

Figure 12. The effect of particle size on the removal efficiency by physical separation of fine particles (Ko et al., 2006)

$$C_j = \sum_i m_i \times C_i = m_1 \times C_1 + m_2 \times C_2 + \cdots + m_i \times C_i \quad \text{Equation (11)}$$

Table 1. Separation technology of fine particles in soil-solution mixture (USEPA, 1995)

Property	Size separation screening	Hydrodynamic separation (Classification)	Density (gravity) separation	Froth flotation	Magnetic separation
Basic principle	Various diameter openings allow passage of particles with different effective size	Different due to density differences	Separation due to density differences	Particles are attracted to bubbles due to their surface properties	Magnetic susceptibility
Advantage	High-throughput continuous processing with simple, inexpensive equipment			Very effective for fine particles	Can recover a wide variety of materials when high gradient fields are used
Limitation	Screens can plug; fine screen is fragile; dry screening produces dust	Difficult when high proportions of clay, silt and humic materials are present		Particulate must be present at low concentration	High capital and operating cost
Typical implementation	Screens, sieves or trammels (wet or dry)	Clarifier, elutriator, hydrocyclone	Shaking table, spiral concentrator, jig	Air flotation column or cells	Electromagnets, magnetic filters

C_j is the adjusted bulk concentration of the contaminant, C_i is the measured concentration from each soil particle fraction, and m_i is the mass ratio of each soil particle fraction. The removal efficiencies were computed from the selected size fraction except for the particle fraction less than the specific separation size. Then, from the calculated results, the overall removal efficiencies are enhanced as the size of physical separation increases from 64 to 92 % after the separation of specific under-size particle. Fine particles are highly enriched with recalcitrant contaminants; hence, their separation from treated soil can reduce the concentration of contaminants. It is effective to remove the fine particles in the contaminated river deposit for the remedial work. The contaminant will be treated with fine particles, and then the remediation becomes simple and effective process. For a full-scale design, the performance of a soil washing system is also typically measured by the volume reduction attained (Ko et al., 2005). This reduction can be calculated by weighing the separated fine particles and feed soil (See Eqn. 12).

$$\text{Performance by volume reduction (\%)} = 1 - \left(\frac{S_i - S_c}{S_i}\right) = 1 - \frac{S_f}{S_i} \qquad \text{Equation (12)}$$

where S_i, S_c, and S_f represent the weight of the initial feed soil, the clean soil, and the fine particle fraction for the physical separation, respectively. The soil volume reduction is proportional to the treatment cost of fine particles. The low amount of fine particles for physical separation reduces their treatment cost. The objective of the process is to treat the entire volume of contaminated soil. Hence, the preferred size of fine particles is determined by minimizing the separation size and attaining the guideline of contamination. In this case study, about 10%, that is, 149μm is critical, considering the cost-effectiveness for the post-treatment of produced fine sludge (Figure 12). Oversize and sand fractions separated from the fines will be treated with the main full-scale washing scrubber and returned to the site as clean backfill. Concurrently, the fines will be separated and either dewatered into a sludge cake or further treated using other extraction methods. For the physical separation of fine particles, practical examples of separation facility are size screening, density separation (jig type), hydrodynamic separation (hydrocyclone), hydrodynamic separation (clarifier type), froth flotation and magnetic separation (USEPA, 1995). Classically, the mining ore extraction or soil washing technologies were water-based treatment system, and their separations have adopted these separation technologies (Table 1). For the remedial design of this case study, the screen seems to be suitable for the physical separation, because the wet river deposit, composed dominantly with sand and gravel, can be applied for the separation of the exact particle size, minimizing the dust to the air.

4.3 Surfactant Washing of Contaminated Soil

Soil-water interfacial phenomena with surfactant are also related with colloidal control. The stability of colloidal structure of surfactant, that is, colloidal micelle is significant for the enhanced removal efficiency of contaminant. The soils contaminated with petroleum or both petroleum and metal has been treated with surfactant solution for their removal (Doong et al., 1998; Mulligan et al., 2001; Khodadoust et al., 2005; Ko et al., 2005). In this study, the composite contaminants from soil were washed with nonionic surfactant (Tween 80). The properties of contaminated soil are shown in Table 2. The contaminants are consisted with TPH (total petroleum hydrocarbon), Pb, Zn, Ni, and F as the composite contaminant with both organic and inorganic contaminants. In soil washing technology, it is difficult to choose the suitable washing solution for the composite contaminants. Recent studies of soil washing show the feasibility of surfactant additives for the enhancement of composite contaminants from soil and sediment (Table 2). It indicates that the surfactant washing has been verified for the removal of the organic contaminants, and then it also shows the feasible efficiency for the composite contaminants. In the surfactant washing, surfactant molecules form the colloidal micelles in washing solution. Particularly the formation of colloidal micelles can effectively remove the organic contaminants through enhanced solubilization and mobilization of contaminants, but the formation condition should be examined for the optimization of colloidal formation of micelle. In addition, inorganic contaminants can be also desorbed by hydrophobic interaction of surfactant, organic complexation and solubilization (Mulligan et al., 2001). In this study, for the favorable washing condition, the surfactant and acid was chosen as Tween 80 and hydrochloric acid, respectively.

Table 2. Investigation of soil washing using surfactant and additives

Soil	Experiment	Washing agents	Contaminants	Implication	Reference
Artificially contaminated kaolinite	Batch	Tween 80 Igepal CA-720 EDTA Cyclodextrins Organic acid Sequential extraction	Phenanthrene Ni	Sequential use of surfactant and organic acid (Tween 80 and citric acid)	Khodadoust et al. [2004a]
Field soil	Batch	Tween 80 Igepal CA-720 EDTA / DTPA Sequential extraction	Phenanthrene Pb Zn	Tween 80 + EDTA or Citric acid +Tween 80	Khodadoust et al. [2004b]
Sediment	Batch	Surfactin from *Bacillus subtilis* Rhamnolipids from *Pseudomonas aeruginosa* Sophorolipid from *Torulopsis bombicola*	Cu Zn	Chracteritic removal of organically-bound copper, and carbonate and oxide-bound zinc	Mulligan et al.[2001]
Field soil	Batch	Surfactin from *Bacillus subtilis*	Oil Cu		Mulligan et al.[1999]
Field soil	Batch	Surfactant (Witconol 1206)	PCBs Pb	High removal efficiency of both contaminants	Sheets and Bergquist [1999
Artificially contaminated kaolinite	Batch	Cationic/anionic/nonionic surfactant with EDTA and acid solution	Cd Pb Zn	effective washing using cationic surfactant under acidic condition or anionic/nonionic surfactant through the addition of EDTA	Doong et al. [1998]
Field soil	Column	Flushing sequentially with saline and surfactant/cosurfactant/water	tetraethyl Pb Gasoline	surfactant/cosurfactant/water produced the microemulsifications for the immobile or residual contaminants	Ouyang et al.[1996]
Field soil	Batch and column	Surfactant(Dowfax 8390) or surfactant + diphenyl carbazide	Cr	3.7-5.7 times greater than surfactant without the complexing agent	Nivas et al.[1996]

In this study, the soil washing of composite soil was performed by a pilot-scale facility (Figure 13). The pilot-scale equipment used in this study consisted of five principle parts; soil washing scrubber (drum-type), vibrating screen, screw feeder, high-pressure air supply and ceramic filter system as shown in Figure 13. The screw feeder was able to transfer contaminated soil into the washing scrubber simultaneously mixing washing solution with soil:solution ratio of 1:10. The washing scrubber has a drum-type cylinder with the inner screw blade to help soils move forward. While being treated, treated soils and washing solution were shaken with a strong mechanical turbulence by a high-pressure air supply. The compartment extracted contaminants by enhancing contact surface area for the extraction reaction and reducing soil particle size. The rotating velocity of washing scrubber can be controlled by rpm (revolutions per minute). Moreover, after treated soils and washing solution pass through the washing scrubber, they are separated and screened by the vibrating screen with 200 mesh (0.074 mm) sieving size. In the ceramic filter system, the waste washing solution was treated with settling the fine particles and adsorbing dissolved inorganics and recycled for new washing solution with the addition of surfactant and or acid solution. All parts of the soil washing equipment were made of stainless steel.

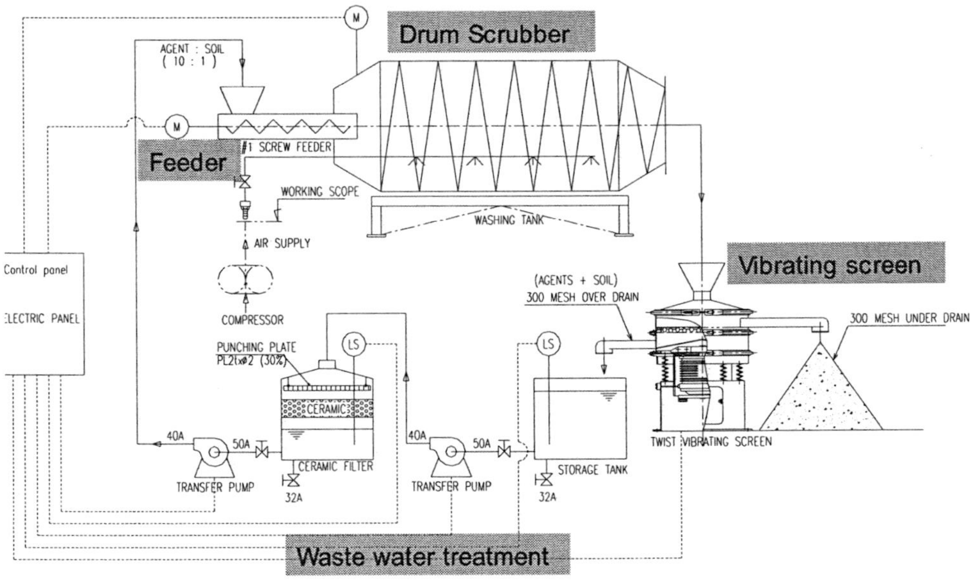

Figure 13. Pilot-scale soil washing facility (2 ton/hr) (Ko et al., 2005)

Surfactant is chosen in terms of the critical micelle concentration (CMC) and cost-effectiveness (Rosen, 1989; Deshpande et al., 1999). The CMC of surfactant represents the condition for the formation of colloidal micelle above its CMC. In addition to its cost, considering the enhanced removal efficiency by the colloidal micelle, low CMC is able to effectively form CMC by minimizing the consumption of surfactant. Therefore, recent studies and applications use the nonionic surfactant, for example, Tween 80 (Deshpande et al., 1999). Figure 14 show the consumption of surfactant by its adsorption onto soil particles. By lowering the surface tension of surfactant through its addition, colloidal micelles form in aqueous washing solution, and then the adsorbed organic and inorganic contaminants can be desorbed simply or through surfactant-metal complexation (Sun et al., 1995). Further the solubility of contaminants increases in the aqueous washing solution. Comparing the surface tension of surfactant in aqueous solution, the soil particles consume the surfactant with coating its surface and desorbing the contaminants, and then the CMC concentration shift to the more amount of surfactant addition (Figure 14).

In terms of colloidal stability, charged surfactant act as a stabilizer by the charge compensation due to the strong adsorption of surfactant on the soil particle (Rosen, 1989). For nonionic surfactant, like Tween 80, the steric forces are the main stabilization mechanism. With increasing surfactant concentration, the coagulation concentration of colloid move to higher background ion concentration, i.e. pH and ionic strength, and then lead to the fast aggregation. It indicates that the steric stabilization can be rationalized with an additional short-range repulsive component to the interparticle force. The additional repulsion results from the overlap of the adsorbed surfactant functional group on the colloidal surfaces. Below the CMC, charged surfaces, covered with surfactants, typically interact by classical DLVO-forces down to distances of about 10 nm. At shorter distances steric overlap forces lead to stronger repulsion. In addition, depletion forces may also become significant in surfactant systems as a different mechanism. This force become only relevant above the

CMC, as the depletion force will be generated by the colloidal micelles (Edwards et al., 1994). In this regard, for the stabilization of colloidal micelles dispersed in the soil-solution system, it is favorable to use surfactants at concentrations above the CMC.

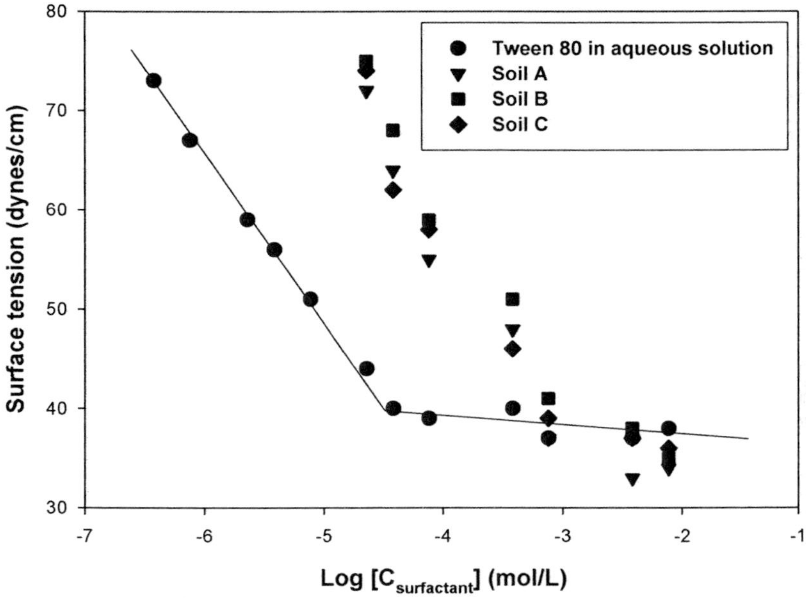

Figure 14. Measurement of surface tension with respect to the concentration of nonionic surfactant in aqueous solution and soil solution.

Table 3. Removal efficiency of composite contaminants from soil by a pilot-scale soil washing

Washing solution *	Reaction pH **	Removal efficiency (%)				
		TPH	Pb	Zn	Ni	F
Acid washing	3.3	25	21	25	29	45
Tween 80 washing	7.8	27	15	4	37	47
Tween 80 washing at pH 3	3.1	48	7	15	14	9
Tween 80 washing + Acid washing	3.8	35	46	73	89	91
Acid washing + Tween 80 washing	4.6	41	19	52	45	88

* soil(g) : solution(mL) ratio = 1:10 and washing duration = 30 min

** Reaction pH is the solution pH after the washing.

The single washing solution, like acid or Tween 80 washing, shows very low removal efficiencies from 4 to 47 % (Table 3). Further the simultaneous washing, that is, Tween 80 washing with an adjusted pH 3, also has low efficiency for various contaminants. However, the sequential washing with acid and Tween 80 or vice versa, evidently enhances the removal efficiency (from 19 to 91 %). In acidic condition, nonionic surfactant easily is adsorbed onto soil particles, resulting into the consumption of the surfactant with less formation of colloidal micelles. In this regard, the acid washing had better be separated in the surfactant washing, such as sequential washing rather than simultaneous washing. Therefore, the consumption of surfactant also increases at an acidic pH in addition to its adsorption onto soil particles. The stability of colloidal micelles should be increased through solubilization and mobilization of contaminants for the enhancement of removal efficiency. During the surfactant washing process, the stability of colloidal micelle should be monitored and controlled.

5. Concluding Remark

The colloidal particles themselves are generated by contaminants, and enhance the activity of contaminants by chemical and physical perturbation. In particular, the colloidal particles are used for the environmental remediation due to the highly reactive surface and controlled colloidal mobility. It is evidently possible to use the colloidal particles with the control technology of colloidal particles.

In the colloid barrier formation, the reactive colloidal materials, like nano-scale iron particles, are favorably deposited on the subsurface environment. In these regards, the chemical and physical conditions of the colloidal deposition will be able to enhance the colloidal stability and coagulation for the migration and coating, respectively. Polymerized colloid surface is kept with stability in aqueous solution. On the contrary, multivalent metal cations, like Ca^{2+}, reduce the colloidal stability, and then lead to the short-range barrier formation.

In general, the colloidal particles are enriched with high concentration of contaminants due to the reactive surfaces. In the environmental remediation system, like soil washing, wastewater treatment and so on, the physical separation of colloidal particles can reduce the volume of contaminated soil, and lead to the lowering the level of contamination in bulk contaminated soil. The environment-friendly remediation, like this water-based separation system, should be ecologically less toxic with no chemicals for the extraction or stabilization of contaminants.

Many surfactant washings have been applied for soil contaminated with hydrocarbon despite the expensiveness of various surfactants. Nowadays nonionic surfactant, due to its low amount consumption and a low level of CMC, should be controlled in the surfactant-based remediation system. In this regard, the colloidal micelles of nonionic surfactant are adsorbed onto soil particles and dispersed in the mixed soil-solution system. The detachment of hydrocarbon and metal is related with the surfactant concentration above the CMC, leading the high stability of micelle.

References

Alber, M., Estuar. Coast. *Shelf Sci.* 2000, 50, 805-816.

Alexander, S.C.; Alexander, E.C. Jr. *Minnesota Acad. Sci. J.* 1989, 55, 48-52.

Benner, S.G.; Blowes, D.W.; Ptacek, C.J., 1997, Int. Cont. Tech. Conference St. Petersburg, FL, Proceeding, 56.

Buckau, G.; Artinger, R.; Geyer, S.; Wolf, W.; Fritz, P.; Kim, J.I. *Appl. Geochem.* 2000, 15, 819-832.

Cantrell, K.J.; Kaplan, D.I. *J. Environ. Eng.-ASCE* 1997, 123(5), 499-505.

Cantrell, K.J.; Kaplan, D.I.; Gilmore, T.J. *J. Environ. Eng.-ASCE* 1997, 123(8), 786-791.
Cantrell, R.W.; Kaplan, D.I.; Wietsma, T. W. *J. Hazard. Mater.* 1995, 42, 201-212.
Cha, J.; Lee, J.; Lee, M. *Econ. Environ. Geol.*, 2003, 36(6), 457-467.
Degueldre, C.; Triay, I.; Kim, J. I.; Vilks, P.; Laaksoharju, M.; Miekeley, N. *Appl. Geochem.* 2000, 15, 819-832.
Derjaguin, B.V.; Landau, L. *Acta Phys. Chim.* URSS. 1941, 14:633–662.
Doong, R. A.; Wu, Y. W.; Lei, W. G. *Wat. Sci. Technol.* 1998, 37, 65-71.
Durmusoglu, E.; Corapcioglu, M. Y. *J. Environ. Eng.-ASCE.* 2000, 126, 833-841.
Edwards, D. A.; Liu, Z.; Luthy, R. G. *J. Environ. Eng.-ASCE.* 1994, 120, 23-41.
Elimelech, M. J. *Colloid Interface Sci.* 1991, 146, 337– 351.
Elimelech, M., Nagai, M., Ko, C.-H., Ryan, J.N. *Environ. Sci. Technol.* 2000, 34, 2143–2148.
Federal Interagency Stream Restoration Working Group (FISRWG). Stream corridor restoration: principles, processes, and practices.1998.
Jain, A.; Raven, K. P.; Loeppert, R. H. *Environ. Sci. Technol.* 1999, 33, 1179-1184.
Kanel, S.R.; Manning, B.; Charlet, L.; Choi, H. *Environ. Sci. Technol.* 2005, 39(5), 1291-1298.
Khodadoust, A.P.; Reddy, K.R.; Maturi, K. *Environ. Eng. Sci.* 2004b, 21(6), 691-704.
Khodadoust, A.P.; Reddy, K.R.; Maturi, K. J. *Environ. Eng. Sci.* 2004a, 21, (6), 691-704.
Khodadoust. A. P.; Reddy, K. R.; Maturi, K. *J. Hazard. Mater.* 2005, 117, 15-24.
Ko, I. J. Hazard Mater. 2006 (In submission)
Ko, I. W.; Lee, S.W.; Kim, J. Y.; Kim, K. W.; Lee, C.H. *Journal of KoSSGE.* 2004, 9(1), 63-69
Ko, I.; Jang, Y.Y.; Lee, C.H.; Kim, K.W. *J. Hazard. Mater.* 2005, A127, 1-13.
Ko, I.; Kim, J.Y.; Kim, K.W. *Colloid. Surf.* A. 2004, 234, 43–50.
Ko, I.W.; Kim, J.Y.; Kim, K.W. Colloid barrier formations by nanoscale hematite particles. Proceeding of 4th International Symposium on Advanced Environmental Monitoring. 2002, PW10.
Ko, I.W.; Kim, J.Y; Ahn, J.S.; Kim, K.W. Korea Patent No. 2003, 10-0476064.
Ko, I.W.; Lee, C.H.; Choi, H.J.; Ahn, Y.; Kim, J.G.; Kim, K.W. Effect of nonionic surfactant on the extraction of inorganic and organic contaminants from soil. Kossage International Symposium on Soil & Groundwater Environment. 2005, 145-146.
Lackovic, J.A.; Nikolaidis, N.P.; Dobbs, G.. *Environ. Eng. Sci.* 2000, 17, 29-39.
Lee, J.Y.; Choi, J.C.; Yi, M.J.; Kim, J.W.; Cheon, J Y.; Choi, Y K.; Choi, M.J.; Lee, K.K. *Water Air Soil Pollut.* 2005, 165, 167-185.
Lee, M.; Lee, J.; Cha, J.; Choi, J.; Lee, *J. Econ. Environ. Geol.* 2004, 37(1), 121-131.
McCarthy, J.F. *Physics and Chemistry of the Earth.* 1998, 23(2), 179-185.
Mulligan, C.N.; Yong, R.N.; Gibbs, B. F. *Eng. Geol.* 2001, 60, 371-380.
Mulligan, C.N.; Yong, R.N.; Gibbs, B.F.; James, S.; Bennett, H.P.J. *Environ. Sci. Technol.* 1999, 33, 3812-3820.
Nivas, B.T.; Shiau, B.J.; Sabatini, D.A.; Harwell, J.H. *Water Res.* 1996, 30(3), 511-520.
Ouyang, Y.; Mansell, R.S.; Rhue, R.D. J. *Hazard Mater.* 1996, 46, 23–35.
Prostein, R.F. *Physicochemical hydrodynamics;* John Wiley & Sons, Inc.: Hoboken, NJ. 2003; pp 109-274.
Rosen, M.J., 1989, *Surfactant and interfacial phenomena* (2nd ed.) New York: Wiley.
Roth, D.A.; Taylor, H.E.; Domagalski, J.; Dileanis, P.; Peart, D.B.; Antweiler, R.C.; Alpers, C.N. *Environ. Contam. Toxicol.* 2001, 40, 161-172.

Roy, S.B.; Dzomback, D.A. *Environ. Sci. Technol.* 1977, 31, 656-673.
Ryan, J. N.; Elimelech, M. *Colloid. Surf.* A 1996, 107, 1-56.
Sen, T. K.; Khilar, K. C. Adv. *Colloid Interf. Sci.* 2006, 119, 71-96.
Sen, T. K.; Mahajan, S. P.; Khilar, K. C., AlChE J. 2002, 48, 2366-2385.
Sheets, R.G., Bergquist, B.A. *J. Hazard Mater.* 1999, 66(1-2):137-50.
Smedley, P.S.; Kinniburgh, D.G. *Appl. Geochem.* 2002, 17, 517.
Stumm, W. Chemistry of the Solid-Water Interface: Processes at the Mineral-Water and Particle-Water Interface in Natural Systems. Wiley-Interscience. 1992.
Sun, S.; Inskeep, W.P.; Boyd, S. A. *Environ. Sci. Technol.* 1995, 29, 903-913.
Thomas, J.M.; Ward, C.H. *Environ. Sci. Technol.* 1989, 23, 760-766.
Tufenkji, N.; Ryan J.N.; Elimelech, M. *Environ. Sci. Technol.* 2002, 36, 422-428.
USEPA. Contaminants and Remedial. Options at Selected Metal-Contaminated Sites. EPA/540/R-95/512. 1995.
USEPA. Innovative treatment technologies: Annual Status Report. EPA 542-R-96-010, 1996.
USEPA. *Permeable reactive barrier technologies for contaminant remediation.* USEPA 600-R-98-125, 1998.
Verwey, E.J.W.; Overbeek, J.T.G. *Theory of Stability of Lyophobic Colloids.* Elsevier Publishing Company, Amsterdam, The Netherlands, 1948.
Yao, K. M.; Habibian, T.; O'Melia, C. R. Environ. *Sci. Technol.* 1971, 5, 1105-1112.

In: Trends in Hazardous Materials Research
Editor: Edward C. Booking, pp. 137-164
ISBN 1-60021-335-9

Chapter 5

IN SITU TECHNOLOGIES FOR HAZARDOUS MATERIALS CONTAMINATED SOILS AND GROUNDWATER: THE CASE OF MTBE

L. di Palma[1]*, C. Alimonti[1], D. Lausdei[2] and R. Mecozzi[1]
[1]Università di Roma "La Sapienza", Dip. Ingegneria Chimica, Roma, Italy
[2]Golder Associates s.r.l., Roma, Italy

ABSTRACT

In this chapter, experimental tests of different techniques for contaminated soils and groundwater remediation are presented and discussed. The contaminant of concern is the MtBE, a common antiknock additive in the production of unleaded fuel. MtBE's great water solubility, low partitioning into soils and low tendency to be degraded by microbial activity, are responsible for a considerable mobility that allows rapid dispersion and strong persistency in groundwater.

To remove or/and destroy this contaminant, three *in situ* technologies are evaluated and compared: Air Sparging (AS), soil flushing (ISSF), and chemical oxidation (ISCO). AS, which mainly involves the stripping of the contaminant by injected air in the subsurface, is currently one of the most commonly used, due to the simplicity of the plant installation, operation and maintenance and, consequently, to its low associated costs. Other systems, like ISSF and ISCO can be also successfully implemented in MtBE contaminated soils and groundwater treating. Both ISSF and ISCO technologies deal with the chemical oxidation of the pollutant, with the difference that for ISSF the oxidant does not come into contact with the soil matrix. In the experimentation presented in this chapter, Fenton oxidation was performed in either technologies used.

For each technology after the description of the basic principles and environmental applications, the results of lab scale experiments conducted on samples of saturated soil artificially contaminated, are discussed.

* corresponding author: luca.dipalma@uniroma1.it, tel: +39-06-44585571; fax: +39-06-44585622

AS tests were performed both in a cylindrical column (small scale test) and in a large tank of 1 m^3 (intermediate scale test). These two testing set-ups were used to study the optimal air flow rate for MtBE transfer to the gas phase and the influence of the geometry around the injection well.

ISSF tests were performed by flushing the soil with an aqueous solution of a surfactant agent (ethanol). The pH, the ethanol concentration and the flow rate of this extracting solution were optimized to achieve the complete removal of MtBE from the matrix. The recovered solution was then oxidized with Fenton's reagent to destroy the pollutant. The effect of ethanol concentration towards MtBE oxidation was also evaluated, considering the non selectivity of the oxidant.

ISCO tests were performed by injecting Fenton's reagent directly into the soil, both at the natural pH of soil and under acidic conditions.

INTRODUCTION

Methyl *ter* butyl ether (MtBE) is a volatile, flammable, uncolored transparent liquid with a strong hydrocarbon's smell.

This product, which has been synthesized since the early 70^{ies} (as an antiknock additive for unleaded fuel, instead of tetraethyl lead), had a dramatic increase of production during the 90^{ies}, when in some areas of US it was required the use of oxygenates in common gasoline formulations to improve air quality, by improving the combustion of the volatile compounds, reducing the heat of vaporisation and lowering the boiling point of gasoline.

The principal chemical and physical properties of MtBE are:

- Molecular weight (g/mol): 88.15
- Relative vapor density (air=1): 3.0
- Relative liquid density (water=1): 0.744
- Vapor tension (mmHg): 245-276 at 25°C
- Henry's constant : 0.02-0.05
- Water solubility (mg/l): 43000-54000 at 25°C
- Boiling Point (°C): 55.2
- Melting Point (°C): -109
- log Koc: 1.035-1.091
- log Kow: 1.24

The polarity of the Ether group in the molecule, enables it to have any sort of hydrogen bond with water.

Despite the MtBE was mainly introduced for environmental purposes, in the last decade the groundwater contamination by this product due to fuel spills or leakage from underground storage facilities became a high sensible issue. When it reaches the soil, this pollutant partially volatilizes and partially is adsorbed. Unfortunately, due to its great water solubility and low biodegradability, it has a great mobility that allows a rapid dispersion and contamination of the groundwater, especially after rainfalls. It is therefore very persistent in the environment, and the plume generated by MtBE contamination is generally of a greater extension than that of other pollutants that may be present in the fuel.

As a consequence, the need exists to remove the MtBE from the contaminated environments.

The technologies commonly used for the remediation of contaminated soil and groundwater are generally classified as:

- in-site (or *in situ*), when the contaminated environmental matrix (soil or groundwater) is treated directly in place, without pulling it out of the subsoil;
- on-site, when the contaminated matrix is treated on the surface, directly above the place where it has been extracted;
- off-site (or *ex situ*), when the contaminated matrix is treated away from the site where it has been extracted.

Generally *in situ* technologies are cheaper and safer than the *ex situ* ones, because excavation and management of huge quantities of material are not required. As a consequence, the related risk of pollutant spread in the environment is also reduced.

In this chapter the application to the remediation of soil and groundwater contaminated by MtBE of a physical treatment, the Air Sparging (AS) technology, and two chemical treatments, soil flushing (ISSF), and chemical oxidation (ISCO), is described and experimentally evaluated. After a general description of the principles of each technology, a summary of experimental research at lab-scale and intermediate pilot scale are presented and discussed.

AIR SPARGING

Air Sparging (AS) is an *in situ* technology designed to remove volatile organic compounds (VOCs) from the saturated zone through the injection of a gas under pressure (generally air) from one or more wells (specifically designed for injection) below the contaminated plume. AS has been used to address a broad range of volatile and semi-volatile groundwater and soil contaminants, including gasoline and other fuels, associated components such as BTEX or MtBE, and chlorinated solvents (Miller R.R., 1996).

The target of the air injection is to volatilize groundwater contaminants and to promote biodegradation in saturated and unsaturated zone, by increasing subsurface oxygen concentration. Volatilized vapours migrate into vadose zone where they are extracted via vacuum, generally by a soil vapour extraction (SVE) system (Figure 1).

The injected air travels upwards through buoyancy as discrete bubbles or in channels. This produces the mass transfer of the dissolved VOCs from the liquid phase into the vapour phase through the air-water interfacial surface created in the soil pore space (stripping process). AS applicability is restricted to unconfined aquifers where the injected air, extracting organic vapours, can migrate to the unsaturated zone through the water surface.

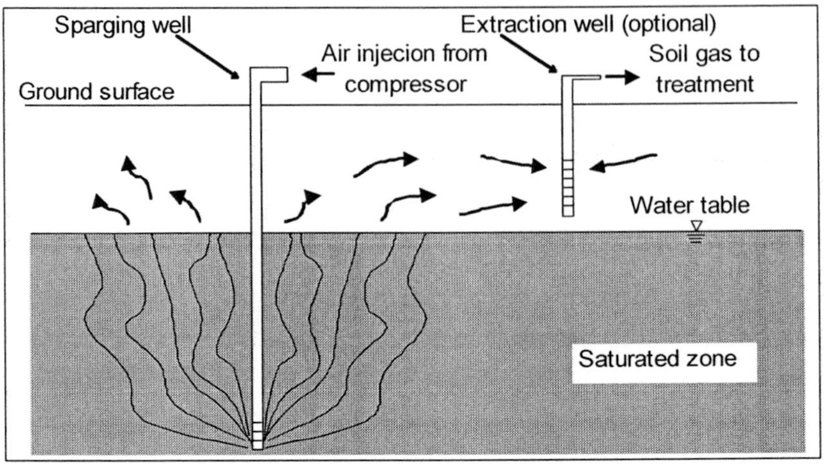

Figure 1. Scheme of an Air Sparging systems coupled with a SVE well for the organic vapors recovery

The removal effectiveness of AS technology highly depends on the geological characteristics of the soil as well as on the physico-chemical properties of the polluting compounds (Burns and Zhang, 2001). In effect, AS effectiveness relies on the fact that the airflow could reach all the volume of aquifer to treat. Thus, the most suitable aquifers to be treated by AS technology consist of permeable rock or soils ranging from medium to highly permeable (aquifers with values of the *hydraulic conductivity* within the range: $10^{-4}<K<10^{-2}$ m/s). With respect to the physical-chemical properties of the pollutants, it is mandatory that they allow a quick mass transfer into the vapour phase within the air volumes (bubbles or channels); hence, the most suitable chemicals are the ones classified as volatile and semi-volatile (Lausdei, 2003).

In situ Soil Flushing

In situ Soil Flushing performs the removal of organic or inorganic pollutants from soil, through the infiltration of an extracting fluid in the subsurface and its successive extraction together with the groundwater.

The flushing fluid is typically an aqueous solution and may contain many additives, such as chelating agents or surfactants that, due to their particular molecular structure, can help the extraction of the contaminant from the aqueous phase. Surfactants in particular, have a hydrophilic head and an oleophilic tail, which allow them to alter the properties of organic-water interfaces (Jafvert 1996; Roote 1998). The retention of contaminants by soil is affected by many factors. Among them the most important are: the soil nature, the pH, the soil cation-exchange capacity and the eventual simultaneous presence of different organic pollutants. Hydraulic conductivity (K) is the principal physical parameter that can determine the feasibility of a soil flushing process (Freeman and Harris, 1995). In effect, as soils with low permeability ($K<1.0\cdot10^{-5}$ cm/s) naturally inhibit the permeation of fluids, soil flushing proves to be effective only for those classified as permeable ($K>1.0\cdot10^{-3}$ cm/s) or, to a lesser extent, as slightly permeable ($1.0\cdot10^{-5}<K<1.0\cdot10^{-3}$ cm/s).

In situ soil flushing technologies must be combined with the successive treatment of the extracted solution (Pump & Treat). The mixture of the injected fluid and contaminant (the extracted solution), along with ground water, is collected through extraction wells and sent to the successive chemical treatment (Freeman and Harris 1995; Di Palma et al. 2003). This is done to destroy the pollutants or to remove them from the liquid, allowing their recycle in the process and the discharging of all treated water, which can diminish the overall treatment costs.

The flushing technology avoids the direct contact of the soil with the oxidant. However it produces a large amount of effluent to be treated and needs a lot of time to achieve the results. In addition, though the contact time of the flushing solution with the soil is generally low (on the contrary of the washing processes), the extracted solution will probably contain a substantial amount of soil constituents other than the target compound. These are mostly from natural organic matter, and are strongly susceptive to the successive oxidation. Furthermore, since the surfactant itself can undergo oxidation, the process needs a previous separation stage to recover it. This permits a considerable reduction in chemical reagents consumption. To reduce the extracted solution's volume, membrane filtration, liquid extraction or evaporation processes are necessary, thus increasing the overall cost of the treatment.

The *in situ* flushing technology (extraction, followed by oxidation), has been already supported by decades of experimental laboratory, pilot and field tests. Therefore, it is commonly adopted in the remediation of contaminated soils. In the remediation of heavy metals polluted soil, the effectiveness of EDTA (ethylenediaminetetraacetic acid, E-H_4) as a chelating agent in flushing treatments has been widely demonstrated (Beckett, 1989; Tejowulan, 1998; Martin and Allen, 1996). Using EDTA presents some advantages in comparison with other chelating agents: it has a low degree of biodegradability in soil and groundwater, coupled with a high capacity of complexing heavy metals (Sun et al., 2001). Conversely, the high consumption and cost of EDTA, strongly affects the economic feasibility of a soil flushing remediation (Chang, 1995; Di Palma et al., 2003).

In the remediation of organic pollutants contaminated soils, the addition of flushing additives, as surfactants, to the extracting solution has also proven to be highly effective (Beckett, 1989; Tejowulan, 1998; Liu et al., 1995; Chang et al., 2000).

The implementation of the flushing technology involves the design and installation of a system of injection and extraction wells.

IN SITU CHEMICAL OXIDATION

The chemical oxidation of contaminated soils or groundwater can be performed through *ex situ* or *in situ* technologies.

The *ex situ* Chemical Oxidation performs the mixing of the oxidizing agent with soil or groundwater in a vessel. Excavation and transportation of the soil are therefore necessary: those operations not only largely affect the overall cost of the treatment, but also may imply additional risks of environmental contamination (Ghassemi 1988). The oxidizing agent can be an aqueous solution, or less commonly, a gas (ozone). The most frequently used oxidizing agents for the treatment of organic contaminants are: ozone, hydrogen peroxide and potassium permanganate.

In the past few years (US EPA 1998; Siegrist 1998; Yin and Allen, 1999; ITRC 2000; Chen et al. 2001; Quan et al. 2003), chemical oxidation processes have been successfully introduced in the remediation of contaminated soils. In recent times, *in situ* chemical oxidation technology (direct oxidation) has become to be implemented due to its potential advantages, both as a unique remediation technology, and as a pretreatment step to enhance biodegradation. Its implementation is particularly useful in case of toxic or recalcitrant compounds because it's aggressive and non selective.

In situ chemical oxidation involves the direct injection of chemical oxidants into the vadose zone or groundwater (Fig. 2).

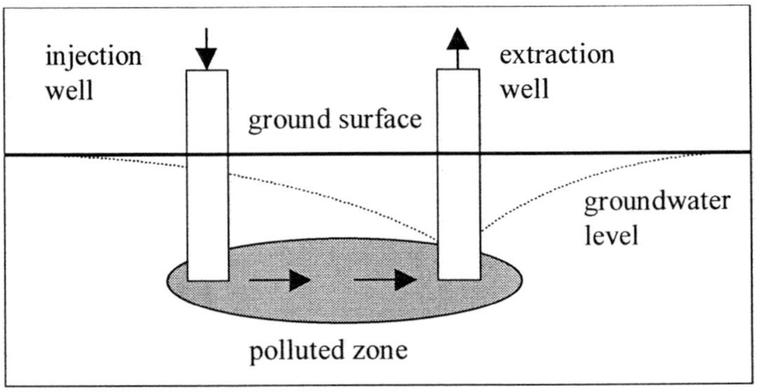

Figure 2. Scheme of the In situ Chemical Oxidation process

The objective of the treatment can be the total degradation of the contaminant or simply its transformation in an oxidized product which can undergo to a successive biological degradation (Ghassemi, 1988). The performance of the treatment is greatly affected by the distribution of the contamination, and by soil characteristics such as pH, permeability, and the presence of other matter susceptive to oxidation (ITRC 2001; Tyre et al. 1991; Tarr and Lindsay 1998).

The main advantage of this emerging technology towards the *ex situ* one, is that soil excavation and mobilization are avoided, and consequently the risks of release of contaminants into the wider environment are reduced (Di Palma, 2005).

PHYSICAL TREATMENT

Experimental Tests of MtBE Remediation by Air Sparging

Laboratory Scale Experiments

In this section the results of experimentation on the MtBE removal from a contaminated aquifer, performed by Air Sparging (AS), are presented and discussed. The objective was to evaluate the effectiveness of AS in MtBE removal from a saturated porous media. Beside this, tests were performed also to evaluate and compare two types of air diffusers, each with a different orifice size.

The laboratory apparatus used to perform experimental tests consisted of the following components, as shown in Figure 3:

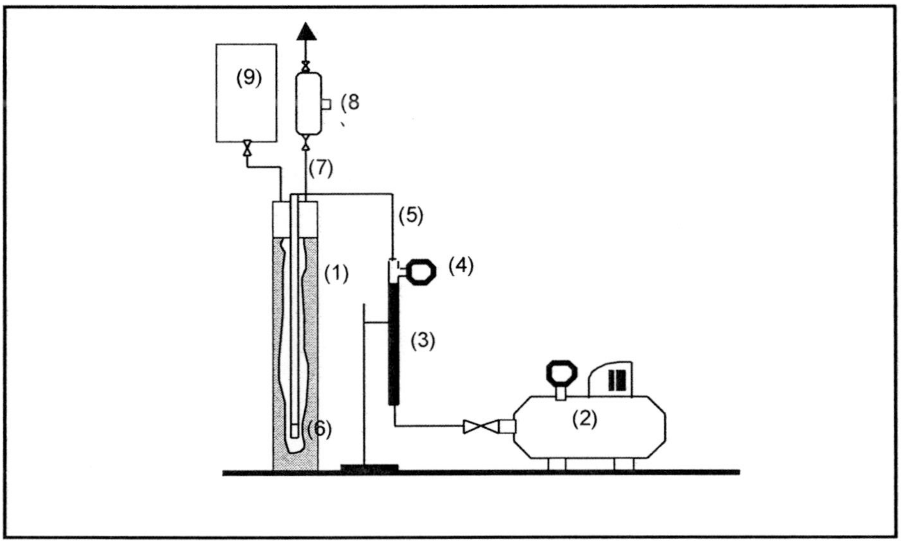

Figure 3. Laboratory experimental apparatus and related flow diagram

(1) plexiglas column (11 x 4.5 x 100 cm), with a top wooden cap, packed with a saturated porous media;
(2) air compressor, to provide pressurized air to the system;
(3) flowmeter, to measure the injected airflow;
(4) manometer, to measure the injected air pressure;
(5) "Rilsan" line, to join the sparge point to the air compressor; and
(6) sparge point, to inject air within the saturated media.
(7) vapour recovery line, to drive vapours out of the column;
(8) vapour sampler (250 ml), to withdraw vapour samples coming out of the column; and
(9) saturation tank, to prepare and store the contaminant solution.

The column was packed with highly uniform gravel and glass beads (idealized porous media), in order to easily replicate identical conditions for each test.

The first sampling point, conventionally called A, was placed at 50 cm above the bottom of the column, the second point, conventionally called B, was placed at the bottom of the column below the zone reached by injected air.

The sparge point was placed within the column, just at 10 cm above the bottom, to allow the injected airflow to have enough space for its developing upwards. In table 1 the main characteristics of the two diffusers tested are reported.

Table 1. Diffusers' geometrical properties

Diffuser no.	Inside diameter (mm)	Outside diameter (mm)	Orifice size (μm)
1	4	8	<1.6
2	4	8	41÷100

The column was saturated with a solution of MtBE in water, at the concentration of 10 mg/l. The solution was prepared inside the saturation tank and then poured within the column through nylon piping. Using a peristaltic pump, three continuous cycles of saturation and drainage were completed to uniformly spread the solution within the saturated media.

The removal tests began with air injection start-up, done right after the initial sampling of solution, and were pursued with several other scheduled samplings of solution.

Results of the two removal tests are shown in Figure 4. The reported concentrations of MtBE are normalized with their initial concentrations, in order to compare the results of both diffusers in terms of removal percentage.

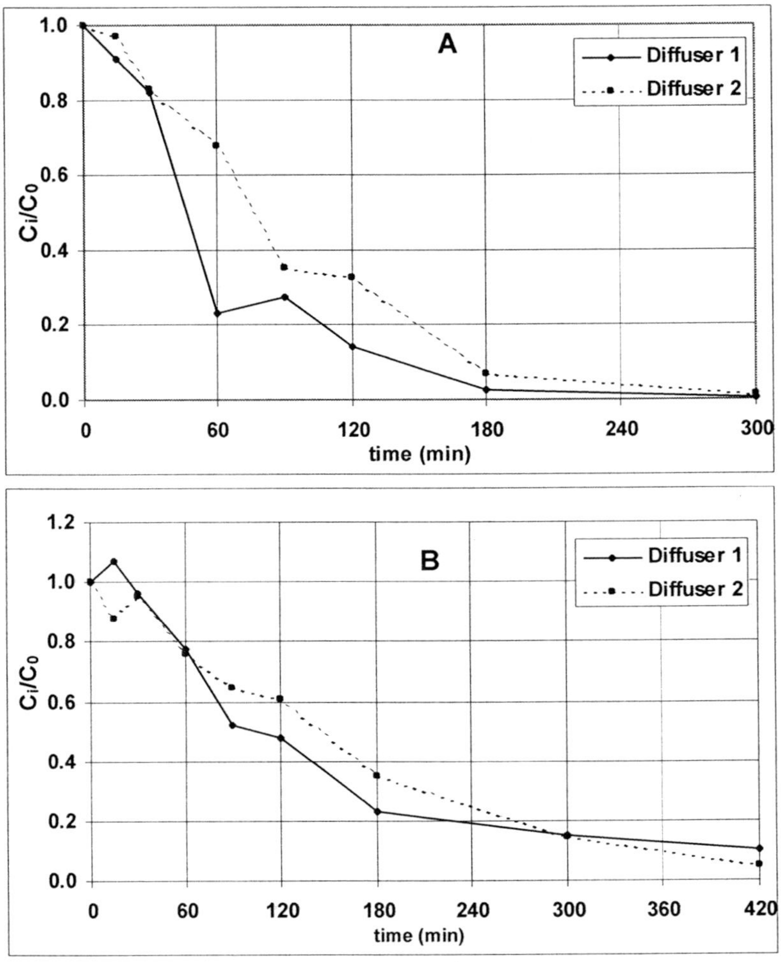

Figure 4. MtBE concentrations within the pore water as a function of time at sampling points A and B.

Comparing the results of the tests obtained using diffusers 1 and 2, it can be noticed that:

- with both diffusers, the residual concentration of MtBE decreases more rapidly at the point A because this is directly impacted by the injected airflow and therefore the stripping process is stronger than at point B; and
- at both sampling points, the concentration of MtBE decreases more rapidly when air is injected using diffuser 1, the one which produced the highest percentage of smaller size bubbles as evidenced during a previous image analysis tests.

In both cases, the results showed that all the MtBE initially dissolved in the pore water solution, was stripped by the air flow injected within the saturated media and passed in the vapour phase. Therefore the treatment proved to be effective.

Intermediate Scale Experiments

MtBE removal from a contaminated aquifer by AS was also performed in a intermediate scale test. This was carried out in the experimental apparatus described in figure 5, consisting in a steel tank (dimensions 1 x 1 x 1,2 m) containing 1 m^3 of soil (Alimonti et al., 2006). Figure 6 shows the sampling points distribution within the saturated porous media.

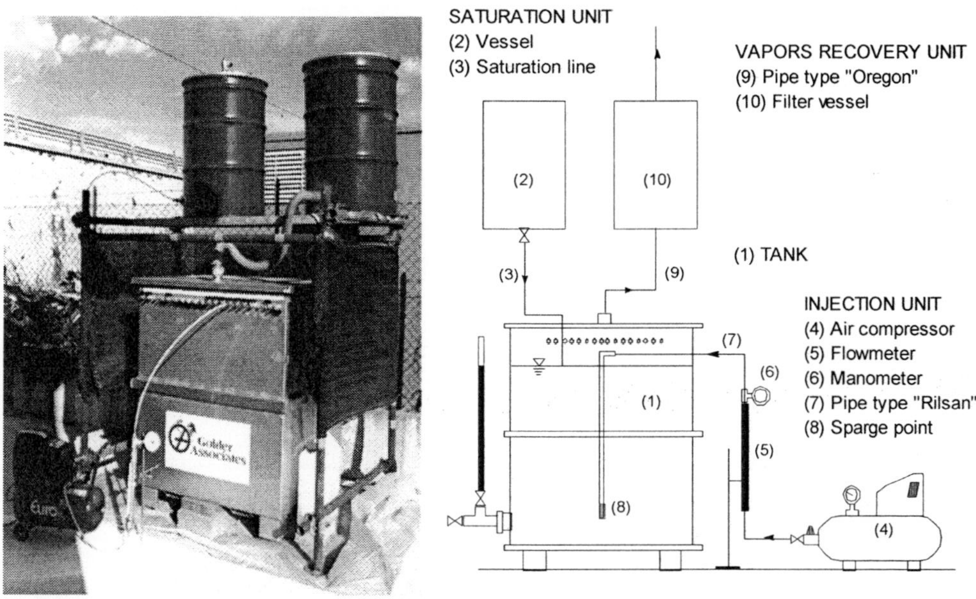

Figure 5. Experimental apparatus and its related flowsheet.

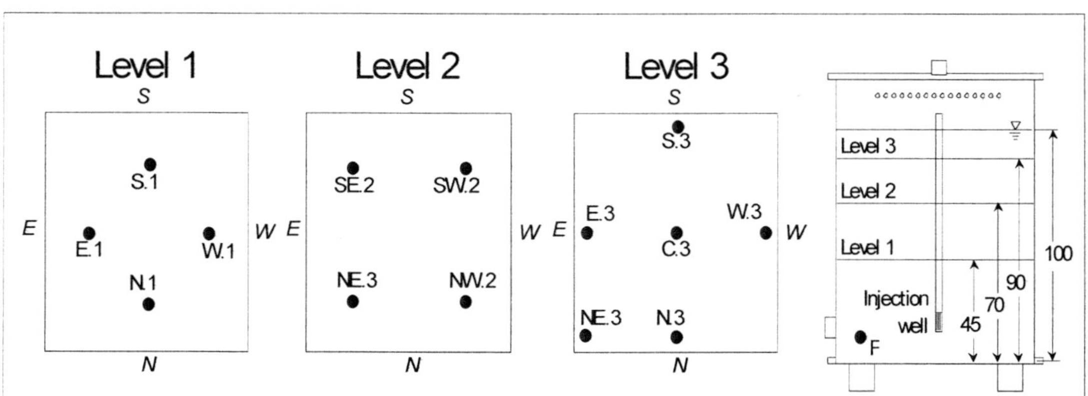

Figure 6. Sampling points distribution within the saturated porous media (all elevations are expressed in cm).

The trends of MtBE concentration observed in the experiments, are illustrated in Figure 7.

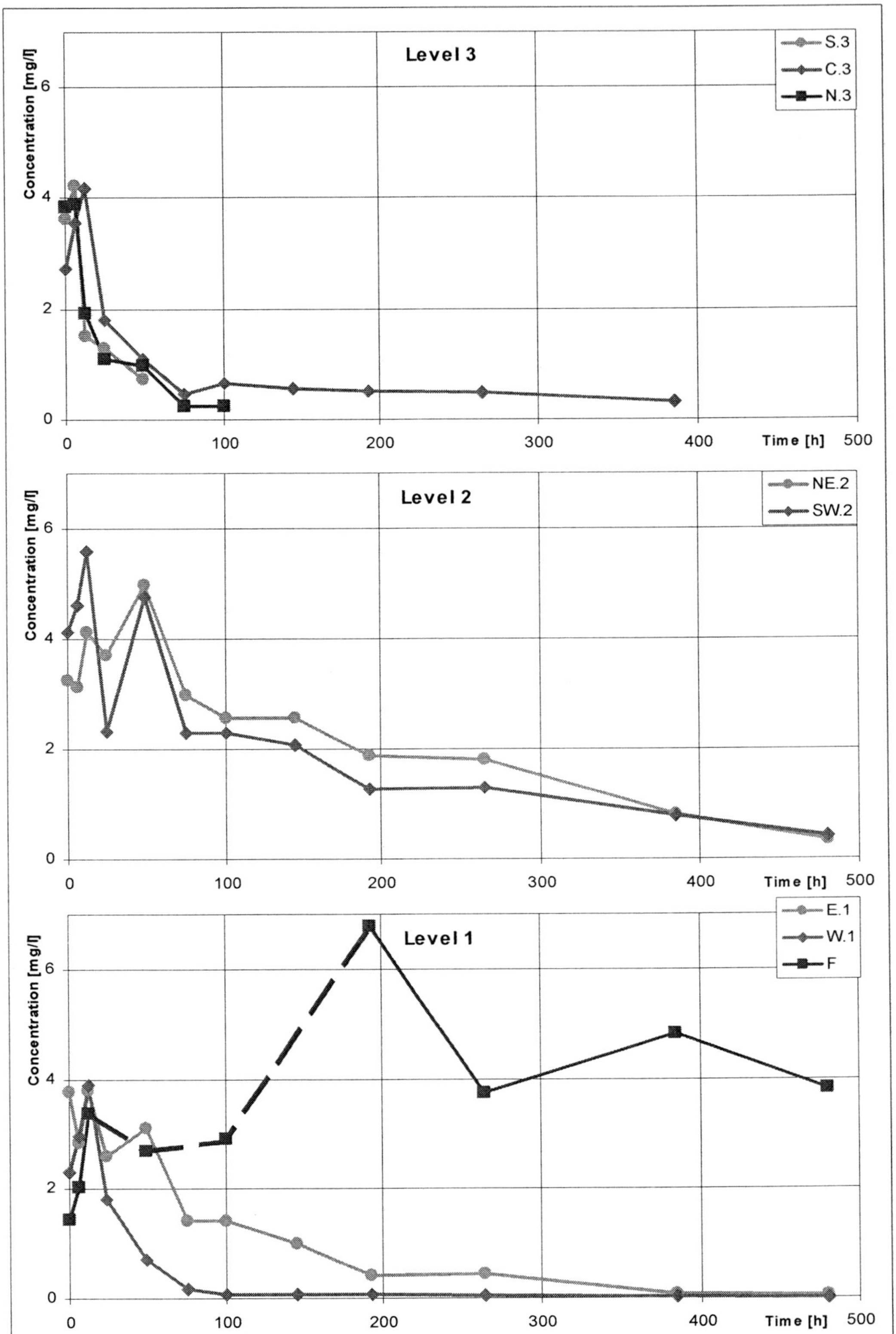

Figure 7. MtBE concentration measured during the removal test.

A general effectiveness of the AS technology in MtBE removal from the contaminated saturated media was observed whereas the media had been directly reached by the injected airflow, MtBE's concentration trend, after an initial fluctuation, shows a strong tendency to attenuate. The fluctuations of the concentration occur at the early stage of air injection, generally within the first two days of operation. After this lapse of time, the concentrations gradually decrease and there is an asymptotic tendency to zero.

Among the three main contaminant removal mechanisms (stripping, direct volatilization and aerobic biodegradation) generally involved in air sparging treatments,(Suthersan S. S., 1999), the experiment highlighted the action of stripping as the mechanism of MtBE removal.

Assuming *stripping* as the dominant mass removal mechanism, Henry's law provides a basic explanation of the trend and the distribution of MtBE concentration within the saturated media.

Since the partitioning of the dissolved contaminant mass occurs at the discrete air-water interface created by the air injection (bubbles and/or air channels), lower MtBE concentrations were expected in the immediate vicinity of the air paths, within the saturated media. Furthermore, to replenish the mass loss around the air paths, mass transfer by diffusion occurred from the water located far from air-water interfaces. Therefore, it is likely that the density of air paths played a significant role in mass transfer effectiveness, as well as the residence time of the injected air traveling through these discrete paths. The dissolved MtBE concentrations, measured at different times and sampling points, consequently rely upon local availability of interfacial surface area and upon MtBE's concentration in the vapor phase within the air paths.

CHEMICAL TREATMENTS

Principles of Chemical Oxidation with Fenton's Reagent

In case of organic contamination, hydrogen peroxide is one of the most used chemical oxidants because it is environmentally safe and easy to handle. Its high oxidizing power may be further increased through the combination with UV radiation (Maillard et al. 1992; Venkatadri and Peters 1993) or transition metal salts (Lu et al. 1997; Weiss 1935; Zepp et al. 1992). The significant increase in hydrogen peroxide's oxidizing capacity is due to the formation of the hydroxyl radical which is highly unstable but has an oxidation power ($E°=$ 2.8 V), lower only to that observed for fluorine, as shown in Table 2.

The most common environmental application of hydrogen peroxide is the Fenton's reagent, an aqueous mixture of H_2O_2 and Fe^{2+} (Fenton, 1894). Under acidic condition, the reaction between iron and hydrogen peroxide generates hydroxyl radicals, which are strong enough to oxidize most organic and inorganic compounds in a non-selectively way. The following equations (Mohanty and Wei 1993) describe the reaction mechanism of the Fenton's system:

$$Fe^{2+} + H_2O_2 \rightarrow Fe^{3+} + HO^{\bullet} + OH^- \qquad (1)$$

The initiation reaction (1) is the reaction that leads to the hydroxyl radical generation, and it is the process limiting step (Mohanty and Wei 1993).

Table 2. Oxidation powers of oxidizing agents

Oxidizing agent	Oxidation reaction	E^0 (V)
Oxygen	$O_2 + 4\,H^+ + 4\,e^- \Leftrightarrow 2\,H_2O$	+ 1.229
Dichromate	$Cr_2O_7^{2-} + 14\,H^+ + 6\,e^- \Leftrightarrow 2\,Cr^{3+} + 7\,H_2O$	+ 1.33
Hypochlorite	$HOCl + H^+ + 2\,e^- \Leftrightarrow Cl^- + H_2O$	+ 1.495
Permanganate	$MnO_4^- + 8\,H^+ + 5\,e^- \Leftrightarrow Mn^{2+} + 4\,H_2O$	+ 1.51
Hydrogen peroxide	$H_2O_2 + 2\,H^+ + 2\,e^- \Leftrightarrow 2\,H_2O$	+ 1.776
Ozone	$O_3 + 2\,H^+ + 2\,e^- \Leftrightarrow O_2 + H_2O$	+ 2.07
Hydroxyl radical	$HO^\bullet + H^+ + e^- \Leftrightarrow H_2O$	+ 2.8
Fluorine	$F_2 + 2\,e^- \Leftrightarrow 2\,F^-$	+ 2.866

The substrate can react with the oxidants through reactions (2-4).

$$RH + HO^\bullet \rightarrow H_2O + R^\bullet \quad (2)$$
$$R^\bullet + HO^\bullet \rightarrow ROH \quad (3)$$
$$R^\bullet + H_2O_2 \rightarrow ROH + HO^\bullet \quad (4)$$

A lot of secondary reactions occur simultaneously to reaction (1):

$$Fe^{3+} + HO_2^\bullet \rightarrow Fe^{2+} + H^+ + O_2 \quad (5)$$
$$Fe^{3+} + H_2O_2 \rightarrow Fe^{2+} + HO_2^\bullet + H^+ \quad (6)$$
$$Fe^{2+} + HO^\bullet \rightarrow Fe^{3+} + OH^- \quad (7)$$
$$H_2O_2 + HO^\bullet \rightarrow H_2O + HO_2^\bullet \quad (8)$$
$$Fe^{2+} + HO_2^\bullet \rightarrow Fe^{3+} + HO_2^- \quad (9)$$

These reactions (5-9) deal with the cycling of Fe^{3+} to Fe^{2+}, with the production of hydroperoxyl radicals, whose oxidizing power is a lot less than the one of the parent compounds (E°= 0,3 V), and with the quenching of $HO^\bullet$ by Fe^{2+} and H_2O_2.

The presence of the substrate can also help the regeneration of Fe^{2+} through reaction (10) or its depletion through reaction (11).

$$R^\bullet + Fe^{3+} \rightarrow R^+ + Fe^{2+} \quad (10)$$
$$R^\bullet + Fe^{2+} \rightarrow R^- + Fe^{3+} \quad (11)$$

Finally, the radical chain reactions may also be ended by the reactions (12-13)

$$2\,HO^\bullet \rightarrow H_2O_2 \quad (12)$$
$$2\,R^\bullet \rightarrow \text{dimer} \quad (13)$$

The combined system of hydrogen peroxide coupled with a bivalent iron salt, results in the great number of reactions above described. As shown, numerous competing reactions manage to scavenge the hydroxyl radicals, thus decreasing the system's oxidizing power: the production of hydroxyl radical ($HO^•$) is always accomplished by the production of hydroperoxyl radicals ($HO_2^•$), Fe^{3+} and O_2.

A particular attention must therefore be given in finding the correct dosage for the reactants. In particular an excess in hydrogen peroxide leads to the formation of hydroperoxyl radicals $HO_2^•$ (reaction 8), which are competitive with $HO^•$ in the oxidation process, and show a certain reactivity towards organic compounds, even if they exhibit a very low oxidation power. The $HO_2^•$ radicals may also oxidize Fe^{2+} ions (reaction 9), and, by subtracting them from the reaction site, may cause a deficit in $HO^•$ production.

An additional, but non negligible consideration, can be made about the need of establishing acidic conditions to optimize Fenton's reaction effectiveness.

The overall Fenton reaction process can be simplified in the following reaction (Walling, 1975):

$$Fe^{2+} + H_2O_2 + 2\,H^+ \rightarrow Fe^{3+} + 2\,H_2O \qquad (14)$$

This reaction (14) shows that acidic conditions are needed to ensure hydroxyl radical production. In fact, in acid solutions, the energy required to allow electron transfer from Fe^{2+} to H_2O_2 is minimal, and at the same time, peroxide decomposition due to the following reactions (15) and (16), is negligible:

$$H_2O_2 \rightarrow H^+ + HOO^- \qquad (15)$$

$$H_2O_2 \rightarrow H_2O + 0.5\,O_2 \qquad (16)$$

Conversely both these reactions are enhanced at neutral or basic conditions and in the presence of large amounts of Fe^{2+}.

Nevertheless extreme acid conditions may also reduce the oxidation effectiveness, because of the consumption of hydroxyl radicals through reaction (17):

$$HO^• + H^+ + e^- \rightarrow H_2O \qquad (17)$$

Application of Fenton's Treatment to Soil and Groundwater Remediation

The use of the Fenton's reagent has proven to be effective for the oxidation of a wide range of well known organic pollutants such as: BTEX (Lou and Lee, 1995), p-chlorophenol (Kwon et al. 1999), trihalomethanes (Tang and Tassos 1997), aromatic amines (Casero et al. 1997), ethylene glycol (McGinnis et al. 2001), nitrophenols (Goi and Trapido 2002), surfactants (Kitis 1999), pesticides (Scott et al. 1995; Pignatello et al. 1995), a few commercial azodyes and textile effluents (Kang and Chang, 1997; Szpyrkowicz et al. 2001; Kang et al. 2002) as well as wastewater and landfill leachates (Bae et al. 1997; Kang and Hwang 2000). On the contrary, Fenton's reaction has been rarely used for the treatment of inorganic pollutants like arsenic, in drinking water (Balarama Krishna et al. 2001).

It must be stressed that Fenton's reaction is the simplest way of producing hydroxyl radicals because no special reactants or special apparatus are required. Furthermore, iron and hydrogen peroxide are relatively safe chemicals (ITRC 2000). Consequently the whole system appears to be easy to handle, environmentally safe and cost effective.

The possibility of applying Fenton's process to contaminated soils was first demonstrated by Tyre et al. (1991), followed by Watts et al. (1993) in batch lab-scale experiments, by Ravikumar and Gurol (1994) with sand-packed column tests and by Kakarla and Watts (1997) with soil-packed column tests.

On the basis of these considerations, in many of the above cited studies, Fenton's reagent optimal conditions have been assessed in a quite restricted pH range, between 2 and 4, while at pH higher than pH=6, a very low effectiveness has been observed (Baciocchi et al. 2003). In order to achieve low pH conditions, soils may need to be acidified. The soil buffering capacity becomes then an important factor that must be considered in any feasibility study involving *in situ* Fenton's oxidation. Highly alkaline soils (containing limestone) may require a substantial acid addition to bring the pH into the optimal range, thus increasing the cost of the process. A pH adjustment may also be required during Fenton's treatment, due to the formation of H^+ ions from reaction (5) and (15).

In several works, the possibility of using the soil's naturally occurring iron as a catalyst, possibly without pH adjustment, in the so-called Fenton-like process, was also successfully investigated (Baciocchi et al. 2003).

In most of the experimental studies on the oxidation of organic compounds, the reaction times required to achieve a quite complete degradation were between 5 minutes and 2 hours (Arnold et al. 1995). However, reaction times from some hours even to some days were needed in case of complex substrates (Chen et al. 2001).

The efficiency of Fenton's oxidation is strongly affected by operative conditions (Kang and Hwang, 2000).

Fenton oxidation's kinetic follows generally a first order rate equation. However, there is a main factor which cause results to deviate from this trend. One of the limiting steps of the process is the regeneration of ferrous ions. As a consequence, Fenton's reaction development along time has been divided in two phases. First a quick substrate oxidation is generally observed, due to the initial rapid formation of hydroxyl radicals. The following decrease in the oxidation rate can be associated with the presence of large amounts of Fe^{3+} not converted to Fe^{2+}. The unfavorable Fe^{3+}/ H_2O_2 ratio is then responsible for the slow oxidation of the remaining substrate.

There is a limit in the value of the Fe^{2+}/H_2O_2 ratio: the difficulty arises from the regeneration of ferrous ions. In effect, an accumulation of Fe^{3+} ions during the oxidation process is caused by reaction (1). This happens because the Fe^{2+} regeneration reactions (5) and (6) are slower. In addition, the Fe^{2+} concentration should be high enough to activate hydrogen peroxide's decomposition through reaction (1) but at the same time low enough to limit the undesirable reactions of Fe^{2+} (7) and (9).

As a result an optimal Fe^{2+}/H_2O_2 ratio has to be determined for each specific application. As an example, Arnold et al. (1995) found that Fenton oxidation of atrazine was greatly affected by the Fe^{2+}/H_2O_2 ratio. A ratio of 1:1 was found to be optimal, since increasing either the H_2O_2 concentration or the Fe^{2+} one, strongly stimulated the undesirable reactions.

Another parameter affecting the overall cost of the process is the H_2O_2 vs. contaminant ratio. A first order kinetic with respect to the concentration of Fenton's reagent , has been widely observed (Gallard and De Laat 2000). Consequently it would be expected that increasing the peroxide concentration would improve oxidation's effectiveness. However, excessive peroxide dosage can act as a scavenger of hydroxyl radicals according to reaction (8). It must be stressed that, according to reaction (1), an increase in peroxide concentration must always correspond to an increase in Fe^{2+} concentration. In wastewater treatment, the H_2O_2/contaminant ratio is typically maintained in the range between 1/1 and 5/1 (w/w). For soil treatment this ratio can increase up to 50/1, due to the oxidant demand of soil natural matter, contaminant hydrophobicity (expressed as: $\log K_{OW}$, where K_{OW} is the octanol-water partition coefficient), and hydrogen peroxide instability (Quan et al. 2003). In some commercial applications of Fenton's reagent, a mixture of 5-50% wt. H_2O_2 was applied (Gates and Siegrist 1995; Cresap and Burke 2000; ITRC 2001). Some H_2O_2 mixtures contain therefore a large quantity of water. There is an advantage in applying peroxide at low concentrations, due to a resulting better control of heat and gas generation. This permits to reduce the resulting potential stripping of the contaminant. A concentration of 10% is generally recommended for several applications (ITRC 2001).

Experiments on MtBE Extraction by Soil Flushing and Fenton's Treatment of the Extracted Solution

In this section the results of an experimentation of MtBE extraction from a contaminated saturated soil, followed by an oxidation treatment of the extracted solution, are presented and discussed.

The experiments were performed on a soil artificially contaminated with MtBE. The soil was a sandy-loamy soil, whose composition is shown in Table 3 while selected chemical and physical characteristics are shown in Table 4.

Table 3. Composition of the soil used in soil flushing experiments

Component	Weight %	Particle size, mm
Silt	38.75	0-0.02
Clay	7.5	0.02-0.355
Sand	35	0.355-1.7
Gravel	18.75	1.7-9.5

Table 4. Characteristics of the soil used

pH	8.1
Organic carbon, %	3.2
Permeability, cm/s	$3.21 \cdot 10^{-3}$
Porosity, %	46
Humidity, g/kg	24.5
Bulk density, g/ml	1.30
Absolute density, g/ml	2.30

The soil was contaminated by flushing an aqueous solution of MtBE. The solution was flushed into the soil and continuously recirculated, until an homogeneous contamination and a residual level of MtBE in the saturated soil between 100 and 200 mg/l, were achieved.

The extraction tests were performed by flushing the soil with an aqueous solution of ethanol: two series of experimental tests were performed, at two different initial level of MtBE in the soil (80 and 100 mg/l). For each series of test, two flow rates of the extractant solution were tested (70 and 100 ml/h, corresponding to 3.4 and 4.8 m^3/m^2h respectively). The experiments were performed at the controlled temperature of 4°C, to avoid MtBE volatilization.

Several intervals of time were selected, each one corresponding to the flushing of a volume of solution equal to two times the value of the pore volume (PV) of the soil. The collected solution after extraction, was sampled and subjected to the analyses to determine MtBE and total organic carbon (TOC) content.

Figure 7 show the extraction yield observed in the four series of experimental tests.

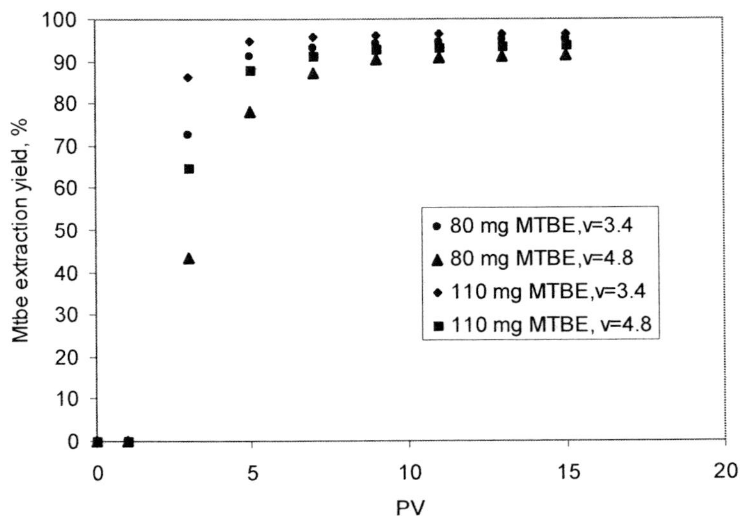

Figure 8. Experimental results of MtBE extraction by soil flushing

As confirmed by Figure 8, an increase in MtBE extraction occurred when increasing volumes of solution were injected. At the same time, the extraction yield was higher when the initial amount of MtBE was higher and the contact time between the flushing solution and the soil was lower. A fast extraction was also observed: after the flushing of a volume of 5 PV of solution, an extraction yield of 91.1% and 94.8% was achieved for the initial concentration of 80 and 110 mg/l respectively.

The following table 5 provides the concentration of MtBE (C_e) and TOC in the extracted solution, together with the values of the flushed and recovered solutions' volumes (test with C_i=110 mg /200 g soil, v =70 ml/h).

The data reported in the above reported Figure 7 and Table 5, show that the relative amount of MtBE and the quantity of natural organic matter (NOM) extracted from the soil (measured as TOC), depended from the experimental conditions. Assuming that MtBE was not converted during the extraction, a mass balance was performed for each of the four experimental condition: the results are shown in Table 6.

Table 5. Experimental results of MtBE extraction by soil flushing (C_i=110 mg /200 g soil, v =70 ml/h)

PV	Volume in (ml)	Volume out (ml)	TOC (mg/l)	C_e (mg/l)	Mtbe extraction yield (%)
1	69	/	/	/	/
3	207	128	743	741	86.2
5	345	260	93.0	72.0	94.8
7	483	398	42.2	5.58	95.6
9	621	536	31.0	4.14	96.0
11	759	670	20.2	2.17	96.3
13	897	808	12.5	0.82	96.4
15	1035	946	0.91	0.71	96.5

Table 6. Mass balances for the flushing experiments

MtBE initial amount (mg/200 g soil)	Flushing flow rate (ml/h)	MtBE extracted (mg)	NOM extracted (mg TOC)
80	70	76	68
80	100	72	56
110	70	106	50
110	100	103	34

MtBE extraction was not significantly dependent on the flushing solution's flow rate, while, this parameter greatly affected the extraction of organic matter from the soil due to its lower water solubility. In addition, higher amounts of MtBE lowered the quantity of NOM extracted. This could be explained considering the higher affinity of MtBE for the extraction agent.

The amount of MtBE volatilized could be considered negligible, due to the operating temperature of 4°C.

Mechanism of MtBE Oxidation with Fenton's Reagent

The solution recovered after the percolation of 5 PV of the flushing solution was then subjected to the oxidation treatment with Fenton's reagent. The initial MtBE concentration was 100 mg/l, and NOM concentration was in the range between 30 and 100 mg/l of TOC, depending upon the operating conditions.

The advantage of using chemical oxidation to treat a solution of water contaminated with MtBE, is that this compound has a very low biodegradability. Chemical oxidation can reduce the duration of the treatment and can eventually provide more biodegradable intermediates.

The radical attack on the MtBE molecule occurs on the side with less steric hindrance i.e. on the methyl group:

$$(CH_3)_3\text{-C-O-}CH_3 + HO^\circ \rightarrow H_2O + (CH_3)_3\text{-C-O-}CH_2^\circ$$

$$(CH_3)_3\text{-C-O-}CH_2^\circ + O_2 \rightarrow (CH_3)_3\text{-C-O-}CH_2\text{-O-O}^\circ$$

$$(CH_3)_3\text{-C-O-}CH_2\text{-O-O}^\circ + (CH_3)_3\text{-C-O-}CH_3 \rightarrow$$

$$\rightarrow (CH_3)_3\text{-C-O-}CH_2\text{-O-OH} + (CH_3)_3\text{-C-O-}CH_2^\circ$$

$$(CH_3)_3\text{-C-O-}CH_2\text{-O-OH} \rightarrow \underset{\text{TBF}}{(CH_3)_3\text{-C-O-CHO}} + H_2O$$

Acid hydrolysis

$$\underset{\text{TBF}}{(CH_3)_3\text{—C—O—CHO}} \leftrightarrow \underset{\text{TBA}}{(CH_3)_3\text{C—OH}} + \underset{\text{formic acid}}{HCOOH} \quad (HCOOH \uparrow CO_2)$$

$$\underset{\text{TBA}}{(CH_3)_3\text{—C—OH}} \leftrightarrow (CH_3)_3\text{—}C^+ + H_2O \rightarrow \underset{\text{Isobutene}}{CH_3\text{—C(}CH_2\text{)—}CH_3} + H^+$$

$$\underset{\text{Isobutene}}{CH_3\text{—C(}CH_2\text{)—}CH_3} + O_2 \rightarrow \underset{\text{Acetone}}{CH_3\text{—CO—}CH_3} + \underset{\text{formaldeide}}{H\text{—CHO}} \longrightarrow CO_2$$

This firstly produces TBF (tertiary-Butyl formate), an esther that successively is involved in a hydrolysis equilibrium because of the acidic environment. The hydrolysis products are TBA (tertiary-Butyl-alcohol) and formic acid. The latter one is then easily transformed in CO_2. TBA then can be oxidized and transformed in acetone, which represents the most abundant and persistent by-product. TBA can also be produced from a direct oxidation of TBF. Finally the oxidation process leads to the formation of methanol and, allowing longer reaction times, to the final conversion to CO_2.

Fenton Oxidation of the Extracted Solution

The experimental tests were performed in sealed vials, on samples of the flushing solution or of a reference solution prepared dissolving in tap water the same amount of MtBE extracted. This was done to study the effects of NOM on the process.

The solutions were acidified with HCl down to pH=3.5 (in the optimal range for Fenton treatment, as shown in the previous sections). Preliminary tests were carried out to evaluate the extent of volatilization in the operating conditions chosen for the oxidation treatment. Results of these tests showed that from a 100 mg/l aqueous solution of MtBE the 8% of the compound volatilized, while in the presence of 80 mg/l dissolved organic matter form soil this amount was reduced to about 3%.

After pH adjustment, the solution was oxidized using hydrogen peroxide coupled with Fe^{2+} (as $FeSO_4 \cdot 7H_2O$), under a constant agitation at ambient temperature. After two hours reaction time, the residual iron was precipitated and the liquid phase was then vacuum filtered and analyzed.

Table 7 shows the results obtained in three series of experimental tests, performed at three different Fe^{2+} concentration. The aim was to evaluate, for each test, the optimal H_2O_2/Fe^{2+} ratio. The concentrations of MtBE (after 5 min reaction time) and of all the main byproducts of its oxidation (as described in the previous section) are reported.

Table 7. Influence of reagent concentration on Fenton's oxidation of MtBE

Fe^{2+} (mM)	H_2O_2/ Fe^{2+}	MTBE (mg/l)	TBA (mg/l)	TBF (mg/l)	Methanol (mg/l)	Acetone (mg/l)
4.5	1	52.2	70.3	5.20	1.90	10.3
	1.6	10.2	61.0	10.8	13.1	20.7
	2	9.13	50.5	12.3	11.6	25.3
	3	11.5	52.6	12.9	ND	36.5
10	1.5	0.97	30	ND	9.30	40.7
	2	ND	14.8	ND	9.00	49.1
	2.5	1.20	18.5	10.2	10.6	53.0
	3	0.76	ND	18.3	8.60	44.7
	3.5	1.92	36.2	ND	14.4	50.7
	4	3.40	17.8	36.2	2.60	48.4
20	1	1.36	25.3	22	10.1	59.0
	2	6.64	22.9	46.4	11.5	51.0
	3	12.6	28.5	45.6	16.8	46.8

Results show that, as expected, there is a limit to the increase of the H_2O_2/Fe^{2+} ratio: the difficulty is in the regeneration of the ferrous ions. An accumulation of Fe^{3+} ions during the oxidation progress is in fact caused by reaction (1), that occurs faster than the Fe^{2+} regeneration reactions (5) and (6). Fe^{2+} concentration should be high enough to activate hydrogen peroxide but at the same time, low enough to limit undesirable reactions of Fe^{2+} (7) and (9).

In the investigated range the higher oxidation rate was observed when Fe^{2+} concentration was 10 mM and $H_2O_2/Fe^{2+} = 2$. For this ratio, a maximum MtBE conversion was in fact generally observed.

Since, as shown in the previous sections, Fenton reaction is strongly affected by pH, another sequence of tests was performed to evaluate the influence of the initial solution's pH on MtBE conversion. It must be stressed that, considering the great buffering capacity of the soil, the extraction of soil constituents such as carbonates and NOM, produced a non negligible acid consumption during the acidification prior to the addition of Fenton's reagents.

The results of these tests are shown in table 8, for an initial concentration of 190 mg/l MtBE and a NOM concentration of 30 mg/l.

Table 8. Influence of reaction pH on Fenton's oxidation of Mtbe

pH	MtBE oxidation yield (%)
2	99.0
3.5	99.3
4	99.2
5	99.0
7.5	97.2

MtBE oxidation yield was always very high, even at circum neutral pH conditions, and this is a promising aspect in the prospective of an application of this process, for the possibility to reduce acid consumption. As expected, the higher conversion rate was observed at an initial pH of 3.5, in the common optimal range for Fenton oxidation treatment.

To obtain some information about the reaction kinetics, Table 9 reports the results obtained after the first 2 minutes. Reactions were carried out at the previously determined optimal operating conditions ($[Fe^{2+}]$= 10 mM, H_2O_2/Fe^{2+}= 2, pH=3.5).

Results confirmed that the final product of MtBE conversion was acetone.

Table 9. Kinetic of MtBE degradation by Fenton reagent (pH=3.5)

Time (min)	MTBE (mg/l)	TBA (mg/l)	TBF (mg/l)	Methanol (mg/l)	Acetone (mg/l)
0	187.3	0	0	0	0
0.083	146.7	0	0	0	3.06
0.25	30.1	57.86	0	3.01	25.36
0.5	7.43	47.42	0	3.22	42.61
0.75	3.5	40.46	26.18	0.62	47
1	1.57	20.05	0	0.61	46.1
5	0	19.47	0	0	52.88

Finally, Table 10 shows the results of tests performed at the same conditions than Table 9, except for the initial pH of 5.5.

Table 10. Kinetic of MtBE degradation by Fenton reagent (pH=5.5)

Time (min)	MTBE (mg/l)	TBA (mg/l)	TBF (mg/l)	Methanol (mg/l)	Acetone (mg/l)
0	178.5	0	0	0	0
0.25	26.86	55.58	7.62	15.73	30.66
0.5	14.42	53.89	0	15.57	34.87
1	2.74	25.37	24.76	11.33	46.08
5	0.81	18.82	9.92	9.16	53.21

The results of the two tests are in accordance: in particular it can be observed that TBF was more stable in less acidic conditions.

In situ chemical Oxidation of MtBE in Saturated Media

Lab Scale Experiments

Figure 9 shows the results of preliminary experiments of MtBE chemical oxidation in water.

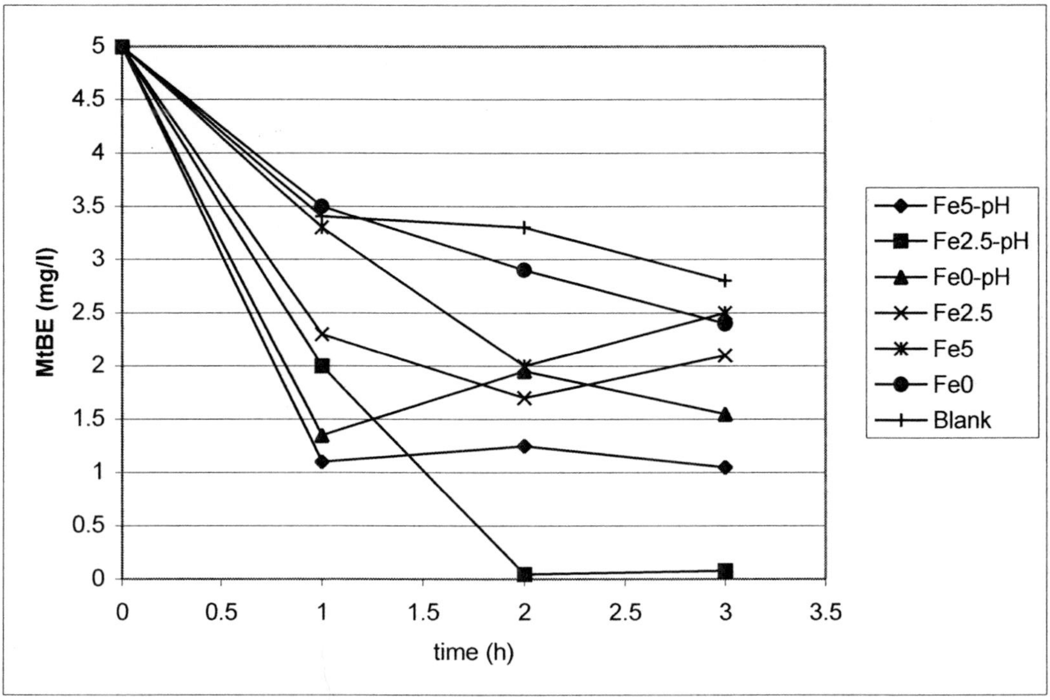

Figure 9. Batch removal tests in water

The batch tests in water were performed in 250-ml beakers, filled with 50 ml of MtBE solution to achieve an initial concentration of 5 mg/l.

The theoretical amount of H_2O_2 required to oxidize MtBE was estimated on a stoichiometric basis (5.78 g of H_2O_2 for every 1.00 g of MtBE). Due to the presence in soil and groundwater of other contaminants and naturally occurring organic compounds in soil or groundwater, the quantity of oxidant required at field conditions is often five to seven time greater (Leethem, 2001). In these tests the quantity of H_2O_2 to be added in each beaker (19.48 g) was chosen five times greater than the stoichiometric quantity.

A quantity of 19.48 g of H_2O_2 was added in six of the seven beakers containing the MtBE solution. A parallel blank experiment was carried out adding distilled water instead the oxidant.

Table 11. Test conditions in the seven beakers of the batch removal tests in water

Test	Sample Name	$H_2O_2/MtBE$	$FeSO_4/H_2O_2$	pH
1	Blank	0	0	5,5
2	Fe0	5	0	5,5
3	Fe5	5	1/5	5,5
4	Fe2.5	5	1/2,5	5,5
5	Fe0-pH	5	0	3
6	Fe5-pH	5	1/5	3
7	Fe2.5-pH	5	1/2,5	3

Ferrous sulfate ($FeSO_4$) was chosen as catalyst of Fenton's oxidation and added in four of the six beakers. Two different ratios of $FeSO_4$ and H_2O_2 solution (1/5 and 1/2,5) were compared to optimize the oxidation. In three of the six beakers also pH was reduced (adding sulfuric acid), to check if acidic condition could enhance the action of H_2O_2 in MtBE removal from water.

Comparing the results of the tests reported in Figure 9 the major evidences are:

- the pH significantly influences the effectiveness of MtBE oxidation;
- the action of removal is quite low in all of conditions tested;
- the highest removal efficiency occurred in the test Fe5 (38.1%), and in the test Fe2.5 when pH was adjusted thus confirming the importance of soil pH in Fenton's treatments.

Intermediate Scale Experiments

Following to the lab scale tests, further experiments were performed at an intermediate scale, to assess the effectiveness of ISCO in MtBE removal from a saturated media.
As well as the above described tests with AS, the experimental set-up consisted of a steel tank (dimensions 1 x 1 x 1,2 m), containing 1 m^3 of soil saturated with a MtBE contaminated solution. In this case the sparging point was substituted by the injection well represented in Figure 10, in order to allow the oxidant solution to spread within the media.

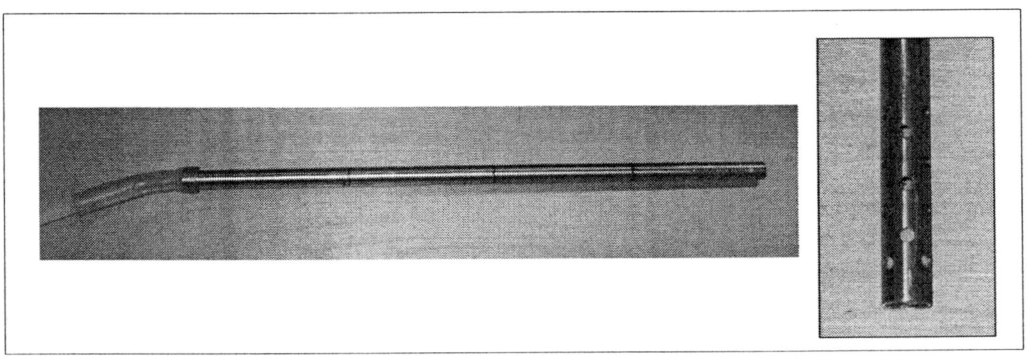

Figure 10. The injection well placed within the saturated media of the tank

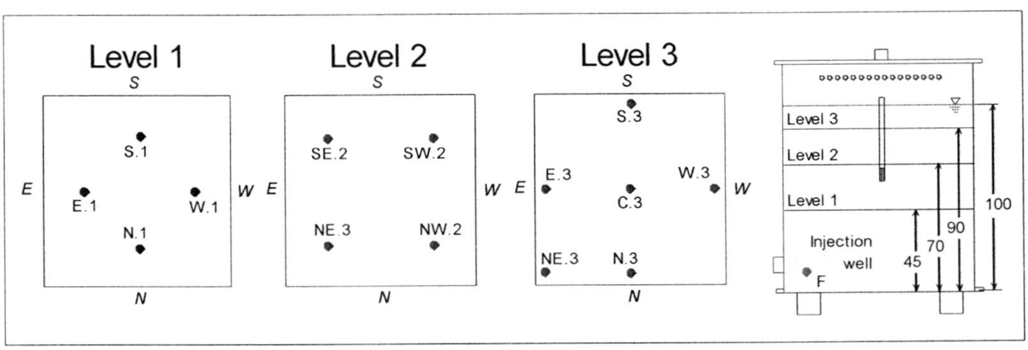

Figure 11. Injection well and sampling points location within the saturated porous media (all elevations are expressed in cm).

The position of the injection wells and the sampling points within the tank is reported in Figure 11.

The saturated media was contaminated with MtBE (6 mg/l). Ferrous sulfate ($FeSO_4$) was then added by mean of a peristaltic pump connected to the injection well, and finally, an H_2O_2 solution was injected. The oxidant concentration was based on MtBE reaction stoichiometry. In order to enhance the efficiency of treatment, a second injection of oxidant, without catalyst, was done 24 hours later than the first one.

Table 14 reports the concentration of MtBE along time in the different sampling points. The chemical analyses carried out on the solution samples also reveal the presence of acetone as the main by-product of the oxidation (not reported in Table 14), thus confirming the degradation pathway generally observed in Fenton oxidation of MtBE. The data of Table 14 have been plotted to better understand the trends of MtBE concentration measured within the contaminated media (Figure 12).

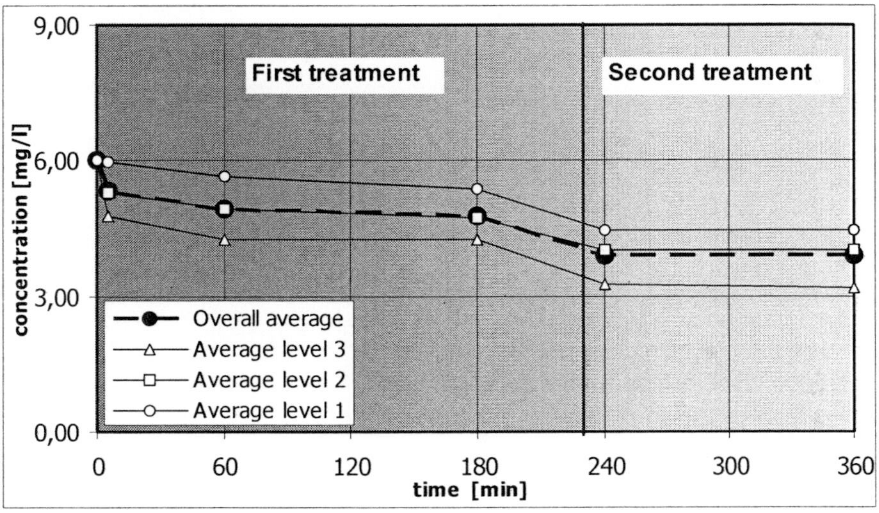

Figure 12. Trends of MtBE concentrations measured during the removal test

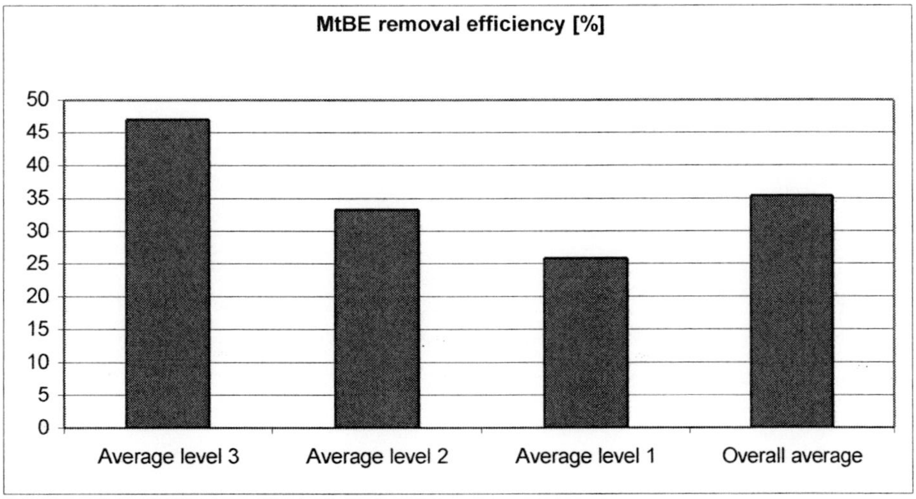

Figure 13. MtBE removal efficiency measured at each sampling points levels

Table 14. Analytical results of MtBE concentrations measured at each sampling point during the removal test

Treatment	First			Second	
Time [min]	0	60	180	0	180
Sampling points	MtBE concentrations [mg/l]				
N III	6,81	5,48	5,50	2,77	2,64
E III	5,94	5,80	5,78	4,12	3,68
S III	6,78	5,52	5,53	4,36	4,25
NE III	4,35	4,57	4,50	5,00	5,43
C III	0,06	0,01	0,05	< 0,01	< 0,01
Average level 3	*4,79*	*4,28*	*4,27*	*3,25*	*3,20*
NW II	4,59	3,88	2,75	3,33	3,05
NE II	4,54	4,36	4,44	4,71	4,09
SW II	4,55	4,4	5,78	2,87	3,90
SE II	7,57	7,18	6,06	5,19	5,10
Average level 2	*5,31*	*4,96*	*4,76*	*4,03*	*4,04*
N I	5,18	4,31	4,27	4,14	4,05
E I	5,90	5,78	5,90	5,42	5,36
S I	6,27	6,30	6,22	4,24	4,16
W I	6,51	6,18	5,05	4,11	4,20
Average level 1	*5,97*	*5,64*	*5,36*	*4,48*	*4,44*
Overall average	5,36	4,96	4,80	3,92	3,89

The observed overall MtBE removal efficiency is presented in Figure 13.
The following conclusions can be drawn on the basis of the results of experiments.

- The results of the intermediate scale test are comparable with those obtained in the laboratory scale tests previous described.
- The overall average removal efficiency was 35,4%. However this data was highly dependent on the sampling level (47.0% in level 3, 33.3% in level 2 and 25.8% in level 1). The decreasing of the removal efficiency with the depth of the sampling levels, was probably due to the tendency of the oxidant solution to move upwards through the saturated media, maybe driven by the generation of O_2 in the solution during oxidation.
- Within the first hour of treatment the higher removal rate was observed, proving that oxidation is a fast reaction.
- Acetone resulted to be the main by-product of MtBE oxidation.
- Differences in MtBE concentrations measured among sampling points belonging to the same level were attributed to the presence of preferential pathways within the media. This indicates that the delivery of the oxidant solution through the porous media plays an important role in the overall removal efficiency of technology.
- As a consequence, the use of a more diluted oxidant solution may favour its distribution and, at the same time, may reduce heat generation thus limiting volatilization.

REFERENCES

Alimonti, C., Lausdei, D., Santoro L.B. (2006) Experimental analysis of the stripping process driven by air sparging systems, in: Contaminated Soil, Sediments and Water: Successes and Challenges, ed. E.J. Calabrese, P.T. Kostecki, J. Dragoon, vol. 10, 163-175, Springer.

Arnold, S. M., Hickey W. J. and Harris R. F. (1995) Degradation of atrazine by Fenton's reagent: condition optimisation and product quantification, *Environ. Sci. Technol.* 28, 2083-2088.

Baciocchi, R., Boni, M.L. and D'Aprile L. (2003) Hydrogen peroxide lifetime as an indicator of the efficiency of 3-chlorophenol Fenton's and Fenton-like oxidation in soils, *J. Hazard. Mater.* B96, 305-329.

Bae J., Kim S. and Chang H. (1997) Treatment of landfill leachates; ammonia removal via nitrification and denitrification and further COD reduction via Fenton's treatment followed by activated sludge, *Water Sci. Technol.* 36, 341-348.

Balarama Krishna M.V., Chandrasekaran K., Karunasagar D. and Arunachalam J. (2001) A combined treatment approach using Fenton's Reagent and zero valent iron for the removal of arsenic from drinking water, *J. Hazard. Mater.* B84, 229-240.

Beckett P.H.T., The use of extractants in studies on trace metals in soils, sewage sludges, and sludge-treated soils, Advances in soil science, 9 (1989), Springer-Verlag New York Inc.

Casero I., Sicilia D., Rubio S. and Pérez-Bendito D. (1997) Chemical Degradation of Aromatic Amines by Fenton's Reagent, *Wat. Res.* 31(8), 1985-1995.

Chang L.Y., A waste minimization study of a chelated copper complex in wastewater-treatability and process analysis, Waste management, 15 (1995) 209.

Chang M.C., Huang C.R., Shu H.Y. (2000) Effects of surfactants on extraction of phenanthrene in spiked sand, *Chemosphere* 41, 1295-1300.

Chen, G., Hoag, G.E., Chedda, P., Nadim, F., Woody, B.A. and Dobbs, G.M. (2001) The mechanism and applicability of *in situ* oxidation of trichloroethylene with Fenton's reagent, *J. Hazard. Mater.* B87, 171-186.

Cresap G.H. and Burke A.L. (2000) In situ chemical oxidation of fuel using hydrogen peroxide: a case study, *Hazard. Ind. Waste*, 32, 31-38.

Di Palma L., Ferrantelli P., Petrucci E. (2003) Experimental study of the remediation of atrazine contaminated soils through soil extraction and subsequent peroxidation, *J. Hazard. Mater.* 99(3), 265-276.

Di Palma, L. (2005) In situ chemical oxidation of contaminated environments with hazardous materials, in: *Soil and Sediment Remediation*, Ed. P. Lens, T. Grotenius, G. Malina and H. Tabuk, pagg. 200-222, IWA Publishing, London, UK, 2005.

Fenton H.J.H. (1894) Oxidation of Tartaric Acid in presence of Iron, *J. Chem. Soc.* 65, 899-910.

Freeman, F. (1975) Possible criteria for distinguishing between cyclic an acyclic activated complexes among cyclic activated complexes in addition reactions, *Chem. Rev.* 75, 439-491.

Freeman H. M. and Harris E. F. (1995) Hazardous waste remediation: innovative treatment technologies, Technomic Publishing Co. Inc., Lancaster, Basel.

Gallard and De Laat (2000) Kinetic modeling of Fe(III)/H2O2 oxidation reactions in dilute aqueous solution using atrazine as a model organic compound, *Wat. Res.*, 34, 3107-3116.

Gates D.D. and Siegrist R.L. (1995) *In situ* chemical oxidation of trichloroethylene using hydrogen peroxide, *J. Environ. Eng.*, 121(9), 639-644.

Ghassemi, M. (1988) Innovative in situ tratment technologiesb for cleanup of contaminated sites, *J. Hazard. Mater.* 17, 189-206.

Goi A. and Trapido M. (2002) Hydrogen peroxide photolysis, Fenton reagent and photo-Fenton for the degradation of nitrophenols: a comparative study, *Chemosphere* 46, 913-922.

Ho, C.L., Shebl, M.A.A. and Watts, R.J. (1995) Development of an injection system for in situ catalyzed peroxide remediation of contaminated soil, *Hazard. Wastes Hazard. Mater.* 12, 15-25.

ITRC (2000) Dense non aqueous pahase liquids (DNAPLs): Review of emerging characterization and remediation technologies, Interstate Technology and Regulatory Cooperation Work Group, USA.

ITRC (2001) Technical and regulatory guidance for *in situ* chemical oxidation of contaminated soil and groundwater, Interstate Technology and Regulatory Cooperation Work Group, USA.

Jafvert C.T. (1996) Surfactants/Cosolvents. Technology evaluation report, EPA Groundwater Remediation Technologies Analysis Center, Pittsburgh, PA.

Kakarla, P.K.C. and Watts, R.J. (1997) Depth of Fenton-like oxidation in remediation of surface soil, *J. Environ. Eng.* 123, 11.

Kang S.F. and Chang H.M. (1997) Coagulation of Textile Secondary Effluents with Fenton's Reagent, *Water Sci. Technol.* 36(12), 215-222.

Kang S.F., Liao C.H. and Chen M.C. (2002) Preoxidation and coagulation of textile wastewater by the Fenton process, *Chemosphere* 46(6), 923-928.

Kang Y.W. and Hwang K.Y. (2000) Effects of reaction conditions on the oxidation efficiency in the Fenton process, *Wat. Res.* 34(10), 2786-2790.

Kitis M., Adams C. D. and Daigger G.T. (1999) The effects of Fenton's Reagent pretreatment on the biodegradability of nonionic surfactants, *Wat. Res.* 33(11), 2561-2568.

Kuo W.G. (1992) Decolorizing dye wastewater with Fenton's reagent, *Wat. Res.* 26(7), 881-886.

Kwon B.G., Lee D.S., Kang N. and Yoon J. (1999) Characteristics of *p*-chlorophenol oxidation by Fenton's reagent, *Wat. Res.* 33(9), 2110-2118.

Lin S.H. and Lo C.C., (1997) Fenton Process for Treatment of Desizing Wastewater, *Wat. Res.* 31(8), 2050-2056.

Liu M., Roy D. and Wang G. (1995) Reactions and transport modeling of surfactants and anthracene in the soil flushing process, *Waste Manage.*, 15(5/6), 423-432.

Lou J.C. and Lee S.S. (1995) Chemical Oxidation of BTX Using Fenton's Reagent, *Hazard. Waste & Hazard. Mater.* 12(2), 185-193.

Lu M.C., Chen J.N. and Huang C.P. (1997) Effects of Inorganic Ions on the Oxidation of Dichlorvos Insecticide with Fenton's Reagent, *Chemosphere* 35(10), 2285-2295.

Maillard C., Guillard C. and Pichat P. (1992) Comparative Effects of the TiO_2-UV, H_2O_2-UV, H_2O_2-Fe^{++} Systems on the Disappearance Rate of the Benzamide and 4-hydroxibenzamide in Water, *Chemosphere* 24(8), 1085-1094.

Martin, S.B. and Allen, A.E., Recycling EDTA after heavy metals extraction. Chemtech, 26 (1996) 23-25.

McGinnis B.D., Adams V.D. and Middlebrooks E. J. (2001) Degradation of ethylene glycol using Fenton's reagent and UV, *Chemosphere* 45(1), 101-108.

Mohanty N.R. and Wei I.W. (1993) Oxidation of 2,4-Dinitrotoluene Using Fenton's Reagent: Reaction Mechanisms and Their Practical Applications, *Hazard. Waste Hazard. Mater.* 10(2), 171-183.

Pearlman, R.S., Yalkowsky, S.H. and Banerjee, S. (1984) Water solubilities of polynuclear aromatic and heteroaromatic compounds, *J. Phys. Chem. Ref. Data* 13(2), 555-62.

Pignatello J.J., Sun Y. (1995) Complete Oxidation of Metolachlor and Methyl Parathion in Water by The Photoassisted Fenton Reaction, *Wat. Res.* 29(8), 1837-1844.

Quan, H.N., Teel, A.L. and Watts, R.J. (2003) Effect of contaminant hydrophobicity on hydrogen peroxide dosage requirements in the Fenton-like treatment of soils, *J. Hazard. Mater.* B102, 277-289.

Ravikumar, J.X. and Gurol M.D. (1994) Chemical oxidation of chlorinated organics by hydrogen peroxide in the presence of sand, *Environ. Sci. Technol.* 28(3), 394-400.

Roote D.S. (1998) Technology status report: *In situ* soil flushing, EPA Groundwater Remediation Technologies Analysis Center, Pittsburgh, PA.

Scott A.M., Hickey W.J. and Harris R.F. (1995) Degradation of Atrazine by Fenton's reagent: condition optimization and product quantification, *Environ. Sci. Technol.* 29, 2083-2089.

Siegrist R.L. (1998) *In situ* chemical oxidation: Technology features and applications, *Conference on Advances in Innovative Ground-water Remediation Technologies, Atlanta, GA, 15 December 1998*, Ground-water Remediation Technology Analysis Center, US EPA Technology Innovative Office.

Sun, B., Zhao, F.J., Lombi E. and McGrath S.P., Leaching of heavy metals from contaminated soils using EDTA. Environmental Pollution, 113 (2001) 111-120.

Szpyrkowicz L., Iuzzolino C. and Kaul S.N. (2001) A Comparative Study on Oxidation of Disperse Dyes by Electrochemical Process, Ozone, Hypochlorite and Fenton Reagent. *Wat. Res.* 35(9), 2129-2136.

Tang W.Z. and Tassos S. (1997) Oxidation Kinetics and Mechanisms of Trihalometanes by Fenton's Reagent. *Wat. Res.* 31(5), 1117-1125.

Tarr M.A. and Lindsay M.E. (1998) Role of dissolved organic matter in Fenton degradation of hydrophobic pollutants, Abstract presented at the 19th Annual Meeting of SETAC, Nov. 15-19, Charlotte, NC, USA.

Tejowulan, R.S., Removal of trace metals from contaminated soils using EDTA incorporating resin trapping techniques, Environmental Pollution, 103 (1998) 135.

Tyre B.W., Watts R.J. and Miller G.C. (1991) Treatment of four biorefractory contaminants in soil using catalyzed hydrogen peroxide, *J. Envron. Qual.*, 20(4), 832-838.

US EPA - Environmental Protection Agency (1998) Field applications of *in situ* remediation technologies: Chemical Oxidation, EPA 542-R-98-008.

US EPA - Environmental Protection Agency (2002) Arsenic treatment technologies for soil, waste and water, *Solid Waste and Emergency Response* EPA 542-R-02-004.

Venkatadri R. and Peters R.W. (1993) Chemical Oxidation Technologies: Ultraviolet Light/Hydrogen Peroxide, Fenton's Reagent and Titanium Dioxide-Assisted Photocatalysis, *Hazard. Waste & Hazard. Mater.* 10(2), 107-149.

Yin Y. and Allen H.E. (1999) *In situ* chemical treatment, Groundwater Remediation Technologies Analysis Center, Technology Evaluation Report, TE-99-01, Pittsburgh, PA, 82 pp.

Walling C. (1975) Fenton's Reagent Rivisited, *Acc. Chem. Research* 8, 125-131.

Watts, R.J., Udell, M.D. and Mansen, R.M. (1993) Use of iron minerals in optimizing the peroxide treatment of contaminated soils, *Water Environ. Res.* 65, 839-844.

Weiss J. (1935) The Catalytic Decomposition of Hydrogen Peroxide on different Metals. *Trans. Farad. Soc.* 3L, 1547-1557.

West, C.C. and Harwell, J.H. (1992) Surfactants and subsurface remediation, *Environ. Sci. Technol.* 26(12), 2324-2329.

Zepp R.G., Faust B.C. and Hoigné J. (1992) Hydroxyl Radical Formation in Aqueous Reactions (pH 3-8) of Iron(II) with Hydrogen Peroxide: the Photo-Fenton Reaction. *Environ. Sci. Technol.* 26, 313-319.

In: Trends in Hazardous Materials Research
Editor: Edward C. Booking, pp. 165-181
ISBN 1-60021-335-9

Chapter 6

REMOVAL OF HEAVY METALS FROM CCA-TREATED WOOD BY ION EXCHANGE MEMBRANE-ASSISTED METHODS

Emiliya Velizarova[a*], Alexandra B. Ribeiro[b], Eduardo P. Mateus[b] and Lisbeth M. Ottosen[c]

[a] Forest Research Institute, Bulgarian Academy of Sciences, Sofia, Bulgaria
[b] Departamento de Ciências e Engenharia do Ambiente, Faculdade de Ciências e Tecnologia, Universidade Nova de Lisboa, Caparica, Portugal
[c] Department of Civil Engineering, Technical University of Denmark, Lyngby, Denmark

ABSTRACT

The most common formulation, which has been used to preserve wood, is chromated copper arsenate (CCA). Based on the average wood service life of 20–50 years, it is expected that enormous amount of CCA-treated wood is now being, or will soon be, removed from service, which raises an environmental issue of how to deal with this wood waste management problem. Both legislation and drawbacks of the traditional disposal method (land-filling) is motivating studies of alternatives for treatment and possible reuse of CCA-treated wood waste. Besides the economic and technological issues, the method of choice is expected to be environmentally beneficial.

This paper is designed to provide an overview of the main membrane-assisted techniques that can be used for removal of heavy metals from CCA-treated wood waste, the main advantage of which is the combination of leaching of the target contaminant(s) by suitable agents, with the complete isolation of the treated material by membranes, thus forming separate compartment(s), in which the contaminant(s) can be accumulated and recovered. By a proper selection of the membrane material and operating conditions, the transfer of the target charged pollutants through the membranes can be regulated. Furthermore, the metal removal degrees can be controlled by appropriate pre-treatment of

* Corresponding author: E. Velizarova. E-mail: velizars@bas.bg or velizars@abv.bg

the wood waste, by adjusting the flow rates of the solutions, contacting the membranes, by introducing electrodes into the compartments, etc.

The authors have studied comprehensively both the dialytic and electrodialytic removal of Cu, Cr and As from CCA treated timber waste. The dialytic method is solely based on the Donnan equilibrium principle while the electrodialytic method uses a low-level direct current as the "cleaning agent", combining the electrokinetic movement of ions in the matrix, with the principle of electrodialysis. The electrodialytic treatment, performed under appropriate direct current conditions, proved to efficiently remediate CCA-treated timber waste in the form of sawdust (Ribeiro et al., 2000a; 2000b; Moreira *et al.*, 2005) or wood chips (Velizarova *et al.*, 2002; 2004; Virkutyte *et al.*, 2005) in laboratory scale, as well as in pilot scale (Pedersen *et al.*, 2005; Christensen *et al.*, 2005).

The principles and the applicability of these ion exchange membrane-assisted techniques for the removal of heavy metals from CCA-treated wood waste is discussed and critically compared with the electrokinetic remediation technique throughout the paper. The design of more efficient and cost-effective operating conditions, especially in long-term operations without or with minimal membrane fouling, reduction of the specific energy consumption requirements and development of efficient recovery or disposal options for the contaminants, in order to prevent their adverse impact on the ecosystems, are the major challenges.

Keywords: CCA treated timber waste; Copper; Chromium; Arsenic; Ion exchange membranes, Dialytic remediation; Electrodialytic remediation

1. Electrokinetic Remediation

1.1. Introduction

Electrokinetic remediation is a technique for removal of contaminants from polluted sites (Lageman *et al.*, 1989; Pamukcu and Wittle, 1992; Acar and Alshawabkeh, 1993; Lageman, 1993; Probstein and Hicks, 1993; Pamukcu and Wittle, 1994; Ottosen, 1994; Trombly, 1994; Ottosen, 1995; Acar *et al.*, 1995; Acar *et al.*, 1996). The U.S.EPA general classification is that it is a physical remediation treatment for phase separation (USEPA, 1989). According to the experiments and pilot-scale studies conducted, metals such as arsenic, lead, chromium, cadmium, copper, uranium, mercury and zinc, as well as polychlorinated biphenyls, phenols, chlorophenols, toluene, trichlorethane, acetic acid and atrazine, are suitable for electrokinetic remediation and recovery (Virkutyte *et al.*, 2002; Ribeiro *et al.*, 2005).

The increasing demand to solve the challenging problems in environmental restoration has stimulated the desire to use electrical fields for subsurface species transport. Species transport mechanisms under electrical fields are now proposed to be employed not only for the in situ and/or ex situ remediation of soils from inorganic and organic species, but also for the injection of nutrients, electron acceptors and other process additives in in situ bioremediation of subsurface deposits, for the injection of surfactants that enhance solubilization and transport, for the injection of grouts in soil stabilization and waste containment, for precipitation of species in migrating plumes, for soil and pore fluid characterization and species extraction using penetrating probes, for diversion systems for contaminant plumes, for leak detection systems in containment barriers and repair of failing containment barriers (Acar *et al.*, 1995; Ribeiro, 1998; Lear *et al.*, 2004; Saichek and Reddy, 2005).

Electrokinetics is not a new phenomena, since it was first discovered in 1809. A brief scheme showing the evolution on the use of electrokinetics is presented in Figure 1, whereas comprehensive ones can be found in Pamukcu and Wittle (1994) or in Bruch and Lewis (1973).

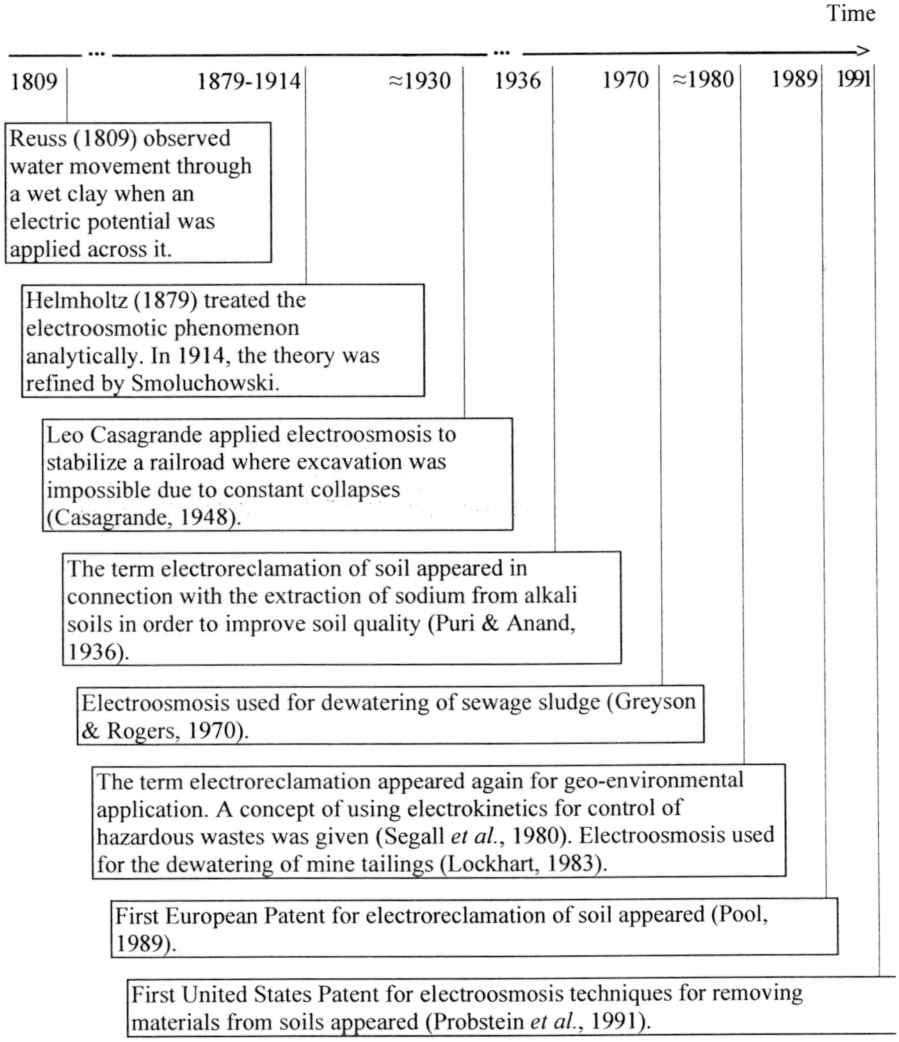

Figure 1. Brief scheme showing the evolution on the use of electrokinetics (Ribeiro, 1998)

1.2. General Principle

Electrokinetic remediation uses low-level direct current producing electric potential drops of the order of volts per centimeter or an electric current density of the order of milliamps per square centimeter applied to the cross sectional area of the soil mass between the electrodes, the electrodes being placed in an open or closed flow arrangement. The main interest in electrokinetic remediation in environmental cleanup operations lies in an attempt to confine and concentrate contaminants close to an electrode. The current state of knowledge of this cleanup technology is critically reviewed by Pamukcu and Wittle (1994), Ottosen, (1995), Yeung and Datla (1995), Acar *et al.* (1995), Ribeiro (1998), Virkutyte *et al.* (2002) or Ribeiro and Rodríguez-Maroto (2006).

Figure 2 presents the general principle of the electrokinetic process. Due to the electric field present when a direct current (dc) is passed between a pair of electrodes, placed in a system containing charged particles (e.g. moist contaminated soil), the pollutant species are driven towards one of the electrodes, from where they may be removed. Three main mechanisms are responsible for this movement: electromigration, electroosmosis and electrophoresis.

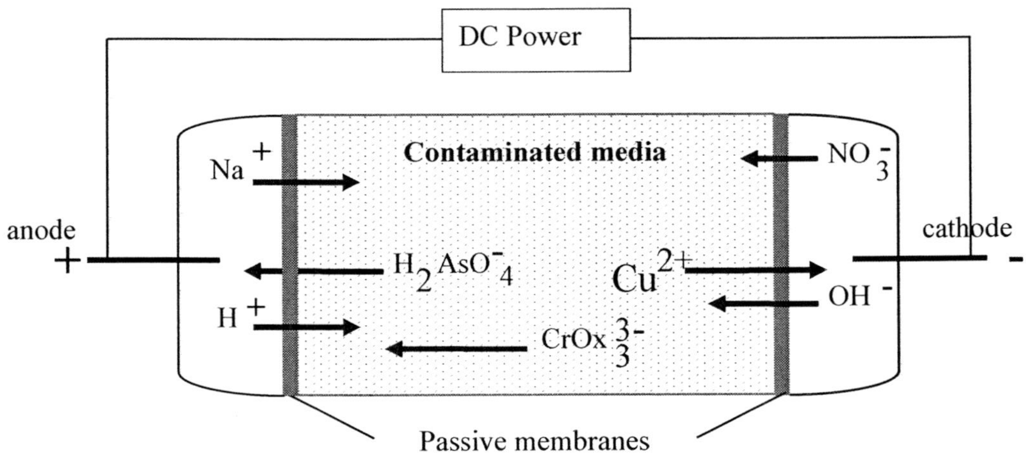

Figure 2. Schematic principle of the electrokinetic process

On inert electrodes, the electrode reactions can be controlled so as to produce H^+ ions at the anode and OH^- ions at the cathode. The electrolysis of water at the electrodes can be described by equations (1) and (2), which is the primary cause of the pH changes in the soil system chemistry during electrokinetics.

$$\text{Anode: } 2H_2O \rightarrow O_2\uparrow + 4H^+ + 4e^- \quad (1)$$

$$\text{Cathode: } 4H_2O + 4e^- \rightarrow 2H_2\uparrow + 4OH^- \quad (2)$$

Also, when the concentration of ions in the electrolytes is increased, other electrode processes may take place at the surface of inert electrodes. Equations (3) and (4) are examples where the Me^{n+} is a metal ion with n positive charges.

$$\text{Cathode: } Me^{n+} + ne^- \longrightarrow Me \quad (3)$$

$$\text{Cathode: } Me(OH)_n + ne^- \longrightarrow Me + nOH^- \quad (4)$$

Metals can precipitate at the cathode surface (equation (3)), also as oxides, hydroxides, carbonates and phosphates, depending on pH, the chemical constituents of the pore water and current flow (Segall and Bruell, 1992).

Several authors present detailed description of the coupled transport processes of fluid, electricity and contaminants during electrokinetics, and associated complicated features generated by various electrochemical reactions (Michell, 1991; 1993; Eykholt and Daniel 1994; Yeung and Datla; 1995; Denisov *et al.*, 1996; Alshawabkeh and Acar, 1996 or Ribeiro and Rodríguez-Maroto, 2006).

When the technique is applied without conditioning (unenhanced electrokinetic remediation), water electrolysis reactions at the electrode compartments (cf. equations 1 and 2) promote the development of an acid and an alkaline front in the soil. The rate at which the hydrogen and hydroxide ions are produced is fixed by the current and occasionally also by competition of other type of electrode processes. However, the mobility of H^+ is about 1.8 times higher than the mobility of OH^-, and the pH fronts meet in the soil at approximately 2/3 of the distance from the anode compartment. An accumulation of contaminants (e.g. heavy metals) appears in this region. Excavation of the soil volume in the accumulation zone may be a way of attaining the soil cleaning target concentrations. However, the excavated soil volume has to be submitted, itself, to a proper treatment and/or disposal, so a transfer of the contaminants always occurs at least point-wise (Ribeiro, 1998).

These limitations may be overcome by the use of ion exchange membranes instead of passive membranes, a concept introduced by the Environmental Electrochemistry Group at the Technical University of Denmark: Electrodialytic remediation.

2. Electrodialytic Remediation

2.1 Introduction

Since 1948, ion exchange membranes have been applied into a variety of processes (Figure 3), where ions have to be concentrated or diluted. In the electrodialytic remediation application, the use of the membranes increase the efficiency of the removal process.

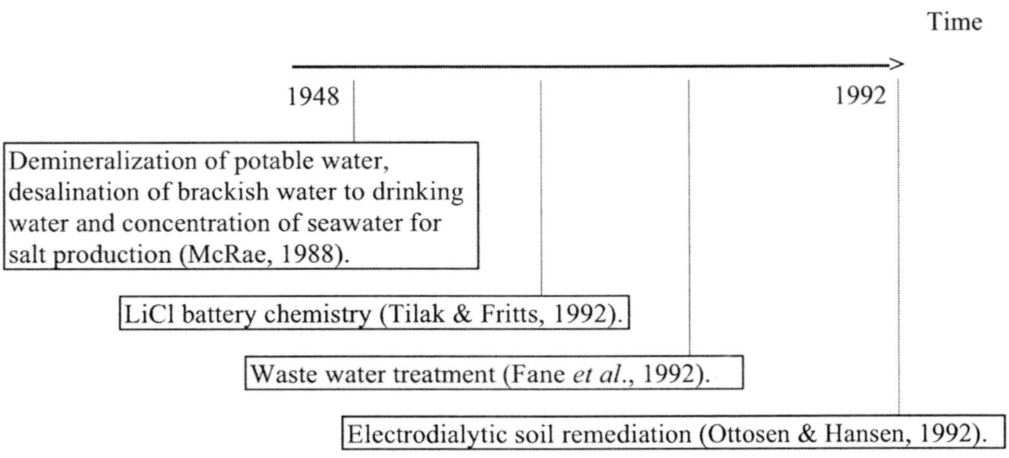

Figure 3. Brief evolution on the use of ion exchange membranes in different processes (Ribeiro, 1998)

2.2. General Principle

Electrodialytic remediation of solid waste products started to develop in 1992 (Ottosen and Hansen, 1992) and was patented in 1995 (PCT/DK95/00209).

Figure 4 presents the general principle of the electrodialytic process. It is similar to the electrokinetic one (Figure 2), differing by the use of ion exchange membranes, instead of passive membranes, which are used to separate the porous contaminated volume from the electrode compartments. A comprehensive overview on the study of practical and theoretical aspects concerning the use of ion exchange membranes in electrodialytic soil remediation can be found in Hansen (1995). During the remediation, an acidic front will develop from the anion exchange membrane towards the cathode. The acidic front is partly due to water splitting at the anion exchange membrane (Ottosen et al. 2000a) and partly due to non-ideal anion exchange membranes that allow a small amount of H^+ ions to pass from the anolyte into the contaminated media. Once produced by water splitting or passed the anion exchange membrane, the H^+ ions are transported by electromigration through the contaminated media. The acidic front aids desorption of most heavy metals from contaminated media.

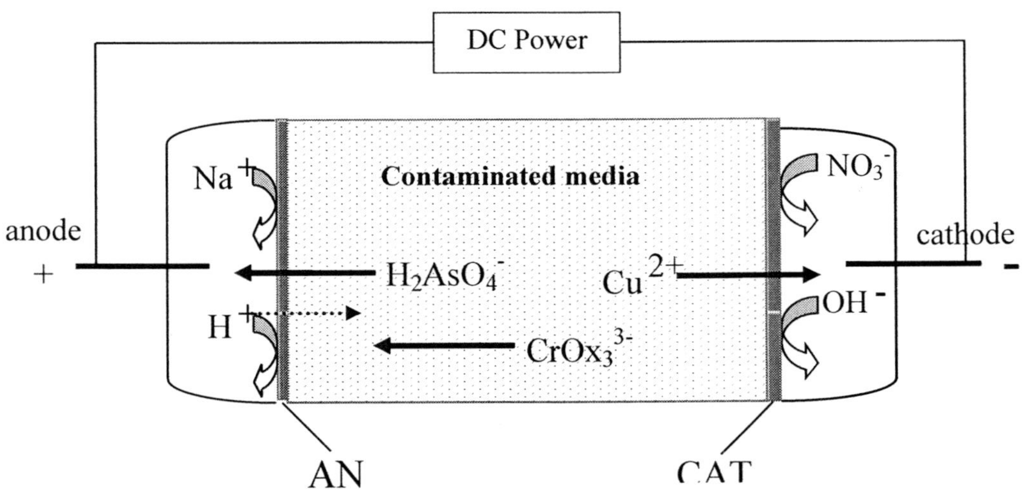

Figure 4. Schematic principle of the electrodialytic process.
AN - anion-exchange membrane; CAT - cation-exchange membrane

2.3. Electrodialytic vs. Electrokinetic Technique. Expected Improvements

The advantages of using ion exchange membranes, in the design presented in Figure 4, may be summarized in the following (Ribeiro, 1998):

– The ion exchange membranes do not allow the ions of the electrolyte solutions, the removed species from the central cell compartment or new ions formed at the electrodes, to pass into the central cell compartment. These species will stay in the electrode compartment to which they were transported, although some may change polarity in certain medium conditions. The membranes prevent energy to be wasted in moving ions between the two electrode compartments;

- it is possible to use any kind of electrodes, inert or reacting ones;
- the control of the pH gradients created in the porous media is facilitated, and the choice of keeping the electrolyte solutions at a certain pH. According to Hansen (1995), precipitation of material or salts in the electrode compartments will decrease the efficiency of the current transport, mainly by increasing the resistance over the cell and a fouling of the membranes.
- Although the porous media acidification favours remediation, when the increase in the hydrogen ion concentration is considered in conjunction with electromigration of a species of interest, a substantial increase in the hydrogen ion transference number may hinder transport of other species. This means that the acidification should be controlled to a limit where the desorption will occur but, at the same time, the acidification should be kept at a minimum, so that the transference number of H^+ ions is not higher than needed. Otherwise, energy would be wasted in carrying the H^+ ions from the anode to the cathode. The desired control of the pH is possible both in the electrokinetic and in the electrodialytic remediation cells; however, the latter case is easier to control, due to the use of the ion exchange membranes (Ribeiro and Rodríguez-Maroto (2006)).
- Membranes are selective for negatively or positively charged ions, which will result in a reduction in power consumption. However, any kind of element, either contaminant or not, is submitted to the same process, as the method is not specifically selective for contaminants.

2.4 Remediation Results Obtained with Electrodialytic Remediation of Different Media

The method was first developed for remediation of heavy metal polluted soil (Ottosen *et al.*, 1997; 2000; 2001; 2002, Ribeiro and Mexia, 1997). The soil was placed in the central compartment (Figure 4) as a water saturated matrix. Very good removal percentages were obtained e.g. the Cu concentration was decreased from1360 mg/kg to 36 mg/kg (corresponding to a removal percentage of 98%) in a soil that was polluted from wood preservation (Ottosen et al. 1997). However, it was found, that in the case of calcareous soils the remediation process proceed very slowly due to the high buffering capacity of these soils and the following slow development of the acidic front and addition of assisting agents may be necessary in some cases (Ottosen *et al.* 2005). One example on how electrodialytic remediation can be successfully obtained in a Cu polluted calcareous soil is by addition of ammonia to the soil prior and during the remediation. Ammonia increases soil pH to above 9 and at this pH Cu is normally not in ionic form and thus not mobile in the electric field but in the presence of ammonia, $[Cu(NH_3)_4]^{2+}$ species are formed and these are mobile in the electric field even at alkaline conditions. A reduction on Cu concentration of 93% was obtained in the actual case (Ottosen *et al.* 1998).

Promising experimental results have been obtained for other matrices than soil (Ottosen, *et al.*, 2003), and one example is Cu mine tailing where a removal percentage of 53% Cu was obtained after sulphuric acid addition prior to the electrodialytic remediation (Hansen *et al.*, 2005; Hansen *et al.,* 2006). It was found, when using the electrodialytic method for decreasing the Cd concentration in fly ash from straw combustion (Hansen *et al.,* 2004), that operational problems occurred due to the reduction of matrix from dissolution of ash particles, though good Cd removal result was obtained and 97% Cd was removed. The operational problems were solved by treating the ash as a stirred suspension instead of a water saturated matrix (Pedersen, 2002). Other contaminated medias have been treated electrodialytically as stirred suspensions. From sewage sludge 69% Cd was removed (Jakobsen *et al.*, 2004). Successful removal of Cu (48%), Zn (80%), Pb (96%) and Cd (98%) has been obtained from harbour sediments (Nystroem *et al.* 2005; 2006). Also different municipal solid waste incineration fly ashes have been subjected to elecrodialytic remediation as a stirred suspension (e.g. Ottosen *et al.*, 2006; Ferreira *et al.*, 2005a; 2005b; Pedersen, 2002). The remediation scheme for these ashes is not as straight forward as for the other matrices treated since these fly ashes contain many different heavy metals and further some of the heavy metals are build into the ash particles. It is possible to remove a large fraction of many heavy metals from the ashes, but the remaining matrix is unstable and will continue leaching small amounts of heavy metals. So for the treatment of these ashes further development is needed.

The electrodialytic method has also shown good removal results for CCA-treated timber waste. The first results were obtained with sawdust wetted with oxalic acid and the best result obtained here was removal of 93% Cu, 95% Cr and 99% As (Ribeiro *et al.*, 2000a; 2000b; Moreira *et al.*, 2005). The electrodialytic method was also tested for timber waste in the form of wood chips (Moreira *et al.*, 2006; Virkutyte *et al.*, 2005; Kristensen *et al.,* 2005; Velizarova *et al.*, 2002; 2004). Again good results were obtained but with slightly lower removal efficiencies as was the case when treating sawdust, and it was found that, in general, the larger wood chips the lower remediation efficiency (Moreira *et al.*, 2006; Kristensen, 2004). Oxalic acid is a very good assisting agent for mobilization of Cu, Cr and As at the same time, but the concentration of oxalic acid was very high to obtain the best result. Following other assisting agents and combinations of different assisting agents was tested (Velizarova *et al.* 2004; Kristensen, 2004). A good combination of assisting agents was found to be a combination of phosphorus acid and oxalic acid during laboratory work (Kristensen, 2004) and it was decided to test the method in pilot scale (Christensen, 2004; Pedersen *et al.*, 2005; Christensen *et al.,* 2005). Also in pilot scale (0.6 m^3 wood chips) good results were obtained and the best removal percentages were 82% Cr, 88% Cu and 96% As.

3. Dialytic Remediation

3.1 Introduction

While electrodialysis has proved to be an efficient method for the removal of target charged pollutants from water streams, it also has several potential limitations, the most important ones being the need of using electrodes and an external power supply, which make the system inherently more complex and relatively expensive, especially when not used for production but for remediation purposes. The second possible drawback is its relatively lower efficiency when dealing with very dilute electrolyte solutions. As a rule of thumb, solutions with an electric conductivity lower than 0.5 mS/cm are considered not suitable for electrodialytic treatment since the limiting current density is rapidly reached under such conditions due to the low availability of charge carriers. These problems led us to investigate, for the first time, the possibility of integrating dialytic treatment for remediation of CCA-treated wood waste. This choice was motivated by the existing theoretical and practical knowledge of using this process in the treatment of water streams, containing a variety of ions and for different purposes such as water softening, recovery of valuable ions and removal of charged contaminants from a number of industrial as well as natural water streams (Davis, 2000). As in the electrodialytic treatment, since the ion exchange membranes act only as separation barriers between solutions pre-treatment of the CCA-treated wood waste is needed to extract the target pollutants from the solid wood into the liquid phase.

3.2. General Principle

One of the most important properties of the ion exchange materials is their permeability to ions of the opposite charge (counter-ions) while excluding ions of the same charge (co-ions) (Helfferich, 1962). The dialytic process, which uses this property is usually referred to as Donnan dialysis and is a process that uses ion exchange membranes without applying electrodes and an external electric potential difference across the membrane (Davis, 2000, Cwirko and Carbonell 1990). It is named after Donnan, who in 1924 described the equilibrium that results when a semipermeable membrane separates two solutions of electrolytes. The ions, which are permeable to the membrane, equilibrate between the two solutions until the so-called Donnan equilibrium is satisfied. This relationship can be derived from thermodynamics, considering that the electrochemical potentials of any permeable ion i must be equal in the two liquid phases, designated as 1 and 2 respectively, at equilibrium and mathematically reads as:

$$\phi_2 - \phi_1 = \frac{RT}{z_i F}\left(\ln a_{i,1} - \ln a_{i,2}\right) = \frac{RT}{F}\ln\left(\frac{a_{i,1}}{a_{i,2}}\right)^{\frac{1}{z_i}} \qquad (5)$$

Where a_i is the activity of the particular ion, z_i its valence, R is the universal gas constant, T is the absolute temperature, F is the Faraday´s constant, and ϕ is the electrical potential. Equation 1 clearly shows the balance of forces. Any electrolyte concentration difference causes an electrical potential difference in the system that acts on all permeable ions. The ratio $\left(\frac{a_{i,1}}{a_{i,2}}\right)^{\frac{1}{z_i}}$, which is sometimes called Donnan constant, at equilibrium is the same for all of the permeable ions (counter-ions) in the system. Furthermore, in dilute solutions, the ion activities ($a_i = \gamma C_i$) are usually approximated by the corresponding molar concentrations C_i, since the ion activity coefficients (γ_i) approach unity.

In Donnan dialysis (opposite to simple dialysis), the separation achieved between the solutions is not lost even if the system is closed to the surrounding, a feature characteristic of equilibrium separation processes such as absorption, extraction and distillation (Ho and Sirkar, 1992). On the other hand, in simple rate-controlled membrane separation processes, such as dialysis, all separation achieved is lost after some time if the system is closed to the surrounding.

The Donnan dialysis type of operation requires the addition of a so-called "driving" counter-ion to the stripping solution(s), which is transported in a direction opposite that of the target ion in order to maintain electroneutrality. As suitable driving ions from an environmental point of view can be considered, e.g. Na^+ when dealing with cationic pollutants and e.g. Cl^- if anionic pollutants have to be removed. Since concentration ratios determine the Donnan equilibrium, not concentration differences, Donnan dialysis allows for transport of charged micropollutants against their concentration gradients that is of utmost importance when dealing with polluted matrixes, such as CCA-treated wood, which contain low levels of polluting ions. Furthermore, the hydraulic residence time can be independently adjusted in the two contacting compartments, to optimise the degree of extraction of the target pollutant. Due to these characteristics, the Donnan dialysis removal of a number of ions, both positively or negatively charged and originating from different sources has received attention (Davis, 2000; Ho and Sirkar, 1992).

3.3. Dialytic vs. Electrodialytic Technique. Comparative Analysis

The experimental design for studying the dialytic process in the case of CCA-treated wood waste can be found in detail, e.g. in Velizarova *et al.* (2002, 2004) and, in brief, consists in cutting a cation and an anion exchange membrane under study in two circles of equal size and placing them between the two chambers of the cell, as presented in Figure 4, however without using electrodes and applying an external power supply. The fluxes of a target ion i to the two solutions, which for simplicity are still referred to as anolyte (for the one contacting the anion-exchange membrane) and catholyte (for the one contacting the cation-exchange membrane) can be calculated, after measuring the increase in time of the target ion concentration in the corresponding solution, using the general differential mass balance equation under transient state conditions and applied to the respective compartment:

$$J_i = \frac{1}{A}\frac{dVC_i}{dt} \qquad (6)$$

where V is the solution volume (equal for both compartments but slightly changing in time due to sampling) and A is the (equal) membrane surface area facing the respective compartment. The removal degree of a target element can be commonly expressed by:

$$\eta_i = \frac{C_{i,in} - C_{i,f}}{C_{i,in}} \qquad (7)$$

where $C_{i,\,in}$ is the initial concentration of a target element in the wood waste, $C_{i,\,f}$ is its final concentration and i refers to Cu, Cr or As respectively.

It has to be noted that CCA-treated wood had not been previously studied as a matrix for the removal of ionic species by dialytic techniques, however, considerable knowledge has been accumulated in cases of removing of any of the target elements from contaminated aqueous sources using a number of ion exchange membranes and data has been reported by many authors, e.g. Kim (1980) Kojima *et al.* (1982), Huang and Wang (1992), Çengeloglu *et al.*, (2003), Velizarov *et al.* (2004), Scindia *et al.* (2005), etc. When dealing with solutions obtained by leaching CCA-treated waste and their subsequent treatment by dialytic techniques, several specific characteristics of these solutions have to be considered in order to develop an adequate ion exchange membrane-assisted technology. Some of the most important are the ionic composition and the charge of the target elements and/or the complexes in which they might be involved. In the following discussion, the use of dialytic technique will be compared to the electrodialytic one for the three elements of interest: As, Cu and Cr. Moreover, the important role of the leaching agent for the process efficiency using both techniques will be shortly outlined.

The mobility of arsenic in water environment is of great concern since it is relatively more soluble than copper and chromium and therefore less likely to be adsorbed (Lebow, 1996). Indeed, our results (Velizarova *et al.* 2002, 2004) showed that even deionized water could be a suitable leaching agent for this element (Table 1).

Table 1. Comparison of the suitability of deionised water against oxalic acid solution (2.5 %) for leaching of Cu, Cr and As from CCA-treated wood waste. Incubation (24 h) of 40 g of wood chips. Liquid to solid (wood chips) ratio = 8.8 (w/w)

Element	Removal, %	
	Deionized water	Oxalic acid
As	17	53
Cu	3	26
Cr	1	60

Under oxidized conditions, which are typical for surface water, arsenic is usually presented as As(V) oxyanion (arsenate), which speciation depends strongly on pH. Under neutral and slightly acidic conditions, the monovalent form $H_2AsO_4^-$ prevails, while in alkaline solutions $HAsO_4^{2-}$ becomes the dominant species. On the other side, As(III), which is known to be much more toxic than As(V) gives rise to arsenite and can be found mostly in groundwater, where reducing condition usually prevail (Smith *et al.*, 2002).

Our results showed that dialytic and electrodialytic treatment are both efficient for arsenic removal from CCA-treated wood waste. Table 2 compares the efficiencies reached by the two techniques under batch operation conditions and using oxalic acid as the leaching pre-treatment agent. As it can be observed, a high removal degree of 81 % was obtained for As, even without applying any external power supply. The practical significance of this result is that if complete removal of the target element is not required by the local legislation, the simpler and less energy consuming dialytic treatment of the CCA-treated wood waste can be the preferred method. Independently of the method used, As was always preferentially accumulated in the anolyte solution confirming its expected presence in the form of negatively charged species.

It is known that copper solubility and transport are not that much dependent on its oxidation state and that Cu(II) tends to be much more stable under oxidized conditions (Lebow, 1996). Our results showed the important role of the type of extracting solution for the subsequent dialytic and/or electrodialytic removal of the copper from CCA-treated wood waste (Table 1). The leaching agent (oxalic acid in this case) had a strong impact on both magnitude and direction of the copper flux, and on its removal degree (Table 2) therefore, its proper selection is crucial for the overall process performance. Our more recent results (Velizarova *et al.*, 2004) showed that other organic acids, such as formic acid could support the efficient removal of Cu and that the use of mixtures of individual organic acids is also possible, however, it does not improve the overall removal efficiency. If these acids are used as leaching agents under electrodialytic conditions, Cu removal from the wood chips as positively as well as negatively charged compounds occur, so, the element is accumulated in both catholyte and anolyte solutions.

Table 2. Comparison of the removal efficiency (Equation 7) ofAs, Cu and Cr reached under dialytic (no external power supply) and electrodialytic (Initial current = 40 mA) conditions

Element	Removal efficiency, %	
	Dialysis	Electrodialysis
As	81	95
Cu	45	80
Cr	72	87

NaCl extracting solution was also found suitable for Cu removal from CCA-treated wood chips. In this case, Cu was almost entirely accumulated in the catholyte due to its presence in a cationic form. If cathode reduction recovery of Cu for its possible re-use is desired, these conditions will be more suitable, compared to the situations of organic acids as extracting solutions, because, in the latter cases, part of Cu is directed to the anode compartment.

Cr appears to be the least mobile of the CCA elements, however, as pointed out by Lebow (1996), little work has been done to determine the valence state of Cr in compounds that leach from treated wood. Our results confirmed the expectation of an extremely low leaching degree of Cr in case of using water and that acidic solutions (such as oxalic acid) proved to be much more powerful leaching agents (Table 1). The currently available knowledge indicates that both trivalent and hexavalent chromium can be found and that the latter form is more solublev and mobile. Therefore, some authors attribute the resistance of trivalent chromium to leaching as a result of its ability to form stable and inert complexes with a wide range of organic and inorganic ligands (McGrath and Smith, 1990). Hexavalent chromium anionic compounds, such as $HCrO_4^-$, $Cr_2O_7^{2-}$ or CrO_4^{2-}, are well soluble and mobile and compounds like $CuOx_2^{2-}$ are rather stable. Presence of other oxalates, such as $CrOx_3^{3-}$, is also possible. Therefore Cr speciation and its possible effect on the process performance appears to be more complex and, thus, more important that the effects of speciation of As or Cu. Regarding the overall removal efficiencies - relatively high values were obtained by both dialytic and electrodialytic techniques (Table 2), showing again their suitability for the treatment of CCA-treated wood waste.

REFERENCES

Acar, Y. B. and Alshawabkeh, A. N. (1993). Principles of electrokinetic remediation. *Environmental Science and Technology*, *27*(13): 2638-2647.

Acar, Y. B.; Gale, R. J.; Alshawabkeh, A. N.; Marks, R. E.; Puppala, S.; Bricka, M. and Parker, R. (1995). Electrokinetic remediation: basics and technology status. *Journal of Hazardous Materials*, *40*: 117-137.

Acar, Y. B.; Rabbi, M. F.; Ozsu, E. E.; Gale, R. J. and Alshawabkeh, A. N. (1996). Enhance soil bioremediation with electric fields. *Chemtech*, April 1996: 40-44.

Alshawabkeh, A. N. and Acar, Y. B. (1996). Electrokinetic remediation. II: Theoretical model. *Journal of Geotechnical Engineering*, *122*(3): 186-197.

Bruch Jr, J.C. and Lewis, R.W. (1973). Electro-osmosis and the contamination of underground fluids. *Journal of Environmental Planning and Pollution Control*, *1*(4): 50-56.

Casagrande, L. (1948). Electro-osmosis in soils. *Geotechnique*, *1*: 159-177.

Çengeloglu, Y.; Tor, A.; Kir, E. and Ersoz, M. (2003). Transport of hexavalent chromium through anion-exchange membranes. *Desalination, 154*: 239-246.

Christensen, I. V. (2004). Electrodialytic remediation of CCA treated waste wood. Ph.D. Thesis, Department of Civil Engineering, Technical University of Denmark.

Christensen, I. V.; Pedersen, A. J.; Ottosen, L. M. and Ribeiro, A. B. (2005?). Electrodialytic remediation of CCA-treated waste wood in a 2 m^3 pilot plant. *Science of the Total Environment* (in press).

Cwirko, E. H. and Carbonell, R. G. (1990). A theoretical analysis of Donnan dialysis across charged porous membranes. *J. Membrane Science, 48:* 155-179.

Davis, T. (2000). Donnan dialysis. In: *Encyclopedia of Separation Science*, Vol. 4, I. D. Wilson (Ed.), Academic Press, London, pp. 1701-1707.

Denisov, G.; Hicks, R. E. and Probstein, R. F. (1996). On the kinetics of charged contaminant removal from soils using electric fields. *Journal of Colloid and Interface Science, 178*: 309-323.

Donnan, F. G. (1924). The theory of membrane equilibria. *Chemical Reviews*, *1*: 73-90.

Eykholt, G. R. and Daniel, D. E. (1994). Impact of system chemistry on electroosmosis in contaminated soil. *Journal of Geotechnical Engineering, 120*(5): 797-815.

Fane, A. G.; Awang, A. R.; Bolko, M.; Macoun, R.; Schofield, R.; Shen, Y. R. and Zha, F. (1992). Metal recovery from wastewater using membranes. *Water Science and Technology, 25*(10): 5-18.

Ferreira, C.; Jensen, P.; Ottosen, L. M.; Ribeiro, A. B. (2005a). Removal of selected heavy metals from MSW fly ash by the electrodialytic process. *Engineering Geology*, *77*(3-4): 339-347.

Ferreira, C.; Ribeiro, A. B. and Ottosen, L. M. (2005b). Effect of major constituents of MSW fly ash during electrodialytic remediation of heavy metals. *Separation Science and Technology*, *40*: 2007–2019.

Greyson, J. and Rogers, H. H. (1970). Dewatering sewage sludge by electroosmosis. *International Associtaion on Water Pollution Research, 1*(II-26): 1-15.

Hansen, H. K. (1995). *Practical and theoretical aspects concerning the use of ion exchange membranes and resins in electrokinetic soil remediation.* Ph. D. Dissertation, The Technical University of Denmark, Denmark, 216 pp.

Hansen, H. K.; Ottosen, L. M.; Villumsen, A. (2004). Electrodialytic removal of cadmium from straw combustion fly ash. J. Chem. *Technol. Biotechno*l., *79*: 789-794.

Hansen, H. K.; Ribeiro, A. B.; Mateus, E. P. and Ottosen, L. M. (2006). Diagnostic analysis of electrodialysis in mine tailing materials. *Electrochemica Acta* (accepted for publication).

Hansen, H. K.; Rojo, A.; Ottosen, L. M. (2005). Electrodialytic remediation of copper mine tailings. *Journal of Hazardous Materials*, *117*: 179-183.

Helfferich, F. (1962). Ion Exchange. McGraw-Hill Inc., New York.

Helmholtz, H. L. F. (1879). Studies of electric boundary layers. *Annln Phys.*, *7*: 337, translated by P. E. Bocque (1951), *Bull. Dep. Eng. Res. Univ. Mich.*, *33*: 5-47.

Ho, W. S. W. and Sirkar, K. K. (1992). Membrane Handbook. Van Nostrand Reinhold, New York.

Huang, T-C. and Wang, J-K. (1992). Separation of cupric and ferric ions with a cation-exchange membrane in the presence of complexing agents. *Separation Science and Technology*, *27*: 1137-1155.

Jakobsen, M. R.; Fritt-Rasmussen, J; Nielsen, S. and Ottosen, L. M. (2004). Electrodialytic removal of cadmium from wastewater sludge. *J. Hazard. Mater.*, *106 B*: 127–32.

Kim, B. M. (1980). Donnan dialysis for removal of chromates and cyanides. *AIChE Symp. Ser.*, *76*: 184-192.

Kojima, T.; Furusaki, S. and Saito, K. (1982). A fundamental study on recovery of copper with a cation exchange membrane. Part 2 – Transfer rate of copper and hydrogen ion through a cation exchange membrane. *Can. J. Chem. Eng.*, *60*: 650-658.

Kristensen, I. V.; Ottosen, L. M.; Ribeiro, A. B. and Villumsen, A. (2005). Electrodialytic removal of Cu, Cr and As from CCA-treated wood. Cap. 22 *In*: E. Lichtfouse, J. Schwarzbauer, D. Robert (Eds.), *Environmental Chemistry, Green Chemistry and Pollutants in Ecosystems,* Springer-Verlag, Berlin, ISBN 3-540-22860-8, pp. 235-242.

Lageman, R. (1993). Electroreclamation. *Environmental Science and Technology*, *27*(13): 2648-2650.

Lageman, R.; Pool, W. and Seffinga, G. (1989). Electro-reclamation: theory and practice. *Chemistry and Industry*, 18 September 1989, pp. 585-590.

Lear, G; Harbottle, M. J.; van der Gast, C. J.; Jackman, S. A.; Knowles, C. J.; Sills, G. and Thompson, I. P. (2004). The effect of electrokinetics on microbial communities. *Soil Science and Biochemistry*, *36*: 1751-1760.

Lebow, S. (1996). Leaching of wood preservative components and their mobility in the environment. Gen. Tech. Rep. FPL-GTR-93. U.S. Department of Agriculture, Forest Service, Forest Products Laboratory, 36 p.

Lockhart, N. C. (1983). Electro-osmotic dewatering of fine tailings from mineral processing. *International Journal of Mineral Processing*, *10*: 131-140.

McGrath, S. P. and Smith, S. (1990). Chromium and Nickel. *In*: Alloway, B. J. (Ed.) Heavy metals in soils. John Wiley and Sons, New York, pp. 125-150.

McRae, W. (1988). Electrodialysis. In: *Process Technologies for Water Treatment.* S. Stucki (Ed.), Plenum Press, New York, pp. 45-63.

Mitchell, J. K. (1991). Conduction phenomena: from theory to geotechnical practice. *Géotechnique*, *41*(3): 299-340.

Mitchell, J. K. (1993). Conduction phenomena (Chapter 12). In: *Fundamentals of soil behavior*. 2nd Ed., John Wiley and Sons, New York, USA, pp. 228-283.

Moreira, E.; Ribeiro, A. B.; Mateus, E. P.; Mexia, J. T. and Ottosen, L. O. (2005). Regressional modeling of electrodialytic removal of Cu, Cr and As from CCA treated timber waste. Application to sawdust. *Wood Sci. Technol.*, *39*(4): 291-309.

Moreira, E.; Ribeiro, A. B.; Mateus, E. P.; Mexia, J. T. and Ottosen, L. O. (2006). Regressional modeling of electrodialytic removal of Cu, Cr and As from CCA treated timber waste: application to wood chips. *Biometrical Letters* (in press).

Nystroem, G. M.; Ottosen, L. O. and Villumsen (2005). Electrodialytic removal of Cu, Zn, Pb, and Cd from harbor sediment: Influence of changing experimental conditions. *Environ. Sci. Technol.*, *39*: 2906-2911.

Nystroem, G. M.; Pedersen, A. J.; Ottosen, L. O. and Villumsen (2006). The use of desorbing agents in electrodialytic remediation of harbour sediment. *Science of the Total Environment*, *357*: 25– 37.

Ottosen, L. M. and Hansen, H. K. (1992). *Electrokinetic cleaning of heavy metal polluted soil.* Internal Report, Department of Physical Chemistry and Department of Geology and Geotechnical Engineering, Technical University of Denmark, Denmark, 9 pp.

Ottosen, L. M. (1994). *Electrokinetic remediation of heavy metal polluted soil.* Miljø- og Energiministeriet Miljøstyrelsen, Arbejdsrapport fra Miljøstyrelsen, Nr. 67, Denmark, 121 pp.

Ottosen, L. M. (1995). *Electrokinetic remediation. Application to soils polluted from wood preservation.* Ph.D. Thesis, Technical University of Denmark, Denmark, 217 pp.

Ottosen, L. M.; Eriksson, T.; Hansen, H. K. and Ribeiro, A. B. (2002). Effects from different types of construction refuse in the soil on electrodialytic remediation. *Journal of Hazardous Materials, 91*(1-3): 205-219.

Ottosen, L. M.; Hansen, H. K.; Hansen, C. B. (2000a). Water splitting at ion-exchange membranes and potential differences in soil during electrodialytic soil remediation. *Journal of Applied Electrochemistry*, *30*: 1199-1207.

Ottosen, L. M.; Hansen, H. K.; Hansen, L.; Kliem, B.; Bech-Nielsen, G.; Villumsen, A. (1998). Electrodialytic soil remediation. Improved conditions and acceleration of the process by addition of desorbing agents to the soil. Contaminated Soil ′98. Edinburgh, May 17-21, 1998, pp. 471-478.

Ottosen, L. M.; Hansen, H. K.; Laursen, S.; Villumsen, A. (1997). Electrodialytic remediation of soil polluted with copper from wood preservation industry. *Environ. Sci. Technol.*, *31*: 1711-1715.

Ottosen, L. M.; Hansen, H. K.; Ribeiro, A. B. and Villumsen, A. (2001). Removal of Cu, Pb and Zn in an applied electric field in calcareous and non-calcareous soils. *Journal of Hazardous Materials, 85*(3): 291-299.

Ottosen, L. M.; Hansen, H. K.; Villumsen, A.; Bech-Nielsen, G. (2000b). Electrodialytic remediation of an arsenic and copper polluted soil. Continous addition of ammonia during the process. *Environmental Technology*, *21*: 1421-1428.

Ottosen, L. M.; Hansen, H. K.; Villumsen, A.; Bech-Nielsen, G. (2000b). Electrodialytic remediation of an arsenic and copper polluted soil. Continous addition of ammonia during the process. *Environmental Technology*, *21*: 1421-1428.

Ottosen, L. M.; Kristensen, I. V.; Pedersen, A. J.; Hansen, H. K.; Villumsen, A. and Ribeiro, A. B. (2003). Electrodialytic removal of heavy metals from different solid waste products. *Sep. Sci. Technol., 38*(6): 1269– 89.

Ottosen,L. M.; Lima, A. T.; Pedersen, A. J. and Ribeiro, A. B. (2006). Electrodialytic extraction of Cu, Pb and Cl from municipal solid waste incineration fly ash suspended in water. *Journal of Chemical Technology and Biotechnology, 81*: 553–559.

Pamukcu, S. and Wittle, J. (1994). Electrokinetically enhanced in situ soil decontamination. In: *Remediation of Hazardous Waste Contaminated Soils*, Donald L. Wise and Debra J. Trantolo (Eds.), Marcel Dekker, Inc., New York, USA, pp. 245-298.

Pamukcu, S. and Wittle, J. K. (1992). Electrokinetic removal of selected heavy metals from soil. *Environmental Progress, 11*(3): 241-250.

Pedersen, A. J. (2002). Evaluation of assisting agents for electrodialytic removal of Cd, Pb, As, Cu and Cr from MSWI fly ash. *J. Hazard. Mater., B95*:185 –98.

Pedersen, A. J.; Kristensen, I. V.; Ottosen, L. M.; Ribeiro, A. B. and Villumsen, A. (2005). Electrodialytic remediation of CCA-treated waste wood in pilot scale. *Engineering Geology, 77*(3-4): 331-338.

Pedersen, A.J. (2002). Electrodialytic removal of heavy metals from fly ashes. Ph.D. Thesis. Department of Civil Engineering, Technical University of Denmark.

Pool, W. (1989). *A process for electroreclamation of soil material, an electric current system for application of the process, and an electrode housing for use in the electric current system*. European patent 0 312 174 A1 from 19 April 1989, and US patent 5,433,829 from 9.7.1989.

Probstein, R. F. and Hicks, R. E. (1993). Removal of contaminants from soils by electric fields. *Science*, 23 April 1993, *260*: 498-503.

Probstein, R. F.; Renaud, P. C. and Shapiro, A. P. (1991). *Electroosmosis techniques for removing materials from soils*. US patent 5,074,986 from 24.12.1991.

Puri, A. N. and Anand, B. (1936). Reclamation of alkali soils by electrodialysis. *Soil Science, 42*: 23-27.

Reuss, F. F. (1809). Sur un nouvel effet de l'electricité galvanique. *Mem. Soc. Imp. Naturalistes Moscow, 2*: 327-337.

Ribeiro, A B. (1998). Use of electrodialytric remediation technique for removal of selected heavy metals and metalloids from soils. Ph.D. Thesis. Technical University of Denmark, 313 pp.

Ribeiro, A. B. and Mexia, J. T. (1997). A dynamic model for the electrokinetic removal of copper from a polluted soil. *J. Haz. Mat. 56:* 257-271.

Ribeiro, A. B.; Mateus, E. P.; Otttosen, L. M. and Bech-Nielsen, G. (2000b). Electrodialytic removal of Cu, Cr and As from chromated, copper, arsenate-treated timber waste. *Environ. Sci. Technol., 34*: 784-788.

Ribeiro, A. B.; Mateus, E. P.; Otttosen, L. M. and Cabrita, R. L. (2000a). Formic, citric and oxalic acids as assisting agents for the electrodialytic removal of Cu, Cr and As from CCA-treated timber waste In: Proceedings of the 11th Annual International Conference on Heavy Metals in the Environment. J. Nriagu, Editor. Contribution No. 1183. University of Michigan, School of Public Health, August 6-10, 2000, Ann Arbor, MI (CD-ROM).

Ribeiro, A. B.; Rodríguez-Maroto, J. M.; Mateus, E. P. and Gomes, H. (2005). Removal of organic contaminants from soils by an electrokinetic process: the case of atrazine. Experimental and modeling. *Chemosphere, 59*: 1229-1239.

Ribeiro, A. B; Rodríguez-Maroto, J. M. (2006). Electroremediation of heavy metal-contaminated soils. Processes and applications. Cap. 18 *In:* M.N.V. Prasad, K.S. Sajwan, Ravi Naidu (Eds.), *Trace elements in the environment: Biogeochemistry, Biotechnology and Bioremediation*, Taylor and Francis, CRC Press, Florida, USA, ISBN 1-56670-685-8, pp. 341-368.

Rojo, A. and Hansen, H. K. (2005). Electrodialytic remediation of copper mine tailings: Sulphuric and citric acid addition. *Separation Science and Technology, 40*: 1947-1956.

Saichek, R. E. and Reddy, K. R. (2005). Electrokinetically enhanced remediation of hydrophobic organic compounds in soils: A review. *Critical Reviews in Environmental Science and Technology, 35*:115-192.

Scindia, Y. M.; Pandey, A. K. and Reddy, A. V. R. (2005). Coupled-diffusion transport of Cr(VI) across anion-exchange membranes prepared by physical and chemical immobilization methods. *Journal of Membrane Science, 249*: 143-152.

Segall, B. A. and Bruell, C. J. (1992). Electroosmotic contaminant -removal processes. *Journal of Environmental Engineering, 118*(1): 84-100.

Segall, B. A.; O'Bannon, C. E. and Matthias, J. A. (1980). Electro-osmosis chemistry and water quality. *Journal of the Geotechnical Engineering Division, 106*(GT10): 1148-1152.

Smith, A. H.; Lopipero, P. A.; Bates, M. N. and Steinmaus, C. M. (2002). Arsenic epidemiology and drinking water standards. *Science, 296*: 2145-2146.

Tilak, B. V. and Fritts, S. D. (1992). Comments on the estimation of caustic current efficiency of ion exchange membrane chlor-alkali cells. *Journal of Applied Electrochemistry, 22*: 675-677.

Trombly, J. (1994). Electrochemical remediation takes to the field. *Environmental Science and Technology, 28*(6): 289A-291A.

USEPA (1989). *Guide to treatment technologies for hazardous wastes at superfund sites.* U. S. Environmental Protection Agency, EPA/540/2-89/052, March 1989.

Velizarov, S.; Crespo, J. G. and Reis, M, A. (2004). Removal of inorganic anions from drinking water supplies by membrane bio/processes. *Reviews in Environmental Science and Bio/Technology, 3*: 361-380.

Velizarova, E.; Ribeiro, A. B. and Ottosen, L. M. (2002). A comparative study on Cu, Cr and As removal from CCA-treated wood waste by dialytic and electrodialytic processes. *J. Hazardous Materials, B94*: 147-160.

Velizarova, E.; Ribeiro, A.B.; Mateus, E. P. and Ottosen, L. M. (2004). Effect of differents extracting solutions on the electrodialytic remediation of CCA-treated wood waste. Part I. Behaviour of Cu and Cr. *J. Hazardous Materials, B107*: 103-113.

Villumsen, A. (1995). Method and apparatus for the decontamination of products containing heavy metals, PCT/DK95/00209, Technical University of Denmark.

Virkutyte, J.; Sillanpaa, M. and Latostenmaa, P. (2002). Electrokinetic soil remediation critical overview. *The Science of the Total Environment, 289*: 97-121.

Virkutyte, J.; Velizarova, E.; Ribeiro, A. B. and Sillanpaa, M. (2005). Copper and chromium electrodialytic migration behaviour in CCA-treated timber waste. *Water, Air, and Soil Pollution, 160*: 27–39.

Yeung, A. T. and Datla, S. (1995). Fundamental formulation of electrokinetic extraction of contaminants from soil. *Canadian Geotechnical Journal, 32*(4): 569-583.

In: Trends in Hazardous Materials Research
Editor: Edward C. Booking, pp. 183-200
ISBN 1-60021-335-9

Chapter 7

FLAMMABILITY CHARACTERISTICS OF ALTERNATIVE GASEOUS FUELS

S. Y. Liao [1], W. D. Shen[1], C. Yuan[1] and D. M. Jiang[2]
[1] College of Chongqing Communication, Shapingba, Chongqing, P.R China
[2]School of Energy and Power Engineering, Xi'an Jiaotong University, P.R China

ABSTRACT

Flammability limits data are essential for a quantitative risk assessment of explosion hazard associated with the use of combustible gas. To obtain the fundamental flammability data for prevention of the hazards in the practical applications, experiments were conducted in a constant volume combustion bomb, for promising alternative gaseous fuels, such natural gas, liquefied petroleum gas, mixtures in air with or without diluent gas. Following the investigation on the flame initiation and its early development, the role of spark energy on the flame initiation of combustible mixtures had been investigated experimentally. Results showed that, a stable flame kernel yields easily with higher spark energy, but generally, spark energy well above the minimum ignition energy, has relatively weak influence on flame propagation, as identified by the fact that the flame propagation speeds become independent upon ignition energy after flame radii greater than a criterion value. The effect of residence time on flammability limit was investigated because of the possibility of preoxidation reactions during the waiting time from mixture adding to spark ignition. On the basis of global chemical kinetics mechanism, a simple method was proposed to determine the preoxidation rate of fuel. With the growing implementation of exhaust gas recirculation in reducing NOx emissions of engine, it is necessary to perform a fundamental research on the flammability characteristics of fuel-air-diluent mixtures. The influences of diluent gas on the flammability region of these alternative gaseous fuels in air were then experimentally explored. It is known that, the flame burning velocity is one of the most relevant parameters of flame chemistry. Thereby, an assumption of critical burning velocity at flammability limit was proposed to approximate the flammability region of these mixtures, as a result, the diluent limits of studied mixture was achieved.

1. INTRODUCTION

Knowledge of the explosion hazards of gaseous fuels is of importance to ensure the safety in industrial and domestic applications that produce or use flammable mixtures. There is no question that the flammability limit is a most widely used index for representing the flame inhibition and the explosion hazard for combustible gases and vapors. In accordance with generally accepted usage, the flammability limits are known as those regions of fuel-air ratio within which flame propagation can be possible and beyond which flame cannot propagate. There are two distinct separate flammability limits for the fuel-air mixture, namely, the leanest fuel limit up to which the flame can propagate is termed as lower flammability limit (LFL), and the richest limit is called as upper flammability limit (UFL).

There is a large array of flammability limits data available for pure fuels in air, such as methane, ethane, propane, butane etc, while there is still scarce for practical multi-composition gaseous fuels, such as natural gas, liquefied petroleum gas, which are being promoted as alternative fuels in practical applications. Meanwhile, the growing implementation of exhaust gas recirculation (EGR) in reducing NOx emissions of engine is also of paramount motivation to perform a fundamental research on the flammability characteristics of fuel-air-diluent mixtures. Although there are so much experimental data on flammability limits for ternary gaseous mixtures of fuel-air-diluent, the diluent gases considered herein were nitrogen gas, carbon dioxide or their mixture, which are obviously different from the real residual gaseous in combustion chamber. Additionally, most of previous studies were conducted at atmospherical conditions. However, for engineering application, the fundamental research on flammability characteristics of fuel-air mixtures with EGR, especially at high temperature and pressure like those of engine combustion, is much more worthwhile and desirable.

In this work, the experiments were made systematically to determine the flammability limits of premixed NG and LPG mixture in air with or without the residual gas in a constant volume combustion bomb, using conventional spark ignition system. Following an investigation of the flame initiation and early development by means of the schlieren technique with a high speed CCD, the role and effect of spark energy on the flame initiation of combustible mixtures had been studied. The flammability limits were measured and the effect of residence time on flammability limit was investigated as well. Because the preoxidation reactions would occur during the waiting time from mixture adding to spark ignition, a simple method was proposed to determine the change of mixture composition on the basis of global chemical reaction mechanism. The influences of EGR on the flammability region of fuel-air-diluent flames were experimentally studied and an assumption of critical burning velocity at flammability limit had been proposed to approximate the flammable region of these mixtures.

2. EXPERIMENTAL DETERMINATION OF FLAMMABILITY LIMITS

Flammability limits are physical-chemical parameters of flammable gases and vapors of flammable liquids, which are related to many factors including the heat losses from the flame by conduction, convection and radiation to the apparatus walls, instabilities in the flame front resulting from buoyant convection, selective diffusion and flame stretch, as well as radical loss or their generation on apparatus walls [1]. Flammability limits of combustible mixtures had been discussed extensively in combustion literatures. A study by Burgess and Hertzberg [2] described the fundamental process that may complicate the measurement of flammability limits in physical apparatus. In the standardized measurements of flammability limits, experiments were usually conducted in the flammability tubes or closed vessels. In most of recent work, it has become desirable to obtain a more fundamental measurement of flammability through the use of an apparatus that minimizes heat losses and allows flame stretch and flame-front instabilities to be characterized. This desire came from the realization that flammability limits are controlled by external (e.g., heat losses or flame stretch) and internal (e.g., preferential diffusion of a limiting component) parameters [3]. Generally, large size of combustion charmer can minimize the wall effects and potentially allow the use of strong igniters to ensure the absence of ignition limitation, so most of the flammability measurements were conducted in the closed chambers in recent years.

Generally, there are several criteria to determine the flammability limits in the experimental measurements of flammability limits. A successful attempt can be determined by one or a combination of the following two criteria: (1) Visualization inspection of the flame kernel development produced by the spark, namely visual criterion. (2) Measurements of the pressure and/or temperature histories in the vessel, where an appropriate pressure or temperature rise criteria can be used to designate flammability rather than the purely visual observation of flame development. The pressure/temperature rise criteria are helpful in the determination of flammability limits in closed vessels. Nevertheless the observation of flame kernel is generally, still widely used, as the observation of flame kernel is directly visualized.

Table 1. Contents of NG considered

Component	CH_4	C_2H_6	C_3H_8	N_2	CO_2	others*
Volume fraction (%)	96.160	1.096	0.136	0.001	2.540	0.067

*The remains will be butane, pentane, sulfureted hydrogen, and water.

Table 2. Contents of LPG considered

Component	C_2H_6	C_3H_6	C_3H_8	C_4H_8	i-C_4H_{10}	n-C_4H_{10}	C_5H_{10}
Volume fraction (%)	0.02	0.92	27.65	1.72	25.68	42.60	1.41

Natural gas and liquefied petroleum gas have been promoted as clean, alternative fuels in many countries. These gaseous fuels selected for the present study is from P. R. China. The detailed contents are listed in tables 1 and 2. The experimental study was conducted in a constant volume combustion bomb, as shown in Figure 1A. The cubic combustion bomb has an inside size of 108x108x135mm, with 1.57 liters in volume. Two sides of this bomb are fixed quartz glasses to make the inside observable, which are to provide the viewing access for the observation of flame growth. The combustible mixture was prepared within the closed vessel by adding gases at the required partial pressures scaled by a mercury manometer whose sensitivity is about 0.13 kPa/mmHg. A thermocouple with accuracy of 1K was used to measure the initial temperature of combustion vessel. The deviation associated with determining flammability limit using this method was controlled within a limited value by accurately scaled the pressure and initial temperature. Two extended stainless steel electrodes were used to form the spark gap at the center of this bomb, to make center ignition. As the early stage development of flame kernel is of importance to explore the initiation dynamics of premixed laminar flames, which is a key to understand the problem of flammability chemistry, the history of the shape and size of the developing flame kernel had been recorded by a REDLAKE HG-100K high-speed CCD camera, operating at 5000 pictures per second with a schlieren optical system, as given in Figure 1B.

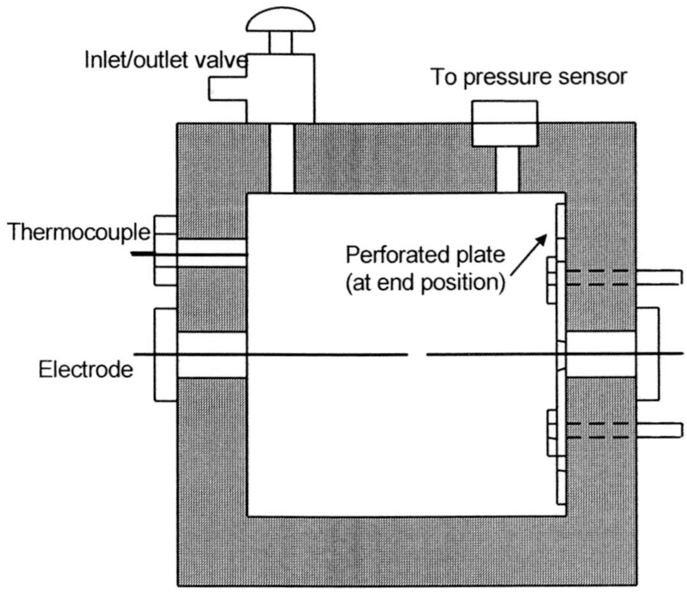

A: Constant volume combustion bomb

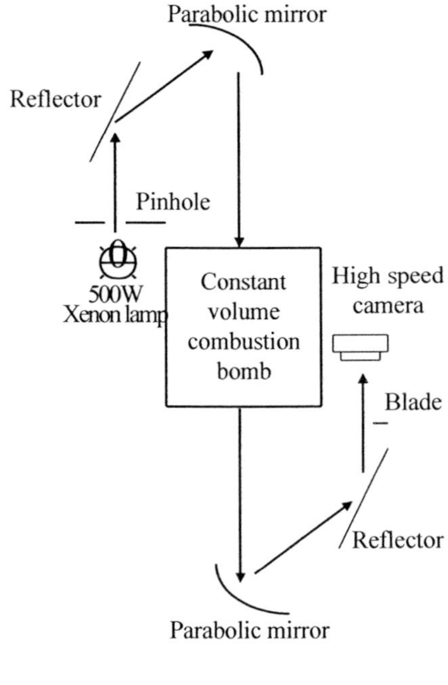

B: Schlieren system

Figure 1. schematic diagram of experimental system

Furthermore, in order to relate the measurements about flammability characteristics to the practical engine conditions, an igniter with a conventional battery-coil ignition system including primary and secondary coils [4], was adopted herein. And based on an assumption of minus losses because of the resistant oscillation in the primary circuit, the spark energy is assumed to be equal to the energy stored in the primary circuit winding in the coil, expressed as

$$Es = 0.5 L_1 I_p^{\ 2}$$

Where Es is the spark energy, L_1 the inductance of the primary circuit winding of the coil, and I_p the charging current in the circuit at the moment of switching off.

As we know, the ignition energy, well above the minimum ignition energy, can lead to very high apparent flame propagation, due to the expansion of the spark plasma and the conductive energy transfer from spark to mixture. Shown in Figure 2 are the cases of flame initiation with different ignition energy. It is obvious that higher spark energy can induce much bigger plasma, which results in a larger flame kernel. After the spark occurs, ignition takes place in the center of the chamber, and the flame propagates spherically to the whole mixture. Shown in Figure 3 is a typical case of flame propagation, it can be seen that the schlieren flame images show a disk whose diameter increases with time. As far as the visualization of flame propagation was concerned, the key aim is to obtain the flame propagating speed. The flame image analysis was then performed to deduce the evolution of flame radius versus time; hence the spatial flame speed can be evaluated as the rate of variation in flame size. In order to characterize the flame propagation, one can plot the evolution of the flame speed in function of the flame radius, as shown in Figure 4. As we know, the role of spark is to initialize a flame, which is to overcome the tendency for the flame to quench because of the high curvature stretch rate and high heat loss to ignition electrode, during the early stage of flame propagation. It is obvious that the flame speeds illustrate a decrease initially, and then increase gradually as the radius increases, which indicates that flame combustion chemistry is weak at the moment of flame kernel initiation, while the heat loss and flame stretch are relatively strong. The effect of ignition energy on the early flame propagation is well plotted in this figure. As we can see that, with higher spark energy, the initial flame speed is faster and the transition stage from flame initiation to stable flame propagation is much shorter. However, the flame propagation speeds become independent of ignition energy after flame radii greater than about 6 mm and this speed becomes the practical flame speed. That is to say, spark energy has relatively weak influence on the whole process of flame propagation.

The data listed in tables 3 and 4 are the results of tests on the quiescent NG and LPG mixtures in air at ambient pressure and temperature. The flammability region of quiescent NG-air mixtures is from 5.0% to 15.6% of NG by volume, corresponding to the equivalence ratio from about 0.486 to 1.707. And that is from 1.84% to 8.54% of LPG, corresponding to the equivalence ratio of about 0.54 and 2.52. In order to validate our experimental measurements, a comparison was made with the available flammability limit data, also shown in Tables 3 and 4. It can be seen that, the result of NG-air is in agreement with those of previous works for pure methane in air, as the concentration of methane in this NG is over

96%. This LPG result falls between those of propane-air and butane-air mixtures, which also indicate that the experimental uncertainty is acceptable.

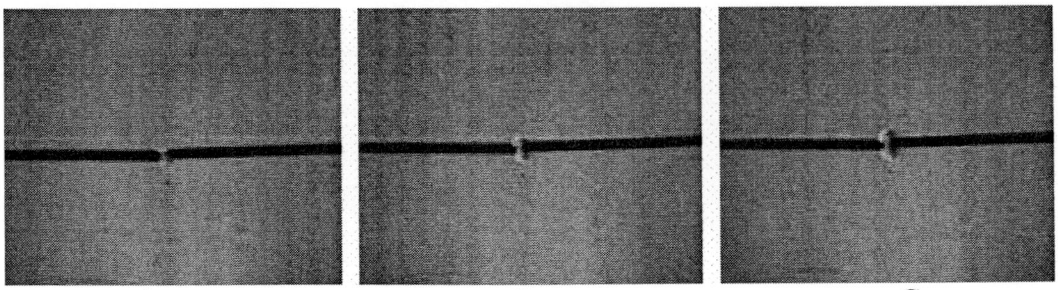

Figure 2. Flame kernel induced by various spark energy for NG-air mixtures with $\phi = 0.6$. A: 25mJ; B: 45mJ; C: 65mJ.

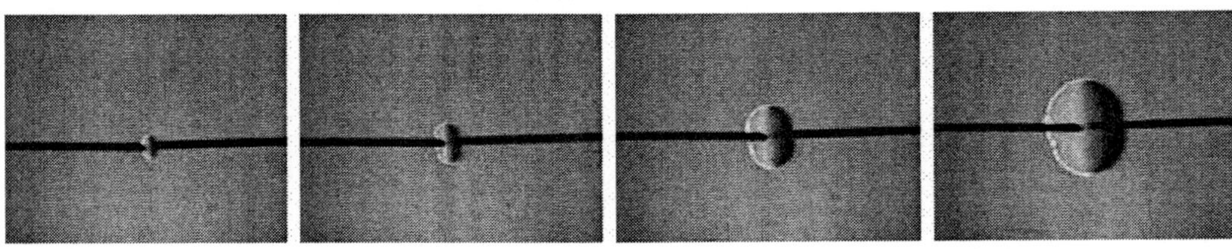

Figure 3. Typical growing schlieren flame kernels for NG-air mixtures with $\phi = 0.6$, the time interval is 4ms.

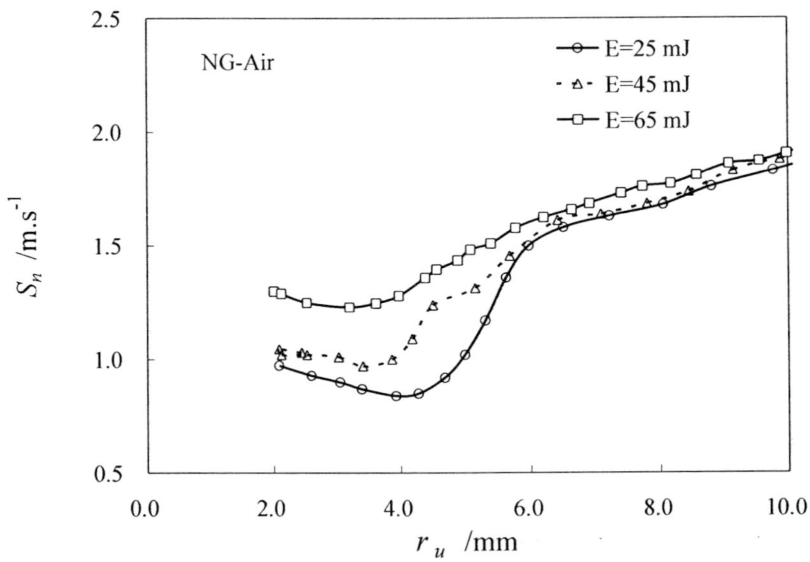

Figure 4. Variations of flame speed S_n with flame radius, r_u for NG-air mixtures with different ignition energy

As to the lower flammability limit of multiple component fuels mixture in air, the LFL value of the mixture can also be estimated by LeChatelier's rule, given as

$$LFL = \frac{100}{\sum(\frac{C_i}{LFL_i})} \quad (1)$$

Where LFL is the mixture lower flammability limit (*vol%*), C_i is the concentration of component i in the gas mixture on an air-free basis (*vol%*), and LFL_i is the lower flammability limit for component i (*vol%*). The estimation of LeChatelier's rule is listed in these tables as well. Generally, acceptable agreement can be found between them.

Table 3. Flammability limit data (in volume %) for methane-air and NG-air mixtures

Mixtures	Test conditions	LFL (Vol.%)	UFL (Vol.%)
NG-air[a]	1.57 L Chamber	5.0	15.6
	LeChatelier's rule	4.98	-
Methane-air	8 L chamber [b]	5.0	-
	20 L chamber[c]	4.9	15.9
	120 L chamber[d]	5.0	15.7
	25.5 m^3 sphere	4.9[e] 5.1 ± 0.1[f]	-
	Flammability tube[g]	5.2	15.1
	Flammability tube[h]	4.9	15.0

[a] Liao et al [5].
[b] Hertzberg & Cashdollar, 1983 [6].
[c] Kenneth L. Cashdollar et al. 2000 [6].
[d] Kenneth L. Cashdollar, 2000, [6].
[e] Burgess et al . 1982 [6].
[f] Furno et al., 1970 and Burgess et al . 1982 [6].
[g] Liao et al, 1996 [7].
[h] Kuchta, 1985[6].

Nowadays, the method of exhaust gas recirculation has being widely used for it has being regarded as an effective way to decrease NOx emissions in combustion engine. Experimental measurements were performed to investigate the flammability limits of diluting gaseous mixtures by adding combustion products or inert gas. Shown in Figure 5 and Figure 6 are the results of flammability limits over wide diluent ranges for the alternative fuels considered herein. It is known that a dependence of flammability limits of various fuels on diluent concentration, which restricts flammability region of ternary gaseous mixtures (fuel, oxidizer and diluent) in the form of a peninsula [9]. And the results obtained by Liao et al. [7] using tubular burner and Coward and Jones [8] using the Bureau of Mines apparatus, are also plotted in this figure. It can be noted that the increase of diluent ratio results in the lessening of flammability region. The reason is known as the addition of dilutions decreases the flame temperature, consequently decreases the burning velocity, so the flammability limits becomes narrow.

Table 4. Flammability limit data for LPG-air mixture

Mixtures	Stoichiometric concentration (vol. %)	Flammability limits (vol. %)
LPG-air	3.39	1.83-8.54
Butane[8]	3.12	1.8~8.4
Propane-air[8]	4.02	2.1-9.5

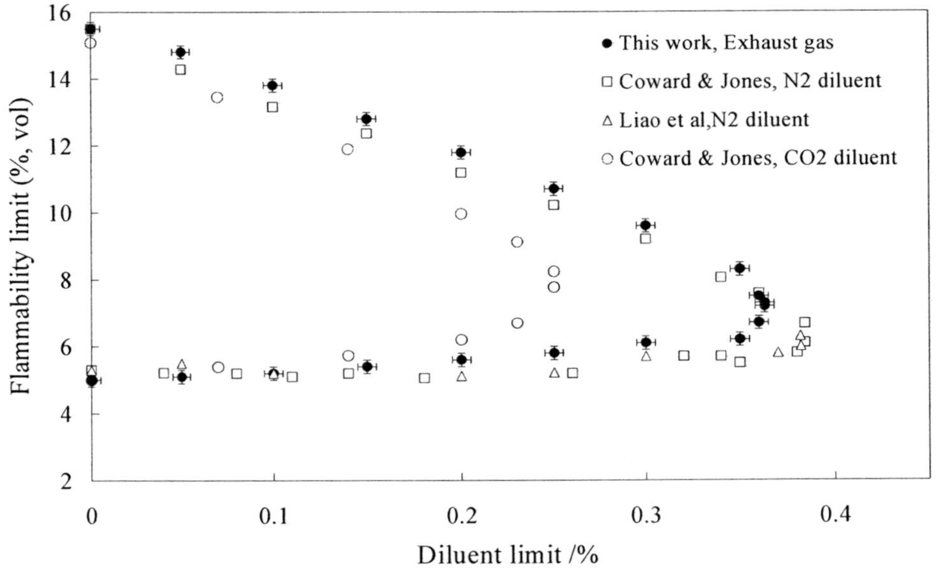

Figure 5. Flammability limits of NG-air mixture diluting by combustion products

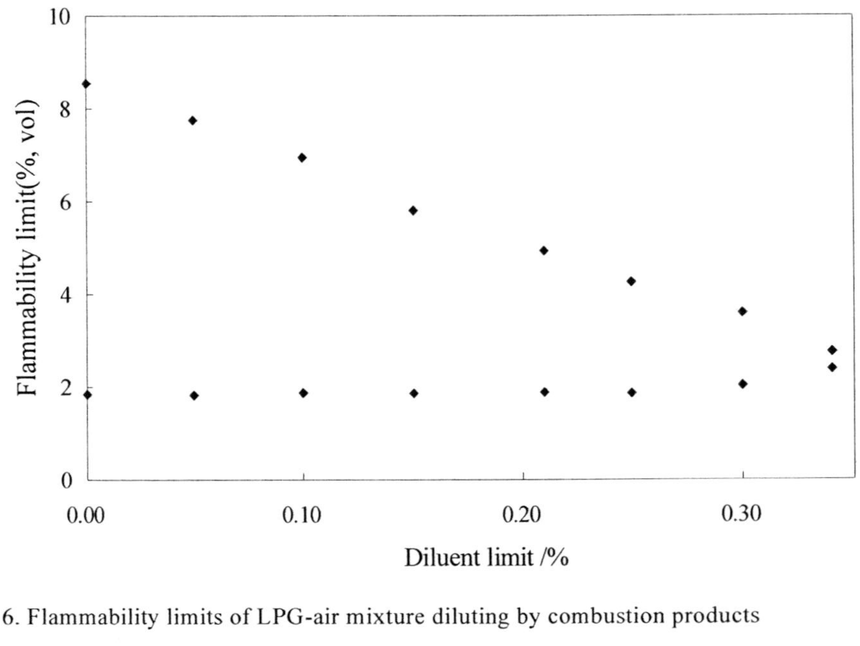

Figure 6. Flammability limits of LPG-air mixture diluting by combustion products

2. Flammability Limit at High Temperature

It is well-known that the preheated temperature has important influence on the flammability limits. Coward and Jones [8] reported that the flammability limits of hydrocarbon-air mixtures show a linear dependence between the limiting concentration and temperature, while certain authors also proposed some theoretical and experimental arguments in flavor of a nonlinear dependence. As we know, the standard measurement of flammability limits is generally, carried out under quiescent condition, however, it would take time to implement the mixing process of gaseous fuel with air. If the quiescent mixture tested is prepared at relatively high temperature, the preoxidation reaction would occur, and there also exists the possibility of surface reactions taking place. That is to say, the change of the mixture composition and state, are possible at temperatures lower than spontaneous ignition temperature, during the preparation of homogenous mixture. Namely, the great difference in the above two opinions maybe can be attributed to the effect of residence (waiting) time before ignition.

It was reported [1] that the rate of a low-temperature reaction is constant in time for given composition and is higher for rich mixture. The upper flammability limits of NG and LPG in air had been investigated in this work over a range of high temperature, and the results are shown in Figure 7, where the residence time of mixture preparation is 10 minutes. Experimental results show that the upper flammability limits slightly increase with the preheated temperature, and an approximate linear relation exists between the temperature and the increasing rich flammability limit. This can be approximately expressed by vanderstraeten rule [10] as

$$UFL(T_1) = UFL(T_0)\left[1 + c(\frac{T_1 - T_0}{100})\right] \tag{2}$$

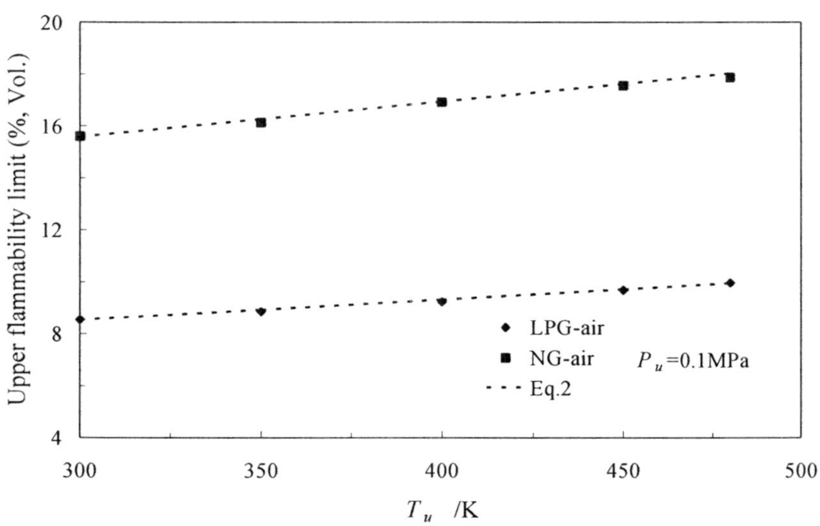

Figure 7. Temperature dependence of the upper flammability limits for NG and LPG in air, where the experiments remaining the residence time of 10 minutes.

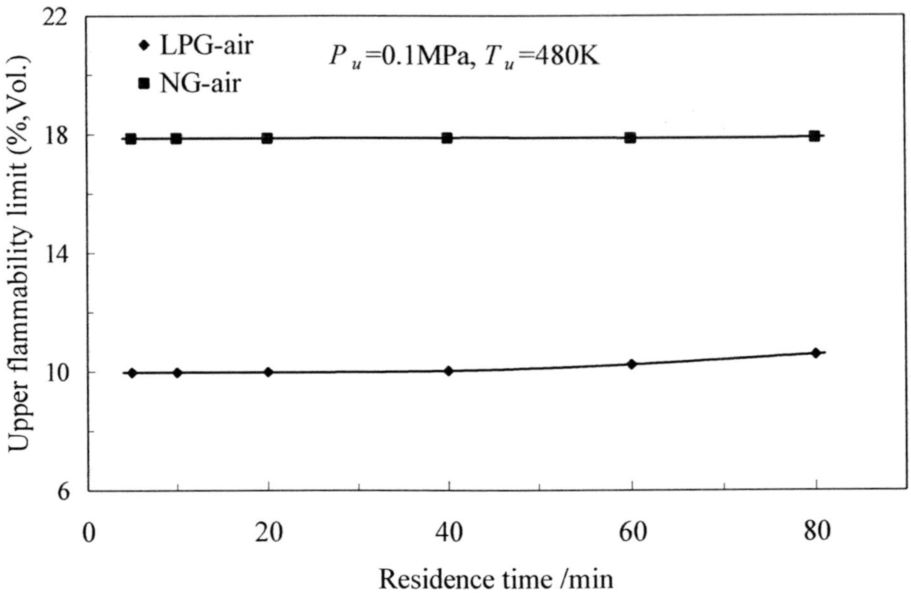

Figure 8. The upper flammability limits as function of the residence time for these NG and LPG in air at 480K.

Here constant *c* indicates the increase in UFL with temperature. For this NG-air, *c* equals to 0.0867, while *c* is 0.0923 for LPG-air.

As mentioned above, the preflame reactions could take place within high temperature mixture, because it is of possibility to change the mixture composition, and consequently to affect the accuracy in measuring flammability limit. The effects of residence time on the upper flammability limits for these two gaseous fuels in air had been studied at 480 K, and the results are plotted in Figure 8. It is shown that the residence time has a weak effect on the upper flammability limits of NG-air mixture at that temperature. While the UFL of LPG-air shows a slight increase with residence time. Wierzba and Ale reported [11] that the rich flammability limits of hydrogen in air determined experimentally in a conventional stainless steel test tube at elevated temperatures were affected very significantly by the time of exposure of the mixture to these elevated temperatures before spark ignition. Bunev [1] investigated the effect of preoxidation on flammability limits of hydrogen and methanol in air. He found that the rich limit of hydrogen-air mixtures was essentially dependent on the residual time for temperatures higher than 513K, while in the temperature range below513K the rich limit was virtually independent of the preoxidation. In previous work, the change in flammability limit with residence time can be attributed to the change of mixture composition because of the low-temperature oxidation reactions. However, few experiment or estimation had even been implemented to analyze to the fraction of fuel composition before ignition, although the possibility of preoxidation of the fuel in the course of preparation of mixture was emphasized. It is known that, detailed chemical kinetic modeling optimized for pre-ignition and auto-ignition conditions was usually used in calculations to evaluate the gas phase reactions. However, for hydrocarbon fuels the elementary reaction schemes are so complex that it is usually not feasible to consider all the chemically reacting species and their reaction rates in analyzing practical combustion system. Thus, in many cases single-step global reaction scheme has received renewed interest in recent year, although it would overestimate the heat release rate of combustion system [12].

The single-step global equation is written as

$$Fuel + \alpha O_2 \rightarrow \beta CO_2 + \gamma H_2O \tag{3}$$

where α, β, γ are stoichiometric coefficients. And the associated global rate expression is expressed as

$$r_{fuel} = \frac{d[fuel]}{dt} = -AT^n p^m \exp(-\frac{E}{RT}).[fuel]^a [O_2]^b \tag{4}$$

Where n, m are generally considered as $n = m = 0$ for many cases, and the global constants a, b can be found in [12].

The calculated results of the extent of fuel consumption during the residence time can be seen in Figure 9. It is shown that the preoxidation rate of NG is nearly close to zero, while that of LPG shows a slightly increase at 480K, which gives the evidence that the NG-air flammability limits have no virtually dependence on the residence time. For comparison, calculations for methanol and methane were implemented as well. It was reported that the flammability limit of methanol can be significantly influenced by the residence time at temperature higher than about 530K [1], which can also be justified by the calculated result, as shown in Figure 9 that at 530K, however the low-temperature oxidation rate of methanol is apparently increased with the increase of residence time. The low-temperature preoxidation of methane is weak, even at 620K, the preoxidated methane has still no obvious increase with the residence time, this is well in agreement with the previous measurements [11].

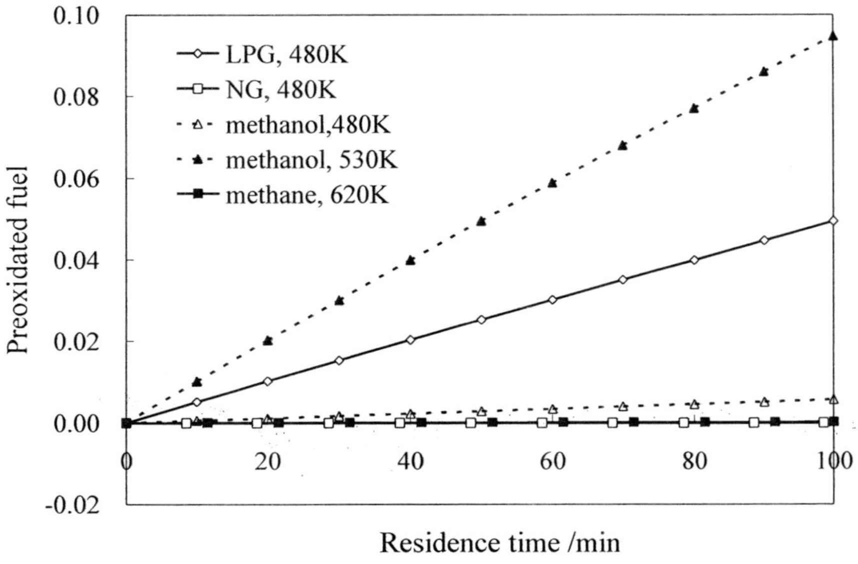

Figure 9. Calculated preoxidation of fuels as a function of the residence time

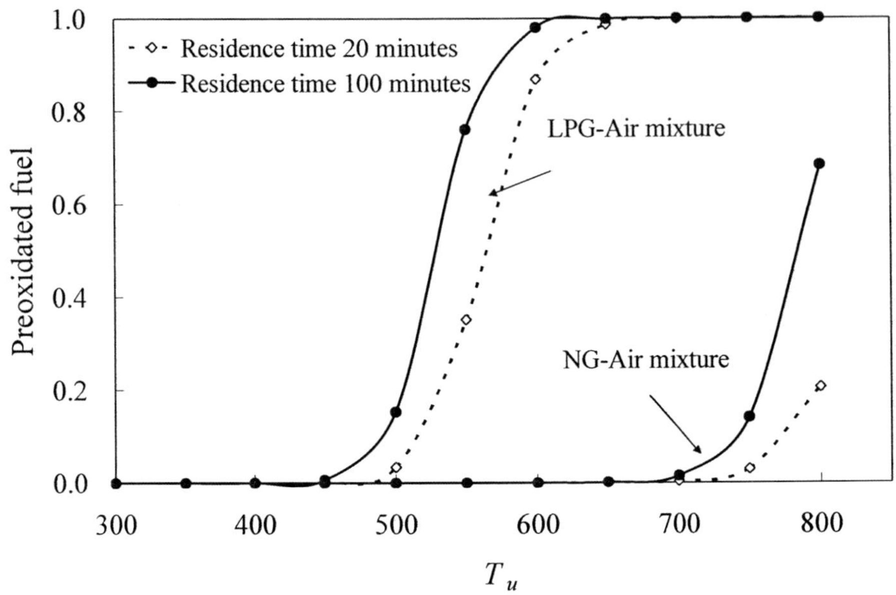

Figure 10. Calculated preoxidation of fuels as a function of the preheated temperature.

Figure 10 gives the preoxidation of fuel with the variation of preheated temperature. It is understandable that increasing temperature can enhance the preflame reactions. The spontaneous ignition temperature of methane is 800K, while those of propane and b-butane are 745K and 638K respectively. In this figure, it can be seen that, the temperature at which the preoxidation is to occur, is dependent upon the type of fuel. As to NG, whose main composition is methane, this temperature is about 700K for a long waiting time of 100 minutes, while to LPG, low-temperature would take place at temperature about 450K. The result also shows that the longer the residence time the more fuel was consumed.

4. Diluent Limits for Alternative Fuels

Although the flammability characteristics of chemical substances had received continuous research in recent years, experimental values of these properties are not always available, and generally they are scarce and expensive to obtain, especially for ternary gaseous mixtures of fuel-air-residual gas. Since experimental values are not available and determining them by experimental means is not practical, a prediction method which is desirably convenient and fast must be used to estimate them. A commonly accepted view is that flames fail to propagate as the burning velocity becomes too low to overcome the dissipation processes during combustion [13]. Burgess and Hertzberg [2] emphasized that at least, burning velocity at the lean limit would tend to be the approximate value for many fuels. Lovachev et al [1] predicted that 5-7 cm/s is the minimum possible flame speed for lean limit hydrocarbon flames, and Huang et al [14] found that the laminar flame speed at the maximum diluent level is on the order of 1-2 cm/s at elevated pressures. Blint [15] calculated laminar burning flame speeds for adiabatic one-dimensional propane/air flames over a range of pressures, initial temperatures, and diluent levels, and an arbitrary flame speed (10 cm/s) was used to determine flammability limit.

Up to now, the laminar flames for fuel-air-diluent mixtures had been well-explored. And these formulas about laminar burning velocity were well available in the literatures. As mentioned above, since it is a commonly accepted view that flame extinction would occur when the burning velocity becomes too low to overcome the dissipation processes during combustion, it is hence desirable to choose a criteria value of flame burning velocity for the quick estimation of flammability limits.

Generally, the burning velocity of laminar flame u_l, were usually explicitly formulated as a function of the initial pressure P_u, initial temperature T_u ,equivalence ratio ϕ and diluent ratio ϕ_r, described as,

$$u_l = f_1(P_u, T_u, \phi, \phi_r) \tag{5}$$

Hence, for given values of ϕ_r, p_u, and T_u, the equivalence ratios at the flammability limits, ϕ_{LFL} and ϕ_{UFL} (corresponding to the upper flammability limit UFL and the lower one, LFL respectively) can be deduced through resolving Eq.5,

$$\phi = f_2(P_u, T_u, u_{l,cr}, \phi_r) \tag{6}$$

where $u_{l,cr}$ is the critical burning velocity, selected from the assumption that below the value the flame cannot propagate. Thereby, the flammability limits of fuel-air-diluent mixture can be obtained by means of the bisection solution to nonlinear functions above. In this study, the burning velocity data of NG-air- diluent and LPG-air-diluent mixtures was derived from [16, 17].

Shown in Figure 11 are the flammability characteristics of methane-air-nitrogen flames. There are three critical burning velocity criteria, i.e. 1cm/s, 5cm/s and 8cm/s, selected to determine the flammability region of mixtures considered. It can be seen that, experimental results fall within the predictions of critical values of 1 cm/s and 8 cm/s. Generally, the whole agreement is obtained by using criterion value of 5 cm/s, and the derivation between the measured data and prediction is within $\pm 10\%$. The diluent limit of 5 cm/s is 0.364, 5.2% derivation against 0.384 of the measurement. However, for rich flames, the prediction underestimates the flammability region, even with a criterion velocity of 1 cm/s, due to the invalidation of fitting burning velocity formula for rich flame. The measurement of Liao et al [7] reported that the UFL of methane-air flame without diluent is about 15% volume fraction, corresponding to the equivalence ratio of 2.04. While the empirical relation on methane-air burning velocity, $u_l = (-150.84\phi^3 + 287.6\phi^2 - 96.327\phi - 1.2924)(1 - 1.208\phi_r^{0.803})$ [18], was validated within the equivalence ratio ranges from about 0.46 to 1.46.

The results of NG-air flames with EGR are presented in Figure 12 [19]. Generally speaking, the measured data are consistent with the prediction of Shebeko method [9] over the diluent ranges, and this still reveals that critical velocity criteria are not suitable to determine the rich flammability limit (UFL). By comparison between experiment and prediction, an appropriate critical velocity was defined to be 5 cm/s herein. The results of LPG-air flames with EGR are presented in Figure 13. In this figure, it appears that the diluent limit just lies around CO stoichiometric curve, which also gives an approximately justification of our work. It is also shown that Shebeko method can give better prediction to the experimental results at both lean and rich flammability limit than does the method based on a critical burning velocity, as shown in Figure 12. However, it is difficult to investigate the effect of temperature and pressure on flammability limit using Shebeko method, as the complex of combustion chemistry at high temperature and pressure conditions.

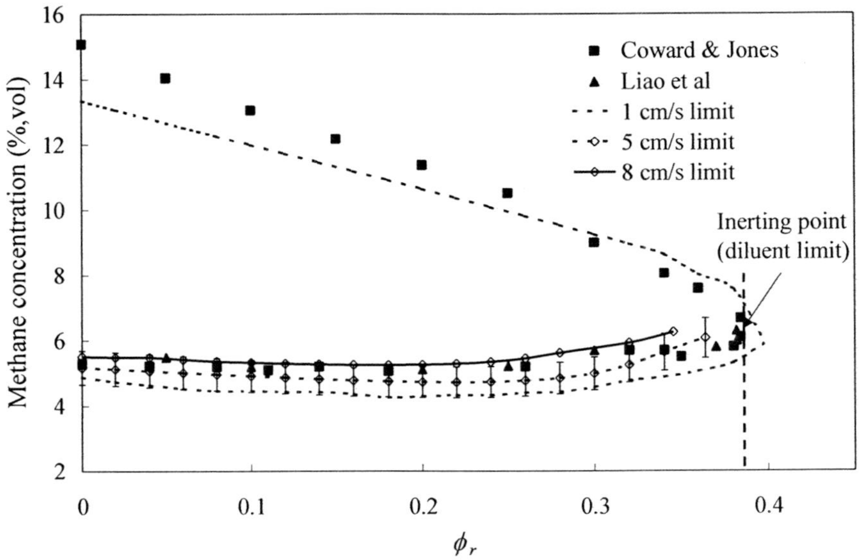

Figure 11. Flammable region of methane-air-nitrogen mixture at 300K and atmospheric pressure

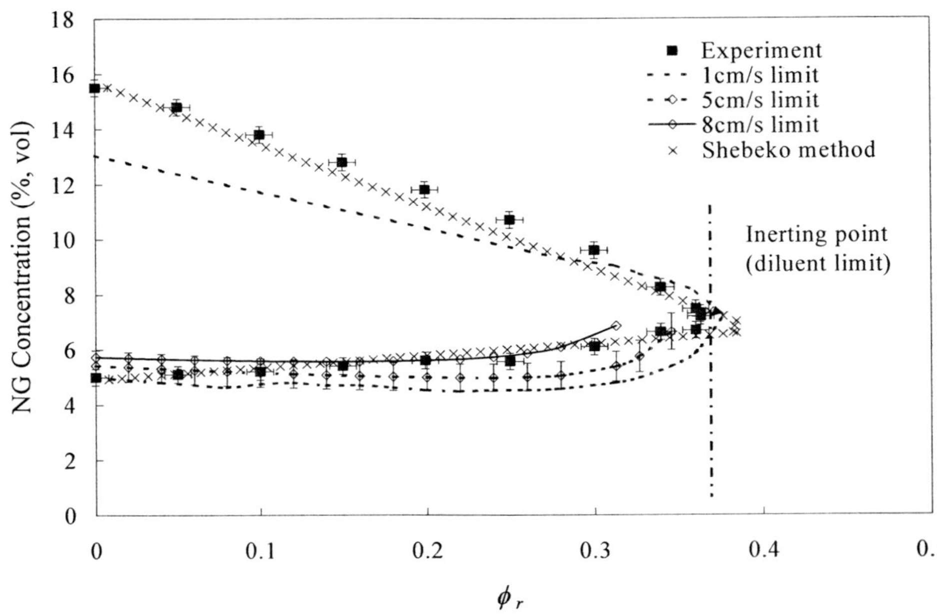

Figure 12. Flammability region of NG-air-diluent mixture at 300K and atmospheric pressure

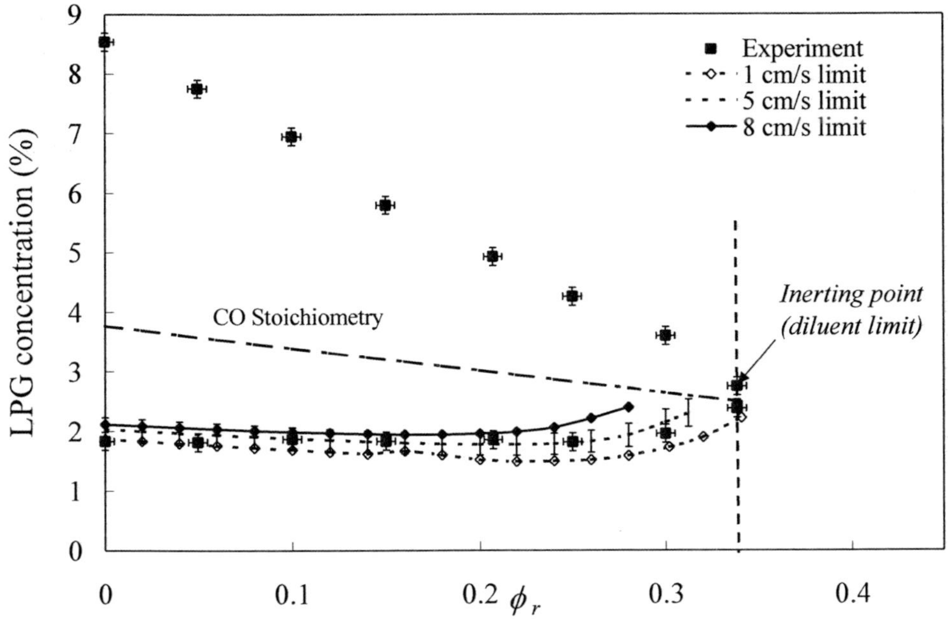

Figure 13. Flammablity characteristics of a LPG-air-diluent mixture at 300K and atmospheric pressure

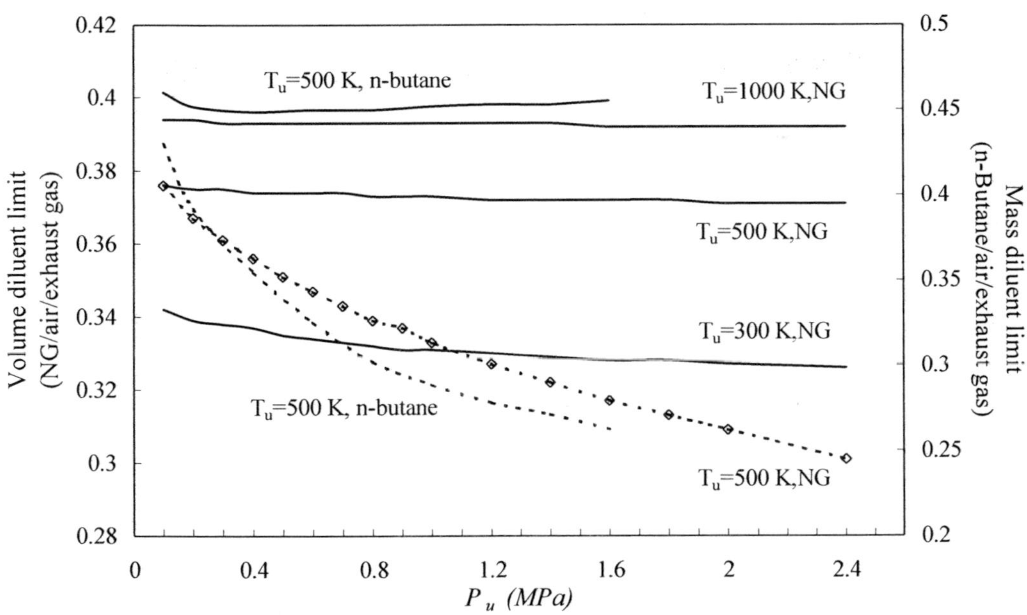

Figure 14. Computed diluent limit of EGR NG-air mixture at various initial temperatures and pressures, dashed curves were derived from fixed flame speed criteria and solid cures from pressure dependence criteria

For method of critical burning velocity, the advantage is that one can approximate the inerting point without an excessive amount of experiments when the burning velocities needed are available. In principle, the pressure and temperature dependences of the flammability characteristics on temperature and pressure could be well established. As reviewed by Lovachev [1], the dependence of burning velocity at the flammability limit on pressure P can be simplified as, $u_{l,\lim} \sim u_{l0} P^{-1/3}$, where subscript 0 denotes the reference conditions (300K and 0.1 MPa). Figures 14 and 15 present the inerting EGR rate for various initial pressures and temperatures. As expected, the inerting value increases with increasing initial temperature. The limits obtained by fixed 5 cm/s critical velocity prediction and pressure dependence criteria are also presented. It shows that the difference of diluent limits between theses two criteria becomes more obvious with the increase of pressure. Similarly, the analogous analysis [14] for n-butane/air/residual gas flames (mass diluent limit) had been reproduced in this figure, a comparable phenomenon were observed as well. It must be noted that, the flame speed of 5 cm/s may be so low and only related to lowest loads or idle conditions of EGR engines. Because the causes of extinction at limits are very complex, it is believed that the further researches and discussions would be needed and useful for better understanding of thermodynamics and chemical kinetics of flammability limits and better improving EGR applications in engines.

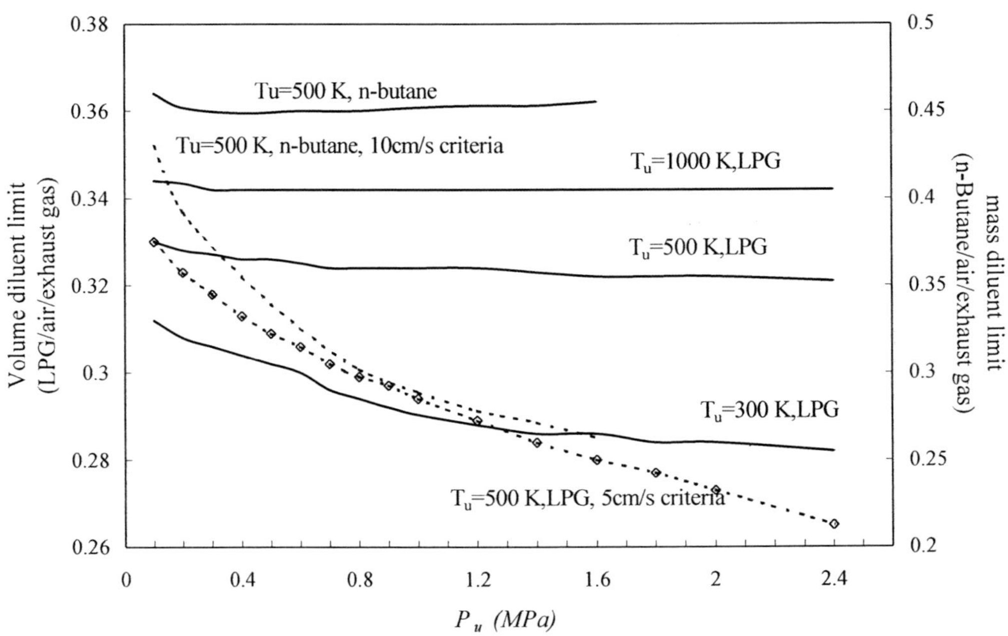

Figure 15. Computed diluent limit of EGR LPG-air mixture at various initial temperatures and pressures, dashed curves were derived from fixed flame speed criteria and solid cures from pressure dependence criteria

5. Conclusion

The flammability limit is a most widely used index for representing the flame inhibition and the explosion hazard for combustible gases and vapors. In order to obtain the flammability characteristics of alternative gaseous fuels, such natural gas and liquefied petroleum gas in air, experiments had been carried out systematically in a constant volume combustion bomb. The results show that the flammability region of quiescent NG-air and LPG-air mixtures is from 5.0% to 15.6% of NG and from 1.84% to 8.54% of LPG by volume respectively. The temperature dependence of rich flammability limits for NG and LPG in air can be formulated as $UFL(T_1) = UFL(T_0)\left[1 + c(\frac{T_1 - T_0}{100})\right]$, where c equals to 0.0867 for NG-air, and 0.0923 for LPG-air. The low-temperature reaction calculation had been implemented to investigate the effect of residence time on the flammability limit, on the basis of single-step chemical reaction mechanism. The influences of EGR on the flammability region of fuel-air-diluent flames were experimentally studied as well. An assumption of critical burning velocity at flammability limit was proposed to approximately determine the flammability region of these mixtures, hence based on this assumption, an estimation of the flammability map for NG/LPG-air-diluent mixtures was obtained by using the empirical formula of burning velocity data. Good agreement of diluent limit between measurements and predictions had been obtained, and this was verified by the calculations reported previously as well.

ACKNOWLEDGMENT

This work is supported by the state key project of fundamental research plan of P. R China (No. 2001CB209208), the key project of NSFC (NO. 50156040), and the NSFC Awarding Fund for Excellent State Key Laboratory (NO.50323001)

REFERENCES

[1] Lovachev, L. A.; Babkin, V. S.; Bunev, V. A.; Výun, A. V.; Krivulin, V. N. and Baratov, A. N. Flammability Limits: An Invited Review. *Combust. Flame* 1973, 20: 259-289

[2] Burgess, D.; Hertzberg, M. The Flammability Limits for Explosions Hazards. Instrument Society of America, Analysis Instrumentation Symposium, Pittsburgh, Pennsylvania, May 12-15, 1974.

[3] Takahashi, A.; Urano, Y.; Tokuhashi K. and Kondo S. Effect of vessel size and shape on experimental flammability limits of gases. *Journal of Hazardous Materials*. A105 (2003) 27-37.

[4] Shen, H. X.; Jiang, D. M. Investigation on the Flame Initiation and Early Development in a Spark Ignition Engine. SAE Paper 922239, 1992.

[5] Liao, S. Y.; Jiang, D. M. Experimental Study of Flammability Limits of Natural Gas-Air Mixture. *J. Hazardous Material,* 2005, B19: 81-84.

[6] Kenneth, G. L.; Isaac, Z. A.; Gregory, G. M.; Thomas, R. A.; Hertzberg, M. Flammability of methane, propane, and hydrogen gases. *Journal of Loss Prevention in the process Industries* 13(2000) 327-340.

[7] Liao, C.; Saito, N.; Saso, Y.; Ogawa. Y. Flammability Limits of Combustible Gases and Vapors Measured by a Tubular Flame Method. *J. Fire Safety.* 1996, 27: 49-68.

[8] Coward, H. F.; Jones. G. W. Limits of flammability of Gases and Vapors, Bureau of Mines Bulletin 503, Washington, DC, 1952.

[9] Shebeko, Yu. N.; Fan, W.; Bolodian, I. A.; Navzenya, V. Yu. An analytical evaluation of flammability limits of gaseous mixtures of combustible-oxidizer-diluent. *J. Fire Safety* 2002, 37: 549-568.

[10] Vanderstraeten, B. Experimental Study of the Pressure and Temperature Dependence on The Upper Flammability Limit of Methane-Air Mixtures. *Journal of Hazardous Material* 56 (1997) 237-246.

[11] Wierzba, B. I.; Ale, B. Rich Flammability Limits of Fuel Mixtures Involving Hydrogen at Elevated Temperatures. *International Journal of Hydrogen Energy* 25 (2000) 75-85.

[12] Borman, G. L.; Ragland, K. W. *Combustion Engineering.* The McGraw-Hill Companies, Inc. USA, 1998.

[13] Macek, A. Flammability limits: a re-examination. Combust. *Sci. Tech,* 1979, 21: 43-52.

[14] Huang, Y.; Sung, C. J.; Eng, J. A. Dilution Limits of n-Butane/air Mixtures under Conditions Relavant to HCCI Combustion. *Combust. Flame* 2004, 136: 457-466.

[15] Blint, R. J. Flammability Limits for Exhaust Gas diluted Flame. *Proc. Combust. Inst.* 22(1988) 1547-1554.

[16] Liao, S.Y.; Jiang, D. M.; Gao, J.; Huang, Z. H. Measurements of Markstein Numbers and Laminar Burning Velocities for Natural Gas-Air Mixtures. *Energy & Fuels* 2004, *18,* 316-326

[17] Liao, S.Y.; Jiang, D. M.; Gao, J.; Huang, Z. H. Measurements of Markstein Numbers and Laminar Burning Velocities for Liquefied Petroleum Gas–Air Mixtures. *Fuel,* 2004, 83 (10), 1281-1288.

[18] Stone, R.; Clarke, A.; Beckwith, P. Correlations for the Laminar Burning Velocity of Methane/Diluent/Air Mixtures Obtained in Free-Fall Experiments. *Combust. Flame* 1998, 114: 546-555.

[19] Liao, S.Y.; Jiang, D. M.; Huang, Z. H; Zeng K. Approximation of Flammability Region for Natural Gas-Air-Diluent Mixture. *Journal of Hazardous Materials ,* 2005 , A125 (2005) 23–28.

In: Trends in Hazardous Materials Research
Editor: Edward C. Booking, pp. 201-217
ISBN 1-60021-335-9

Chapter 8

AGRICULTURAL PESTICIDES POISONING IN SUB-SAHARA AFRICA: PAST RESEARCH AND FUTURE NEEDS

Olurominiyi Ibitayo*
Urban Planning and Environmental Policy, School of Public Affairs, Texas Southern University, TX, USA

ABSTRACT

The current Population Growth Rate (PGR) for Sub-Sahara Africa (3 percent) is by far the highest for any region in the world. This high population growth is particularly ominous regarding food demands if the rapid rate of movement of people from the rural area, the site of food production, to urban areas is considered. In view of the need for self-sufficiency in food production, and the expansion of export agriculture, several countries in Sub-Sahara Africa have been aggressively promoting the use of agricultural inputs such as fertilizers and pesticides.

Agricultural pesticides are, however poisons by design as these chemicals are intended for use in *destroying* any pest that may interfere with the production or the processing of agricultural products. The capacity of pesticides to *destroy* is however not limited to pests but may result in serious risks to human health and to the environment especially. While occupational agricultural pesticide poisoning may be universal, the incidence of such poisoning or contamination is much higher in developing countries.

This paper which provides an overview of past research on the use/misuse of agricultural pesticides in Sub-Sahara Africa includes the following: the upsurge of the use of agricultural pesticides in the region; factors of pesticide poisoning or contamination; adverse health effects of pesticide poisoning including self-reported effects and the extent of these adverse effects, and the adverse effects pesticide contamination on the environment In contrast to the zero-risk approach favored by some researchers, this paper emphasizes the need to reduce agricultural pesticide poisoning in Sub-Sahara Africa through pesticide risk communication and education programs, provision of government

---

* Correspondence address: tel: 1+713-313-7403; fax 1+713-313-7447; *ibitayo_oo@tsu.edu*

subsidy for the purchase of less toxic but usually more expensive pesticides, discontinuation of calendar-based pesticide application, delimiting the importation of highly toxic agricultural pesticides, and reassessing the need for *special* pesticide protective clothing for farmers working in the *hot tropical climate* of Sub-Sahara Africa.

INTRODUCTION

The current Population Growth Rate (PGR) for Sub-Sahara Africa (3 percent) is by far the highest for any region in the world. In spite of some efforts at family planning, the relatively high rate of population growth is likely to continue because as much as 44 percent of the population is less than 15 years of age, compared with 17 percent for Europe (Population Reference Bureau, 2004). This high rate of population growth is particularly ominous regarding food needs if the rapid rate of movement of people from the rural areas, the site of food production, to urban areas is considered. Since the rural-urban migration usually involves the more educated young men, the burden of increased food production is being borne by fewer, older/aging, illiterates or barely literate individuals (Echobichon, 2000).

Under these conditions of high population growth and decreasing younger and educated agriculture labor force, the dire predictions of Malthus of famine, starvation and death seem feasible. The prediction has not occurred especially in developed countries largely because of technological advances in agriculture that has led to substantial increase in agricultural productivity (Bilsborrow and Geores, 1994). The options that are available to sub-Sahara African countries regarding the increased food needs include the extensification of agriculture – bringing new land into cultivation, reducing the fallow period, and the incorporation of agricultural inputs designed to increase productivity and reduce crop loss during storage or transportation. Such inputs include fertilisers, and pesticides – herbicides, insecticides, fungicides, fumigants and others.

The potential for extensification of agriculture is highly constrained in several Sub-Saharan countries because of the limited availability of arable land (Bilsborrow and Geores, 1994). This option is especially less feasible in Zimbabwe and South Africa because of the race-based land ownership system. In South Africa, Blacks account for 75 percent of the population but owns 13 percent of the land thereby resulting in an average of 15 hectares of arable land per Black/subsistence farmer (Schonfield, 1994). The recent effort at redresssing the inequitable distribution of land in Zimbabwe has been vehemently condemned by the international community. Also, bringing new land into cultivation may lead to the cultivation of fragile steep hillsides and semi-arid areas, and result in deforestation, soil degradation and eventual reduced productivity in several Sub-Sahara African countries.

Reducing fallow period - an element of the intensification of agriculture will eventually result in reduced soil fertility. The option being promoted by most developing nations including Sub-Sahara Africa therefore is the increased application of agricultural inputs such as fertilizers and pesticides (Amr, 1989; Karlsson, 2004; Thrupp, 1990; Weselling., McConnell, Partanen, and Hogstedt, 1997). The use of agricultural inputs seems validated by the recent observation that without such chemical inputs, the yield trends over a 41-year period for millet and cassava in four countries in Sub-Sahara Africa are one percent or less (Mortimore, 2003).

Pesticides are substances or a mixture of substances "intended for preventing, destroying, or controlling any pest, including vectors of human or animal disease, unwanted species of plants or animals causing harm during, or otherwise interfering with the production, processing, storage, transport, or marketing of food, agricultural conditions, wood and wood products, or animal feedstuffs, or which may be administered to animals for the control of insects, arachnides or other pest in their bodies" (FAO, 1986 p. 28). Agricultural pesticides include plant growth regulators, defoliants, fruit thinning substances, and substances applied to crops to prevent deterioration during transportation and storage (Al-Saleh, 1994). Pesticides can be classified on the basis of their persistence - a measure of the pesticide's lifetime effectiveness. Non-persistent pesticides usually have lifetime effectiveness of several days to about 12 weeks; moderately persistent ones last one to eighteen months; persistent pesticides which include DDT, aldrin and dieldrin last up to 20 years; and permanent pesticides including mercury and lead last indefinitely (Al-Saleh, 1994). While lifetime effectiveness is a measure of the time period that a pesticide will be effective in destroying a "pest", it is also a measure of how long the pesticide will remain active and potentially adversely affect human health and/or the environment.

This paper which provides an overview of past research on the use/misuse of agricultural pesticides in Sub-Sahara Africa, includes the following: the upsurge of the use of agricultural pesticides in the region; factors of pesticide poisoning or contamination; adverse health effects of pesticide poisoning including self-reported effects and the extent of these adverse effects, and the adverse effects pesticide contamination on the environment In contrast to the zero-risk approach favored by some researchers, this paper emphasizes the need to reduce agricultural pesticide poisoning in Sub-Sahara Africa through pesticide risk communication and education programs, the discontinuation of inflexible, calendar-based application of pesticides, provision of government subsidy for the purchase of less toxic but usually more expensive pesticides, selection of disease resistant hybrids, implementation of more stringent international agreements on exportation of highly toxic pesticides, and reassessing the need for *special* pesticide protective clothing for farmers working in the *hot tropical climate* of Sub-Sahara Africa.

The Green Revolution in Sub-Sahara Africa

The role of pesticides in boosting agricultural productivity is well-documented. The chemical control of insect and fungal pests of potato in the U. S. in the 1940s, the pre and post application of insecticides to rice in Japan in 1953, and the chemical control of boll weevil and boll worm in 1953 and 1970 respectively in El Salvador led to substantial increased yields or vastly reduced losses of the respective crops (Echobichon, 2000). In view of these positive results and the need for self-sufficiency in food production, and the expansion of export agriculture, several countries in Sub-Sahara Africa have been aggressively promoting the use of agricultural inputs such as fertilizers and pesticides. In Nigeria, in the late 1970s, a popular musician waxed a record in support of the country's "Operation Feed the Nation"

(OFN) program, extolled the virtue of farming and enjoined every citizen to grow food crops in their backyards.

Often labeled as "green revolution" such national campaigns promote the use of fertilizers and agricultural pesticides by providing subsidies, and tax discounts or exemption from import and other taxes (Thrupp, 1990). The push for increased use of agrochemicals is buoyed by international aid subsidies that enable these countries to sell pesticides at reduced rates. Consequently, developing nations including Sub-Sahara African countries account for one-third of the US$26 billion world market value for pesticides, and currently represent the fastest growing markets (Amr, 1989; Karlsson, 2004; Weselling., McConnell, Partanen, and Hogstedt, 1997). These national campaigns mirror the tenets of the original Green Revolution - a program developed by Rockefeller and Ford Foundations in the 1940s to alleviate the chronic food shortages and hunger in the Third World through the development of high-yielding crop hybrids (Knight, 1997).

Agricultural Pesticide Poisoning

Pesticides are poisons by design as these chemicals are intended for use in *destroying* any pest that may interfere with the production or the processing of desired products. The capacity of pesticides to *destroy* is however not limited to pests but may result in serious risks to human health and to the environment (Bhopal Working Group 1987). Agricultural pesticide poisoning may be occupational or non-occupational, intentional (as in suicides) or unintentional, and accidental or incidental. On the global scale, occupational poisoning is noted as the most serious problem associated with the use of agricultural pesticides (Goldberg 1985). While occupational agricultural pesticide poisoning may be universal, the incidence of such poisoning or contamination is much higher in developing countries (Al-Saleh 1994; Amr, 1989).

Sources of pesticide poisoning of farmers and farmworkers in Sub-Sahara Africa include lack of regulations regarding the labeling, distribution, use and disposal of pesticides, weak enforcement of existing regulations due at least in part to the strong influence of agricultural and chemical industry interests, and the tendency of the mostly illiterate farmers to ignore label instructions (Akingboungbe, n.d.; Al Saleh, 1994; Clarke, Levy, Spurgeon and Calvert, 1997; Echobichon, 2000; Forget, 1991; Golberg, 1985; Karlsson, 2004; Kimani and Mwanthi, 1995; Thrupp, 1990; Wesselling et al. 1997).

The lack of stringency of agricultural pesticides regulations in Nigeria, and in several other developing countries has been noted in the literature (Akingboungbe n.d; Karlsson, 2004; Kimani and Mwanthi, 1995; Forget, 1991). Kenya's pesticide regulations that banned the use of several toxic pesticides such as ethylene dibromide, chlordane, heptaclor, endrin, parathion and restricted the use of others such as lindane, aldrin, dieldrin and DDT is obviously an exception rather than the rule in Sub-Sahara Africa (Karlsson, 2004). In addition to weak pesticide regulations, the lack of enforcement of the existing regulations has been emphasized (Akingboungbe, nd, Karlsson, 2004; Tijani, 2006; Wesselling et al. 1997). Tijani (2006) noted that while the Nigerian government seems to be aware of the adverse health effects of agricultural pesticides, and has established administrative and legal mandates to curb some of these problems, the enforcement of these mandates is sorely lacking. The potential also exists of transboundary movement or smuggling of pesticides that are banned in one country but available in a neighboring country. Hence, even if restrictions exist on paper these chemicals would still be available in the open market (Wesselling et al 1997). For example, notwithstanding the relative stringency of Kenya's pesticide regulations, toxic agricultural pesticides are sold in small packets in small open shops and in open air markets (Kimani and Mwanthi 1995).

Regarding the level of education of farmers and farmworkers, about one quarter of the households including farmers involved in a pesticide use study in Ginthuri, Kenya were illiterates (Kimani and Mwanthi, 1995). The questionnaire used to obtain information regarding pesticide use by irrigation workers in Ghana was translated into four Ghanaian languages in order to facilitate the respondents' understanding of the elements of the questionnaire (Clarke et al., 1997). One of the adverse effects of low level of education is that the farm workers who took the least pesticide contamination protection measures, such as wearing protective covering were usually those with lower level of general education (Salameh, Baldi, Brochard and Saleh, 2004). In the Ginthuri, Kenya study, many of those that have some formal education tended to ignore the instructions (written in both English and the native language Swahili) depicted on agricultural pesticide container labels (Kimani and Mwanthi, 1995).

The ominous effect of the low level of education is not limited to the inability to read and understand pesticides' instruction labels but includes the lack of general knowledge about pesticides (Kimani and Mwanthi, 1995; Ibitayo, 2006; Koh and Jeyaratnam, 1996; Yassin et al., 2002). Dermal exposure either through clothing or direct contact has been identified as the most important route of exposure to occupational agricultural pesticides in hot climates because volatilization of pesticides as well as the absorption of these chemicals through the skin increases with temperature (Al-Saleh, 1994). However, farmworkers in developing countries generally tend to believe that inhalation (respiratory route) is the most important, most likely because of the obvious odor of pesticides during application (Koh and Jeyaratnam, 1996; Yassin, Abu Mourad and Safi, 2002).

Many of the respondents to the Githunri, Kenya study consider malathion, classified by World Health Organization (WHO) as slightly hazardous, as the most toxic agrochemical, and DDT classified as moderately hazardous as the least toxic. More than half of the respondents to a study regarding pesticide use by farmers in Egypt stated that they were "not sure" whether or not pesticides pollute groundwater, whether there are alternatives to chemical pesticides, and whether pesticides leave residues on plants (Ibitayo, 2006). Unfortunately, one of the effects of the increased migration of young men to urban areas in developing countries is a downward trend in the education levels in rural areas hence the ones left behind and who are responsible for applying pesticides are those with lower levels of education and work skills (Echobichon, 2000).

Another major source of pesticide poisoning in Sub-Sahara Africa is the unsafe use of pesticides, and this has been described as one of the most prevalent occupational hazards faced by agricultural workers in developing countries (Goldberg, 1985). Factors of unsafe use of pesticides include the use of leaking or otherwise faulty equipment, lack of the use of protective gear and clothing during the mixing and application of pesticides, the use of concentrations that exceed the requirements, and "calendar spraying" (Akingboungbe, n.d.; Clarke et al., 1997; Forget, 1991; FASE, 1996; Goldberg, 1985; Ibitayo, 2006; Kimani and

Mwanthi, 1995; Koh and Jeyaratnam, 1996; Salameh et al., 2004; Thrupp, 1990; Williamson, 2004; Wesseling et al., 1997; Yassin et al., 2002).

Domestic utensils and leaking equipment are used extensively for measuring and dispensing pesticides by farm workers in Kenya as poverty precludes the farmers from purchasing equipment that are in good working condition (Kimani and Mwanthi, 1995). Similarly, more than one quarter of the respondent pesticide irrigation workers in Ghana reported that they have never worn any protective clothing such as gloves when mixing or spraying pesticides. Also, the knowledge of the pesticide irrigation workers regarding the use of protective clothing or material varies from "poor" to "moderate" (Clarke et al., 1997). This observation is unfortunate because the use of effective protective clothing drastically reduces skin contamination of pesticides (Nigg, Stamper, and Queen, 1986). Wearing protective covering and washing hands and other exposed skin areas during the mixing and application of pesticides is especially important because absorption through the skin is considered to be the most frequent avenue of agriculture pesticide contamination (Al-Saleh, 1994; Perry, Marbell and Layde, 2000; Yassin et al. 2002).

Regarding "calendar spraying", cocoa farmers in southwestern Nigeria follow an inflexible, specific and scheduled application of lindane (Gammallin 20EC) regardless of pest incidence (Akingboungbe, n.d). Similarly, agricultural credit banks in Costa Rica require farmers to follow inflexible and scheduled application of pesticides in order to qualify for loans (Thrupp 1990). Also, Goldberg (1985) noted a prevalence of pre-set scheduled use of parathion in Central America. Calendar spraying may lead to overuse or unnecessary use of pesticides, increased potential for pesticide poisoning and may also foster neglecting or overlooking other forms of pest control that may be more effective and economical (Thrupp, 1990; Goldberg, 1985).

Pesticide poisoning or contamination may also be accidental and this is not limited to farmers that are involved in the mixing and/or application of the chemicals. Sources of accidental poisoning include drifts from sprayed fields, early reentry to sprayed farms, and eating crops that have recently been sprayed or treated with pesticide (Akingboungbe n.d.; Atuma, 1985; Clarke et al., 1997; Fessenden-Raden, Fitchen and Heath, 1997; Weinger and Lyons, 1992). Drifts from sprayed fields fall directly on the farmers or farm workers during application of pesticides to tree crops such as cocoa (Akinghoungbe, n.d.). Residues of persistent and highly ubiquitous organochlorine pesticide have been observed on leaf vegetables, cow, goat and even antelope meat in Nigerian markets (Atuma (1985). In the irrigation workers' pesticide use study in Ghana, about 40 percent of the workers returned to work in the sprayed field within a few hours of spraying (Clarke et al., 1997).

The general population may also be exposed to "accidental" agricultural pesticide poisoning as vendors may repackage these chemicals in small plastic or paper bags and sell in open air markets next to farm produce and food items (Kimani and Mwanthi, 1995). The effect of accidental poisoning can be extremely ominous. More than 400 people died in Iraq after eating bread that had been prepared from cereals treated with a methylmercury fungicide (Dharir and Clarkson 1985). Also, at least 37 people reportedly died of endosulfan poisoning in northern parts of the Republic of Benin during the 1999/2000 cotton growing season (Akingboungbe n.d). In the late 1950s, a prominent cocoa farmer at Okebode in southwestern Nigeria, and all members of his family were hospitalized after eating a vegetable undergrowth of a cocoa plantation that was recently sprayed with Gammalin 20EC - lindane. Residues of carbofuran pesticide found on several batches of noodles (indomie instant noodles) manufactured in Nigeria may have resulted in 23 reported cases of vomitting and one death (NAFDAC, 2004).

Accidental poisoning also results from the use of empty pesticide containers for storing foodstuffs and water for domestic use and for farm animals, and improper storage of partially used pesticides in family bedrooms, in unlocked cabinets and unmarked bottles or containers that are easily accessible to children. More than three-quarters of the respondent farmers in Ginthuri, Kenya store pesticides close to foodstuffs and in the family bedroom which are usually overcrowded (Kimani and Mwanthi, 1995). The most frequently mentioned storage area for pesticides by irrigation workers in Ghana is the bedroom followed by other rooms in the house. In addition, many of workers have used empty pesticide containers for storing water (Clarke et al., 1997). Other agricultural pesticide use practices that put the general population at risk include cleaning spraying equipment in nearby streams, and throwing empty pesticide containers on garbage sites or along the streets (Amr, 1989; Akingboungbe, nd). Another rarely-mentioned exposure of the general population to agricultural pesticide poisoning is the unscrupulous dumping of Gammalin 20 EC (lindane) into streams to "catch" fish, later sold as "fresh fish" to unwary consumers at Asejire in southwestern Nigeria (Personal observation).

Another factor that contributes to agriculture pesticide poisoning in Sub-Sahara Africa is the importation of pesticides that have been banned or whose use are severely restricted in industrialized countries because of their high levels of toxicity, persistence and bioaccumulation characteristics (Amr, 1989; Echobichon, 2000; Wesselling et al. 1997). As an example, dichlorodiphenyltrichloroethane (DDT) that has been banned for agricultural use in the U.S since 1972 because of its persistence and bioaccumulation attributes was according to U.S Customs' records of 1996, still being shipped from U.S. ports (Echobicon, 2000). While judicious applications of DDT and other organochloride pesticides have resulted in dramatic suppression or eradication of several diseases such as malaria, yellow fever, dengue fever and typhus, these chemicals are environmentally persistent and bioaccumulate, and may pose serious problems to species at the higher levels of the food chain. In addition, organochloride pesticides readily move from the blood stream to the mothers' milk hence these pesticides pose substantial risk especially to babies that are being breast-fed (Forget, 1991). Breast-feeding babies for as long as three years is a common practice in Sub-Sahara Africa.

The 1996 report of Foundation for Advancement in Science and Education (FASE) regarding the export of banned or severely restricted use pesticides from U.S ports is more specific. FASE's (1996) reports based on the analysis of U. S. shipping records show that more than 300 tons of DDT were exported from the U. S. to Peru in 1992. Chlordane that has been banned in at least 47 countries because of its toxicity and persistence, and a restricted use pesticide, alachlor were shipped to developing countries between 1992 and 1994 (FASE, 1996). In addition, "never registered" pesticides that have not been evaluated by the U. S. Environmental Protection Agency (USEPA) for health and environmental risks or whose registration application was rejected because of health and/or environmental concerns may be manufactured in the US as long as they are intended for export and not for use in the country. Nearly 25 million pounds of such "never registered" pesticides including carbosulfan were exported between 1992 and 1994, and presumably to developing countries (FASE, 1996). The report further noted that the estimates of pesticides exported from the U.S. are extremely conservative since the names 74 percent of the pesticides are omitted the shipping manifest.

While one may decry the double standard in U.S. pesticide regulation "any unilateral action by the United States will not stop international trade in hazardous substances but will only replace American hazardous substances with foreign hazardous substances" (Comment, 1981, p.332).

The movement of toxic pesticides has been observed not only at the ports of the exporting countries but also at the "destinations". Toxic organochlorides – DDT, dieldrin, aldrin and lindane were noted to be used extensively in Sub-Sahara African countries - Tanzania, Kenya and Uganda (Wesseling et. al. 1997). Lindane (formulated as Gammalin 20EC) which is used extensively by cocoa farmers in southwestern Nigeria is classified as a potential carcinogen by USEPA, and has been banned in several countries including Denmark, Germany, Holland and Sweden (Akingboungbe, n.d.). The extent of the problem of the importation by developing countries of pesticides that are banned and/or restricted for use in developed countries is probably underestimated. Forget (1991) noted that with the possible exception of Egypt, South Africa and Zimbabwe, African countries often lack a comprehensive system of regulations for importation of pesticides. Similarly, almost half of the African countries that responded to a Food and Agricultural Organisation (FAO) questionnaire indicated that they do not collect any data on the importation of pesticides (FAO, 1996).

The "contribution" of industrialized countries to agricultural pesticide poisoning in developing countries is not limited to exporting highly toxic pesticides alone. As Knight (1997) noted, the large scale use of agricultural pesticides in most developing countries is not geared towards feeding the local population but towards export crops. Quite often, the large farms – tobacco in Zimbabwe, cut flowers in Kenya, and natural rubber in Liberia are owned by companies that are based in developed and industrialized countries.

The adverse effects of "inert ingredients" and impurities in pesticides are rarely considered yet they may pose more serious adverse health effects than the pesticides' active ingredients. Dioxins may be present in as impurities in some herbicides. Also, carbon tetrachloride and chloroform that are often used as inactive ingredients in pesticides pose substantial risks to the liver and the nervous system (Al-Saleh 1994). Finally, pesticide exposure and therefore the potential for poisoning is high in Sub-Sahara Africa because agriculture, the principal basis of the economy, accounts for 80 percent or more of the active labor force in the region (Koh and Jeyaratnam, 1996). Quite often, the labor force on a farm in developing countries including sub-Sahara Africa includes all family members – women, children and elderly (Al-Saleh, 1994). Women especially pregnant or nursing mothers, children and the elderly are particularly susceptible to the hazards of pesticide contamination.

ADVERSE HEALTH AND ENVIRONMENTAL EFFECTS OF PESTICIDE CONTAMINATION

As noted earlier, pesticides are designed to cause damage, albeit to pests but by the same token these chemicals have the capacity to cause damage to human health and the environment. Several studies (Amr, 1989; Clarke et al., 1997; Gomes, Lloyd, Revitt and Basha, 1998; Kimani and Mwanthi, 1995; Koy and Jeyaratnam, 1996; Yassin et al., 2002) noted the following health effects or symptoms of pesticide contamination: eye and skin irritation; nausea/vomiting; irritation of the nose; headache; nervousness; irritability and impaired concentration; weakness; abdominal pain; and blurred vision. The authenticity of the existence and prevalence of these symptoms seem evident as the symptoms are often reported by high percentages of the respondent farmers or agricultural workers. For example, 97.5 percent of the respondents in the Gaza Strip study reported burning sensation in the eyes, while at least 82 percent reported each of the following: watering eyes; breathlessness, chest pain; itching skin; irritation or dizziness (Yassin et al., 2002).

Apart from the self-reported symptoms, other adverse health effects associated with pesticide contamination are non-institutional depression, asthma, nerve damage, blood or liver diseases, and sexual dysfunction (Amr, 1989; Forget, 1991; Goldberg, 1985; Shaw, 2003). In a study carried out in rural Missouri, US, the urine of men with lower sperm concentration and vigor contained significantly higher concentrations of three pesticides, alachlor, atrazine and diazon when compared with urine of men with high sperm concentrations and vitality (Shaw, 2003). This observation is consistent with an earlier report of sexual dysfunction in Egyptian workers employed in pesticide formulation and packaging plants (Amr, 1989). However, in a study of 60 workers at a mosquito-control center in Limpopo Province in South Africa Dalvie et al. (2004) noted a less than normal quality of semen but concluded that there is no clear evidence of the effect of DDT. The research regarding the effect of DDT on semen quality is significant because of the concerns in several parts of Sub-Sahara Africa including South Africa where the pesticide is used for the control of mosquitoes – vectors of malaria.

The effects of pesticide poisoning is not limited to direct adverse human health effects alone as these chemicals pose significant dangers to the environment. Pesticide residues may enter streams through run-offs and lead to extensive fish kill. Also, pesticide residues may bioaccumulate, move through the food chain and eventually ingested by birds, wild animals and domestic livestock (Goldberg, 1985). Methyl bromide which for a long time was used for fumigation of cocoa beans during storage and transportation, and as a soil fumigant for the production of cut flowers is an ozone-depleting substance (Akingboungbe, n.d; Schonfield et al, 1995). Subsequent to the Montreal Protocol however, methyl bromide is being replaced by phosphine for the fumigation of stored cocoa beans (Akingboungbe, n.d.)

However, there is a dearth of information regarding the environmental effects of pesticides in developing countries. Almost all (89%) the developing countries that responded to a FAO questionnaire reported that they do not study the effects of agricultural pesticides on the environment (FAO, 1996). This observation is not surprising because the institutional capacity, human and financial resources for research in subjects such as chemistry, agricultural biology and toxicology that are required for detecting environmental contamination are limited in most countries in Sub-Sahara Africa. In addition, these countries may not be inclined to expend such resources on "low priority" issues such as environmental contamination especially at the expense of increased food production.

Extent of the Adverse Effects of Pesticide Contamination

The problem of adverse health effects of pesticide contamination may be particularly exacerbated in Sub-Sahara Africa because of the high prevalence of infectious diseases and low nutritional status such as protein deficiency. These conditions tend to lessen the citizens' physiological defense against toxic substances and thereby render the farmers, farmworkers and the general population more susceptible to the risks of pesticide poisoning (Amr, 1989; FASE, 1996; Forget, 1991; Karlsson, 2004). Also, the extent of the health problems due to pesticide poisoning may be grossly underestimated due to the absence of proper documentation especially in the rural areas where most agricultural activities take place, misdiagnosis by the attending physicians or health workers, and lack of the necessary expertise in environmental and occupational epidemiology (Amr, 1989; Forget, 1991; Karlsson, 2004; Kimani and Mwanthi, 1995; Wesseling et al., 1997). Any available medical records usually include only the acute cases where the symptoms are immediate and can more easily be traced to the incriminating substance. Often excluded in these records are the chronic health effects of pesticide poisoning or contamination. Chronic cases, the result of exposure to low levels of pesticides over a long time period usually have relatively unobservable symptoms and delayed heath consequences and can therefore not be easily traced to the causal agent or agents (Ibitayo 2000).

Also, the medical records may underrepresent the extent of the problem because many farm workers in Sub-Sahara Africa consider clinical morbidity symptoms of pesticide contamination such as headaches, diziness, neusea and vomitting as minor health problems that will soon pass away or as part of daily activities and therefore not worthy of being reported to any health clinic (Gomes et al. 1998; Karlsson, 2004; Kimani and Mwanthi, 1995; Yassin et al., 2002). In fact, almost 80 percent of the farmers in the Gaza Strip study believe that their body has developed resistance to pesticides (Yassin et al., 2002). Also, many farmers and farm workers in Kenya claimed that there is no evidence of substantive negative effects of pesticide contamination, a belief shared by the Kenyan government (Karlsson, 2004). Similarly, Tijani (2006) reported that marketers (of agricultural pesticides) such as Gammallin 20 claimed that these chemicals do not pose adverse effects to pesticide users, bystanders, and the environment. Unfortunately this erronous information will be passed to those who purchase pesticides from these marketers. As noted earlier, Gammallin 20 has been identified by US EPA as a potential carcinogen, and has been banned in several developed countries.

However, as noted by past research (Gregory, 1992; Ibitayo, 2000; Slovic, 1993; Sokolowska and Tyszka, 1995; Starr, 1969), public perceptions of technological hazard risks are often mediated by the real or perceived benefits of the technology. Public favorable attitude towards diagnostic X-rays is conditioned by the direct and personal benefits derived from this technology (Starr, 1969). The threat of hunger and malnutrition is so real and immediate that people rarely pay any attention to the hazards of pesticide contamination (Atuma, 1985). On the other hand, the wealth that accrues from the sale of cocoa, an export crop, is so attractive that cocoa farmers and farm workers in southwestern Nigeria rarely pay attention to the adverse health hazards of pesticide and pesticide residues.

Agricultural Pesticide Poisoning Reduction Options

In view of the enormous and well-documented adverse health and environmental effects of the use/misuse of agricultural pesticides by farmers and farmworkers in Sub-Sahara Africa, the need arises to develop and implement strategies or a mix of options that would reduce the level of the effects. The suggested strategies include risk communication and pesticide use training programs, the discontinuation of inflexible, calendar-based application of pesticides, and the provision of government subsidy for the less toxic but usually more expensive pesticides. Others include the selection of disease resistant hybrids, the removal and burning of diseased plant parts, especially for tree crops such as cocoa, and the implementation of more stringent international agreements on exportation of highly toxic pesticides.

Most of the past research that investigated the use/misuse of agricultural pesticides in developing countries including Sub-Sahara Africa focused on the attitudes knowledge and behaviors of the farmers and farmworkers. Even Kenya's Pest Control Act of 1984, which conforms in many respects to the guidelines provided by Food and Agricultural Organization (FAO) of the United Nations for pesticide regulation, and touted as one of the most comprehensive pesticide regulation in Africa has been criticized for not providing for the education and training of farmers and farmworkers (Kimani and Mwanthi, 1995). The dissemination of risk information to the public facilitates the citizens' understanding of the hazard events to which the citizens may be exposed, thereby enhancing compliance with the requirements of the risk information (Perry and Nelson 1991). Specifically regarding pesticide risk information, farm workers' accurate knowledge of pesticides is arguably a basic requirements for the prevention of agricultural pesticide poisoning (Koh and Jeyaratnam,1996). The need arises for the provision of educational programs that cover issues such as general knowledge of agricultural insecticides, safe use of the chemicals, proper storage of pesticides, proper disposal of the containers and other pesticide risk information.

However, risk communication must not be developed as a one-way transfer of information process, nor as a single, discrete event that involves only the receiver and sender but as an interactive, complex and dynamic process involving all pesticide stakeholders (Fessenden-Radan et al., 1997; Peterson et al. 1993). The stakeholders should be defined as broadly as possible and should include not just farmers and farmworkers but also agricultural extension workers, labor and environmental workers and health inspectors. Unfortunately some of the comprehensive education programs embarked upon in developing countries have excluded key stakeholders. For example, the training program (Global Safe Use Pilot Project) sponsored by the pesticide industry that claimed to train more than 20,000 farmers in Meru, Kenya neglected casual and permanent workers in the plantation sector – groups that are likely to be heavily exposed to pesticide contamination (Karlsson 2004). Murray and Taylor (2001) noted that while the Global Safe Use Pilot Project otherwise referred to as Safe Use program represents a remarkable and serious effort on the part of industry, the authors argued that the program participants could not have been representative of the target population. This contention is based on the result of the industry-sponsored project evaluation which indicated that 84 percent of the trained farmers reported reading pesticide container labels notwithstanding that 36 percent of the target population can not read or write, and 32 percent do not speak Spanish.

In contrast to the pesticide industry-sponsored program, Weinger and Lyons (1992) described a comprehensive educational program funded by Nicaraguan, Canadian and Norwegian governments and several private voluntary organizations, designed to address pesticide poisoning among farm workers in Nicaragua. The program utilized active "student" participation, focus groups, role playing, and informal discussions to disseminate pesticide poisoning reduction information to a wide variety of audience – farmworkers, storekeepers, pesticide applicators, farmers' cooperatives, farm workers and trainers – educators, physicians, and technicians. More importantly from the perspective of the relationship between knowledge and behavior, the program utilized quantitative measures – medical monitoring and epidemeological investigation to assess program outcome or impact.

The importation of highly toxic pesticides ought to be considered as part of pesticide risk communication to governments, farmers and farm workers in Sub-Sahara Africa. One of the efforts made in this direction is the International Code of Conduct on the Distribution and Use of Pesticides (Pesticide Code) approved by FAO. The Pesticide Code called for uniform international standards regarding the use, distribution, packaging, labelling and distribution of pesticides. However, like other international codes of conduct, the Pesticide Code is voluntary and unenforceable, and according to many developing countries, the pesticide industry routinely violates the provisions and standards of the Code (Applegate, Laitos and Campbell-Mohn, 2000). Another related effort is the initiative by the United Nations Environmental Programme (UNEP) in creating an "alert list" of chemicals that have been banned or severely restricted by ten or more countries. The "list" is sent to all countries and importing countries are required by UNEP to review the "list" and register "Prior Informed Consent (PIC)" decisions to FAO as to whether they are willing to import such pesticides. The PIC decisions are subsequently released to all countries. The PIC concept is an improvement on the Pesticide Code which lacks such a notification mechanism.

Another procedure that needs to be discouraged is the inflexible calendar-based application of pesticides, to be replaced with pest scouting followed by any necessary pesticide application. Akingboungbe (n.d) reported that taken individually, the effects of mirids, psyllids and thrips on cocoa may be relatively low enough as not to adversely affect cocoa yield. The author therefore suggested the establishment of a pest scouting scheme whereby the appropriate threshold level can be determined, and beyond which pesticide application would be economically necessary. In Costa Rica, for example, the scheduling of pesticide application ought to be left to individual farmers on the advice of extension officers rather than agricultural credit banks. Another viable option is to discontinue the use of highly toxic pesticides which incidentally have been in use for a long time and are usually cheaper than the less toxic alternatives. It may therefore be necessary to provide government subsidy for these alternatives, considering that the "benefit" of reduced pesticide poisoning or contamination will outweigh the cost of the subsidy.

The efficacy of protective clothing and equipment in reducing agricultural pesticide poisoning or contamination is well-documented (Nigg et al. 1986; Perry et al. 2000; Yassin et al. 2002; Karlsson, 2004). As a result, several researchers (Kimani and Mwanthi 1995; Clarke et al. 1997; Wesselling et al. 1997; Karlsson, 2004) have emphasized the need for protective clothing for farmers and farm workers in African and other developing countries. However the arguments of these researchers for *special* clothing and equipment suitable for the hot tropical climate needs to be reassessed for at least two reasons; lack of sufficient evidence regarding the inappropriateness of currently available equipment and clothing materials, and the inaccurate perception of a monolithic hot tropical climate for Sub-Sahara African countries.

Kimani and Mwanthi (1995) interviewed 1797 households in Kenya, did not provide any empirical data regarding the use or non-use of protective clothing, yet concluded that "the few who could afford to buy protective clothing claimed such clothing were uncomfortable in the tropical climate" (p. 534). Clarke et al. (1997) noted that 71.8 percent of the respondent irrigation workers in Ghana indicated that they do not use protective clothing because of financial constraints while less than five percent cited discomfort as the reason for non-use of protective clothing, yet the authors concluded that one of the two *predominant* (italics added) reasons given for non-use is that the protective clothing is "uncomfortable to use under the prevailing hot and humid conditions" – a reason given by only five of 123 respondents. Murray and Taylor (2001) flatly concluded that personal protective equipments (which may include gloves, goggles, long pants, respirators and hat) are completely unrealistic for hot tropical climate (of the developing countries)

Table 1. Temperature (in degrees Fahrenheit): Selected Cities in Sub-Sahara Africa

City/Country	A	B	C	D	E	F
Entebbe/Uganda	69	68-71	75-78	60-64		
Dar es Salaam/Tanzania	79	75-83	82-88	67-77	102	50
Arusha/Tanzania	64	59-67	69-84	45-51	102	45
Lusaka/Zambia	70	62-77	71-85	51-69	99	32
Mombassa/Kenya	79	76-83	80-88	71-77	102	50
Johannesburg/South Africa	61	51-69	60-77	42-60	92	19
Cape Town/South Africa	63	55-71	62-78	47-63	100	30
Luanda/Angola	76	69-80	74-86	64-75	98	58
Gaborone/Botswana	70	57-80	69-87	45-72	104	28
Harare/Zimbabwe	67	59-72	69-81	49-63	93	40
Blantyre/Malawi	72	65-78	75-89	55-69	100	46
Kinshasa/Democratic Republic of Congo	79	74-81	78-87	68-75	102	50
Maputo/Mozambique	75	67-81	75-86	59-76	109	28
Windhoek/Namibia	69	57-77	68-84	47-70	102	30
Lagos/Nigeria	81	77-84	79-86	75-82	102	64
Jos/Nigeria	72	69-77	73-79	65-74	97	34
Calabar/Nigeria	80	77-84	77-85	76-83	102	66
Kano/Nigeria	79	75-88	85-101	55-75	114	43
Kumasi/Ghana	79	76-81	78-85	73-78	108	60
Takoradi/Ghana	80	77-83	79-85	74-79	103	59
Moronvia/Liberia	80	77-82	81-89	74-77	115	59

A = Average annual temperature
B = Average monthly temperature (range)
C = Average high monthly temperature (range)
D = Average low monthly temperature (range)
E = Highest temperature recorded
F = Lowest temperature recorded

In contrast to these assertions, there are wide variations in the climate not only of Sub-Sahara Africa as a region but in the climate of the individual countries. As depicted in Table 1, the annual average temperature ranges from a low of 61 degrees F. (Johannesburg, South Africa) to a high of 81 degrees F. in (Lagos, Nigeria). Even the "highest temperatures recorded" which varies from 92 in Johannesburg, South Africa to 114 in Kano, Nigeria are not dissimilar to the highest temperatures recorded in the U. S. Also, most of the "lowest recorded temperature" are 50 degrees or lower. While these do not necessarily represent the "lowest" and "highest" temperatures in Sub-Sahara Africa, these cities were selected from different regions in Sub-Sahara Africa with a particular focus on areas of considerable agricultural production. Regarding variations within individual countries, Kenya can be subdivided into the following four distinct regions on the basis of climate and vegetation: Lake Victoria basin; Rift Valley and highlands; N, NE and Eastern Highlands; and the Coast

CONCLUSION

The adverse health and environmental effects of agricultural pesticide poisoning in Sub-Sahara Africa are real and enormous. However, instead of a zero-risk approach, or dramatising the risks, the need to increase agricultural productivity in Sub-Sahara Africa requires that efforts be geared not necessarily towards a substantial reduction in the use of pesticides per se but towards a mix of options that will maximise safety and efficiency in the use of pesticides. As Williamson (2004), an ardent proponent of reduced use of pesticides by African farmers noted, African countries account for only two percent of global use of pesticides. The obvious disparity between developed nations and African countries regarding the adverse health and environmental effects of agricultural pesticide use does not therefore seem to arise from a lower level of pesticide use in developed nations.

The focus should be on developing comprehensive risk awareness programs not for the farmers and farmworkers but for all stakeholders including the national governments. Such programs should emphasize the chronic health effects of many pesticides so that the farmers will be aware of the existence of relatively unobservable symptoms and delayed heath consequences. The farmers and farm workers would therefore be more able to make informed decisions regarding mixing, application and storage of pesticides, and the disposal pesticide containers. Concomitant with risk awareness, farmers and farm workers should be provided with incentives such as subsidy on protective clothing such as gloves, goggles, respirators and boots. Also, a buy-back program may be instituted for pesticide containers. Many of the countries in Sub-Sahara Africa may need external funds in order to provide these incentive to the farmers.

For maximum effectiveness, the risk awareness programs must involve the national governments of the Sub-Sahara African countries. The importation of highly toxic pesticides that have been banned in industrialized countries, the low level of the stringency of agricultural pesticides regulations and the lack of enforcement of any existing regulations can be addressed mainly by these governments, probably with the assistance of organizations such as FAO and United Nation's World Health Organization (WHO).

Quite often, toxic pesticides can be used if the necessary precautions are taken. For example, methyl parathion, a USEPA class I (extremely hazardous) pesticide is used extensively in the U.S for the control of insect pests especially on cotton. However, it is designated as a restricted-use pesticide which means that it can people who purchase and use (mix, load and spray) the pesticide must be certified applicators (ATSDR, 1995). Also, lindane, a toxicity class II pesticide (less toxic than methyl parathion, but still highly toxic) is registered in the US for treatment of lice.

REFERENCES

Agency for Toxic Substances and Disease Registry (ATSDR). 1995. *Methyl Parathion*, US Department of Health and Human Services, Atlanta, GA

Akingbohungbe A. E., (not dated). Insecticide Usage in Nigerian Cacao Agrosystem: Possible Risks and Amelioration. Department of Plant Sciences, Obafemi Awolowo University, Ile-Ife, Nigeria, 10p.

Al-Saleh I., 1994. Pesticides: A Review Article". *J Environ. Pathology, Toxicology and Oncology*, 13(3), 151-161.

Amr M., 1989. Pesticide Intoxication in Egypt. Abstracts of the Third World Conference on Environment and Health Hazards of Pesticides, Cairo, Egypt, December 11-15, 1989 S-18.

Applegate J., Laitos J., & Campbell-Mohn C., 2000. *The Regulation of Toxic Substances and Hazardous Waste*, Foundation Press, New York.

Atuma S., 1985. Residues of Organochlorine Pesticides in Some Nigerian Food Materials. *Bull. Environ. Contam. Toxicol.* 35, 735-738.

Bhopal Working Group (1987). The Public Health Implications of the Bhopal Disaster, Report of the Program Development Board, American Public Health Association, *American J. of Public Health*, 77, p. 230-236.

Bilsborrow R., & Geores M. K., 2004. Population Change and Agricultural Intensification in Developing Countries. In (Eds.) L. Arizpe, M. P. Stone and D. C. Major, *Population & Environment: Rethinking the Debate*, Boulder: Westview Press p. 171-207.

Clarke E. K., Levy L. S., Spurgeon A, & Calvert L. A. 1997. The Problems Associated with Pesticide Use by Irrigation Workers in Ghana. *Occu. Med.* 47(5), 301-308.

Comment, 1981. Any place but here: A critique of U. S. Hazardous Export Policy, *Brooklyn J. Intl L.*, 7, 329, 331-332.

Dalvie M., Myers, J., Thompson, Mary Lou, Robin, T., Dyer, S., Riebow, J., Molewka, J., Jeebay, M., Millar, R., & Kruger, P. 2004. The long-term effects of DDT exposure on semen fertility and sexual function of malaria vector-control workers in Limpopo Province, South Africa. *Environmental Research* 96(1), 8p.

Dharir H., & Clarkson T., 1985. Toxicological Aspects of the 1972 Methylmercury Poisoning in Iraq. *J. Faculty Med-Baghdad*, 27(2), 7-15.

Echobichon D., 2000. Our Changing Perspectives on Benefits and Risks of Pesticides: A Historical Overview. *NeuroToxicology* 21, 211-218.

Food and Agricultural Organization of the United Nations (FAO). 1996. *Analysis of Governmental Responses to the Second Questionnaire on the State of Implementation of the International Code of Conduct on the Distribution and Use of Pesticides*. Rome: FAO

Food and Agricultural Organization of the United Nations (FAO). 1986. FAO International Code of Conduct on the Distribution and Use of Pesticides. FAO: Rome Foundation for Advancement of Science and Education (FASE), 1996. Exporting Risk: Pesticides Exports from US Ports, 1992-1994, FASE Research Report Spring, 10pp.

Fessenden-Raden J., Fitchen J., & Heath J., 1997. Providing Risk Information in Communities: Factors Influencing What is Heard and Accepted. *Science, Technology & Human Values* 12(3&4), 94-101 (1997).

Forget G., 1991. Pesticides and the Third World. *J. of Toxicol. & Environ. Health* 32, 11-31.

Goldberg K., 1985. Efforts to prevent misuse of pesticides exported to developing countries: Progressing beyond regulation and notification. *Ecological Law Qu arterly*, 12:997, pp. 1025-1051.

Gomes J., Lloyd O., Revitt M., & Bashir M. 1998. Morbidity among farmworkers in a desert country in relation to long-term exposure to pesticides. *Scand J. Work Environ. Health* 24(3): 213-219.

Gregory R., 1992. Implications of Public Perceptions of Hazardous Materials on Risk Management, In D. G. LeVine and A. C. Upton (Eds), *Management of Hazardous Agents: Vol. 2 Social Political and Policy Aspects,*pp. 69-79, Praeger, Westport, Connecticut.

Ibitayo O. O., 2000. Relocated Citizens' Perceptions and Attitudes Regarding Toxic Pesticides, *J. of Contingencies and Crisis Management* 8(3), 141-150.

Ibitayo O. 2006. Egyptian rural farmers' attitudes and behaviors regarding agricultural pesticides: Implications for pesticide risk communication. (Forthcoming), *Risk Analysis.*

Karlsson S. 2004. Agricultural Pesticides in Developing Countries: A Multilevel Governance. *Environment* 45(4), 23-42 (2004).

Kimani V., & Mwanthi M. 1995. Agrochemicals Exposure and Health Implications in Githunri Location, Kenya. *East African Medical J.* 72(8), 531-535.

Knight, D. 1997. Who benefits, who suffers?: The global politics of pesticide use in Brazil. http://zena.secureform.com/Znet/Zmag/articles/jan97Knight.htm accessed 5/11/2006.

Koh D., & Jeyaratnam J., 1996. Pesticides Hazards in Developing Countries. *The Sc. Of the Total Environ* 188 Suppl. 1, S78-S85.

Mortimore, M. 2003. The future of family farms in West Africa: What can we learn from long-term data? In C. Toulmin & B. Gueye (Eds.) *Transformation in West Africa Agricuture and the role of family farms*, CILSS, Ouagadougou, pp. 87-143.

Murray D. L., & Taylor P. L. 2001. Beyond Safe Use: Challenging the International Pesticide Industry's Hazard Reduction Strategy. *Gatekeeper Series* no. 103, a publication of the International Institute for Environment and Development's (IIED) Sustainable Agriculture and Rural Livelihoods Programme, 17p.

National Agency for Food and Drug Control Administration (NAFDAC). Public Notice: Contaminated Indomie Instant Noodles. *This Day*. June 14, 2004;10:20

Nigg, H. N., Stamper, J. H., & Queen, R. M. 1986. Difocol exposure to Florida citrus applicators: Effects of protective clothing. *Arch. Environ. Contam. Toxicol.* 15:121-134).

Perry M, Marbella A, & Layde P. M. 2000. Association of Pesticide Safety Knowledge With Beliefs and Intentions Among Farm Pesticide Operators" *J. of Occupational and Environmental Medicine* 42(2), 187-193.

Perry R., Nelson, L., 1991. Ethnicity and Hazard Information Dissemination, *Environmental Management*, 15(4), 381-387.

Peterson R., & Higley L., 1993. Communicating Pesticide Risks. *Forum*, Winter, 206-211.

Salameh P., Baldi I., Brochard P., & Abi Saleh B., 2004. Pesticides in Lebanon: A Knowledge, Attitude and Practice Survey. *Environmental Research* 94(1), 1-6.

Schonfield, A. 1994. Securing the Future: Environmental Issues as a Priority in South Africa. In E. Kalipeni (Ed.) *Population Growth and Environmental Degradation in Southern Africa*, pp. 179-210, Lynne Rienner Publishers: Boulder, CO.

Schonfield, A., Wamukoya, L., and Glendenig, S. 2005. Under African skies: Methyl bromide use in Africa. *Global Pesticide Campaigner* 5(4), 8p.

Shaw S. 2003. Low Sperm Count Tied to Agricultural Chems: U. of Missouri-Columbia Researcher Calling for Safer Alternatives. Press Release June 18, 2003 *Environmental Health Perspectives,* Journal of the National Institute of Environmental Health Sciences, U. S. Department of Health and Human Services.

Slovic P., 1993. Perceptions of Environmental Hazards: Psychological Perspectives, In Garling T, and Golledge R. G. (Eds) *Behavior and Environment: Psychological and Geographical Approaches*, pp. 223-248, Elsevier Science, New York.

Sokolowska J., & Tyszka T., 1995. Perceptions and Acceptance of Technological and Environmental Risks: Why Are Poor Countries Less Concerned?, *Risk Analysis*, 15(3), 733-743.

Starr C., 1969. Social Benefits VersusTechnological Risks, *Science*, 165, 1232-1238.

Thrupp L. A. 1990. Inappropriate incentives for pesticide use: Agricultural credit requirements in developing countries. *Agriculture & Human Values* 62-69..

Tijani, A. 2006. Pesticide Use and Safety Issues: The Case of Cocoa Farmers in Ondo State, Nigeria. *Journal of Human Ecology*, 19(3) 183-190.

Wesseling C, McConnell, R, Partanen T, & Hogstedt C, 1997. Agricultural Pesticide Use in Developing Countries: Health Effects and Research Needs." *International J. of Health Services* 27 (2), 273-308.

Weinger M., Lyons M., 1992. Problem-solving in the Fields: An Action-oriented Approach to Farmworker Education About Pesticides". *Am. J. of Industrial Med.* 22, 677-690 (1992).

Williamson, Stephanie (2004). Pesticide use in Africa: Doing more harm than good? www.globalpolicy.org/socecon/develop/2004/0702pesticidesafrica.htm. accessed 5/11/2006

Yassin M., Abu Mourad T., Safi J. M. 2002. Knowledge, Attitude, Practice, and Toxicity Symptoms Associated with Pesticide Use Among Farm Workers in Gaza Strip. *Occup. Environ. J.* 59, 387-394.

INDEX

A

B

C

D

E

F

G

H

T

U

V

W

X

Y

Z